Henry G. Freeman

Technisches Taschenwörterbuch

deutsch – englisch

Max Hueber Verlag

Das Werk und seine Teile sind urheberrechtlich geschützt.
Jede Verwertung in anderen als den gesetzlich zugelassenen
Fällen bedarf deshalb der vorherigen schriftlichen
Einwilligung des Verlages.

| 6. 5. | Die letzten Ziffern |
| 1995 94 93 92 | bezeichnen Zahl und Jahr des Druckes. |

Alle Drucke dieser Auflage können, da unverändert,
nebeneinander benutzt werden.
5., neubearbeitete Auflage 1985
© 1965 Max Hueber Verlag, D-8045 Ismaning
Gesamtherstellung: Friedrich Pustet, Regensburg
Printed in Germany
ISBN 3-19-006212-9

VORWORT

Es ist natürliches und legitimes Bedürfnis eines jeden normalen Technikers, einmal *das* Wörterbuch zu Rate zu ziehen, welches

1. ihm mit jeweils 7 von 8 Übersetzungsmöglichkeiten eines Stichwortes keine Kreuzworträtsel vorlegt;
2. ihn nicht zum zeitraubenden Studium der Sprache und der technischen Zusammenhänge zwingt;
3. ihm keine Schiffsladung sachfremder Ausdrücke vorsetzt;
4. ihn nicht mit ewigen Wiederholungen der gleichen Wortverbindung, nur seziert nach Grundbegriffen, langweilt;
5. ihn durch ein optisch hochgezüchtetes Volumen nicht vor Ehrfurcht erstarren läßt;
6. ihm mit einem Phantasiepreis nicht die letzte Mark aus der Tasche zieht.

Dieses kleine Buch mit bescheidenen 13 000 entries, die allerdings die Kernsubstanz der technischen Fachsprache bilden, soll die Lösung sein. Der Benutzer muß allerdings die folgenden Hinweise beachten:

1. Neben-, Unter- und Oberbegriffe des technischen Wortschatzes soll sich der Nachschlagende mit Hilfe des Schlüssels (S. 4) bilden. Dieses lexikographische System reduziert den Umfang des Buches erheblich und damit den Buchpreis.
2. Anstelle endloser Wortzusammensetzungen sind nur die Stammwörter gegeben. Nichts leichter, als mit ihnen jedes gewünschte Kompositum im Nu zusammenzubasteln.
3. Anstelle von 6 oder 7 Übersetzungen eines Stichwortes findet der Nachschlagende meist nur eine, und zwar die gängigste. Das begriffliche Strip-tease eines Wortes mag bei geruhsamer Schreibtischarbeit ein Genuß sein; der Praktiker hat während einer Verhandlung, auf der Baustelle oder unterwegs im Flugzeug oder Zug wenig Sinn dafür.
4. Die Masse der technischen Fachausdrücke ist in der Literatur verankert und genormt. Wozu also überholte Benennungen mit sprachlichem Museumswert aufnehmen? Die Kenntnis der heute gültigen Ausdrucksweise wird vorausgesetzt.
5. Begriffe der Umgangssprache sucht der Benutzer vergebens; es sei denn solche mit fachtechnischer Nebenbedeutung.
6. In diesem Buch wurde die anglo-amerikanische Schreib- und Ausdrucksweise bevorzugt. Ist die insel-englische ebenfalls sehr gängig (z. B. gage, gauge; catalog, catalogue), steht sie an zweiter Stelle mit vorgesetztem *(UK)*.

Wer mit einem solchen Ratgeber umzugehen weiß, wird Freude daran haben. Das Buch wird mit seinen rund 13 000 Grundbegriffen und einem Schlüssel (S. 4) zu weiteren 20 000 Benennungen den Benutzer selten oder nie im Stich lassen. *The proof lies in the pudding.*

Henry G. Freeman

SCHLÜSSEL
zum Gebrauch des Wörterbuches

Der Benutzer dieses Buches benötigt weder Übungen noch Erfahrung, den darin enthaltenen Wortschatz zu vervielfachen. Es sind die folgenden Hinweise zu beachten:

1. Wortverbindungen (Komposita) werden durch einfache Zusammensetzung der Grundbegriffe gebildet.

   Beispiel: Exzenter   = eccentric
               Bewegung   = motion
               Exzenterbewegung   = eccentric motion

2. Substantive (Hauptwörter) bildet man im Englischen durch Anhängen von ... ing an den Stamm des Verbes.

   Beispiel: fräsen   = to mill
               Fräsen   = milling

3. Zahllose Komposita sind in beiden Sprachen durch Vorsetzen eines Bestimmungswortes (Innen~, Außen~, Senkrecht~, Universal~ usw.) enstanden. Für die gesuchte englische Entsprechung sind Bestimmungswort und Grundwort lediglich zu koppeln.

   Beispiel: Senkrecht~   = vertical ~
               Schnitt   = cut
               Senkrechtschnitt   = vertical cut

4. Im technischen Sprachgebrauch ständig wiederkehrende Vorsilben, wie z. B. End~, Eil~, Flach~, Gerad~, Gesamt~ usw., können auch im Englischen dem Grundwort einfach vorgesetzt werden.

   Beispiel: Flach~   = flat~
               Span   = chip
               Flachspan   = flat chip

Bei mehreren Übersetzungsmöglichkeiten einer Vorsilbe sind erklärende Zusätze vorgesehen.

Für die 30 Sekunden, die dem Benutzer mit der Zusammensetzung verlorengehen, ist im Buch Platz für ein Arsenal an neuen Fachausdrücken gewonnen.

Wo immer die einfache Addition der Vorsilben bzw. Bestimmungswörter mit dem Grundwort eine Fehlübersetzung ergibt, wird das Kompositum als Stichwort gesondert angeführt.

Mit diesem System des *do it yourself* hält der Benutzer ein Lexikon mit mehr als 40 000 Fachausdrücken in Händen.

INHALTSVERZEICHNIS – INDEX

	Seite
Vorwort	3
Schüssel zum Gebrauch des Wörterbuches	4
Bibliographie	6
Disposition	8
Unterschiedliche Ausdrucksweise in England und Amerika	9
Unterschiedliche Schreibweise in England und Amerika	10
Die richtige Silbenbetonung englicher Fachwörter	11
Gängige deutsche Kurzzeichen	12
Zum metrischen und englischen Maßsystem	16
Umrechnungsfaktoren	18
Abkürzungen im Wörterverzeichnis (metrol., telev.)	22
Stichwort-Erklärungen (machinery, metallurgy)	23
Von Wörterbüchern	26
Wörterverzeichnis deutsch-englisch	28

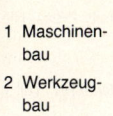

1 Maschinenbau
2 Werkzeugbau
3 Meßtechnik
4 Kraftfahrzeugtechnik
5 Eisen- u. Metallhüttenkunde

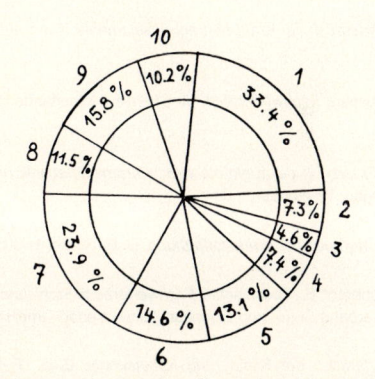

6 Werkstatttechnik
7 Elektrotechnik
8 Bauwesen
9 Grundlagenwissenschaften
10 Werkstoffkunde

Durch Überlagerung der Grundbegriffe ergeben sich mehr als 100%

BIBLIOGRAPHIE

Brinkmann-Schmidt: *Wörterbuch der Datentechnik,* D.-E., E.-D., Brandstetter Verlag

Bucksch, H.: *Wörterbuch für Ingenieurbau und Baumaschinen,* D.-E., E.-D., Bauverlag, Wiesbaden

Bucksch, H.: *Getriebe-Wörterbuch,* D.-E., E.-D., Bauverlag, Wiesbaden

DeVries-Kolb: *Wörterbuch der Chemie und der chemischen Verfahrenstechnik,* D.-E., Verlag Chemie, Weinheim.

DIN-Taschenbuch 4 und 155: *Stahl und Eisen,* Beuth Verlag, Berlin

DIN-Taschenbuch 10 und 43: *Mechanische Verbindungselemente,* Beuth Verlag, Berlin

DIN-Taschenbuch 14: *Spannzeuge,* Beuth Verlag, Berlin

Elsevier's Automobile Dictionary, Elsevier Publishin Co., Amsterdam

EUROTRANS: *Wörterbuch der Kraftübertragungselemente,* Teil 1 und 2, Springer Verlag, Berlin

Fouchier, Billet, Epstein: *Fachwörterbuch für Chemie,* Netherlands University Press, Nijmegen

Goedecke, W.: *Wörterbuch der Elektrotechnik, Fernmeldetechnik und Elektronik,* Teil I, Brandstetter Verlag, Wiesbaden

Goedecke, W.: *Wörterbuch der Werkstoffprüfung,* D.-E.-F., Bd. 1, VDI Verlag, Düsseldorf

Grote, H. und Weichbrodt, E.: *Technisches Fachwörterbuch Stahl- und Eisenbau,* E.-D., D.-E., Wirtschaftsverband Stahl- und Eisenbau, Eilers Verlag GmbH, Bielefeld

Heinrich, G.: *Wörterbuch der Klima- und Kältetechnik,* E.-D., F.-D.-R.-D., Verlag H. Deutsch, Frankfurt

Hyman, C. J.: *Wörterbuch der Physik und verwandter Gebiete,* D.-E., Brandstetter Verlag Wiesbaden

IBM: *Fachausdrücke der Text- und Datenverarbeitung,* E.-D., IBM Deutschland

International Electrotechnical Vocabulary, Group 20, Central Office of the I.E.C.

Katz, M.: *Wörterbuch der Feuerungs- und Heizungstechnik,* Ausg. A: Ölfeuerungen, D.-E.-F., Verlag G. Kopf, Stuttgart

Kleiber, A.: *Schweißtechnik,* VEB Verlag Technik, Berlin

Köhler, E. L.: *Wörterbuch für die Eisen- und Stahlindustrie,* E.-D., E.-D., Springer Verlag, Wien.

Patterson, A. M.: *A German-English Dictionary for Chemists,* John Wiley and Sons, Inc., New York

Rint, C.: *Lexikon der Hochfrequenz-Nachrichten- und Elektrotechnik,* Porta Verlag KG, München

Thompson, R. N. u. Haim, G.: *Welding Dictionary,* Iliffe & Sons, Ltd., London

Van Mansum, C. J.: *Fachwörterbuch der Bautechnik,* 4-sprachig, R. Oldenbourg Verlag, München

Walther, R.: *Technisches Englisch. Zerspanende Werkzeugmaschinen,* E.-D., VEB Verlag, Berlin

Die Wortkartei des Verfassers mit ca. 200 000 Fachausdrücken.

DISPOSITION

Grundlagenwissenschaften: Akustik – Dynamik – Physik – Mechanik – Optik

Allgemeiner Maschinenbau: Maschinenelemente – Werkzeugmaschinen – Maschinen- und Handwerkzeuge – Meßtechnik – Lagerbau – Getriebetechnik – Umformtechnik – Stanztechnik – Schmiedetechnik – Verzahnungstechnik

Eisen- und Metallhüttenkunde: Eisen- und Stahlerzeugung – Stahlverarbeitung – Walzwerkstechnik – Gießereitechnik – Betriebseinrichtungen – Oberflächenbearbeitung – Wärmebehandlungstechnik – Metallographie – Werkstoffprüfung

Elektronik: Starkstromtechnik – Schwachstromtechnik – Fernmeldetechnik – Funktechnik – Lichttechnik – Elektronik – Strahlentechnik

Kraftfahrzeugtechnik: Automobilbau – Krafträder – Fahrräder – Verkehrstechnik

Werkstattechnik: Schweißtechnik – Schmiertechnik – Antriebstechnik – Anstrichtechnik – allgemeine Fertigungstechnik – Technische Zeichnungen – Elektroerosionsverfahren

Bauwesen: Baumaschinen – Hoch- und Tiefbau – Gerüstbau – Betonbau – Straßenbau – Wasserbau – Vermessungswesen

Betriebswissenschaften: Gütekontrolle – Arbeits- und Zeitstudienwesen – Kostenrechnung

Unterschiedliche Ausdrucksweise in UK und US

	UK	US
Abkantpresse	folding press	press brake
Antenne	aerial	antenna
Aufzug	lift	elevator
Ausleger	jib	boom
Auspufftopf	silencer	muffler
Benzin	petrol	benzine, gasoline
Biegepresse	bending press	press brake
Bitumen	bitumen	asphalt
Blockvorwalzen	cogging	blooming
Döpper	snap die	header
Eisenbahn	railway	railroad
Elektronenröhre	valve	tube
Fallprobe	falling weight test	drop test
Flankendurchmesser	effecitve diameter	pitch diameter
Futterautomat	chucking automatic	chucker
Getriebegehäuse	transmission casing	gearbox
Güterzug	goods train	freight train
Hanfseil	cotton rope	manila rope
Kolbenbolzen	gudgeon pin	piston pin, wrist pin
Linde	lime-tree	basswood
Motorhaube	bonnet	engine hood
Pleuel	connecting rod	pitman
Rachenlehre	snap gauge	caliper gage
Schmirgelschleifmaschine	emery grinding machine	sander
Schraubenschlüssel	spanner	wrench
Stahlwerk	steel mill	steel plant
stufenlos regelbar	steplessly variable	infinitely variable
Tellerrand, Kronenrad	crown wheel	rim gear
Thomasroheisen	Thomas pig	basic bessemer pig
Thomasstahl	Thomas steel	basic converter steel
Trapezgewinde	trapezoidal thread	Acme thread
Vorwalzen	cogging	blooming
Walzblock	cogged ingot	bloom
Windschutzscheibe	windscreen	windshied

Weitere Beispiele, die keiner Sprachregel unterliegen:

UK	USA	
ageing	aging	Altern
aluminium	aluminum	Aluminium
briquette	briquet	Brikett
carburetter	carburetor	Vergaser
catalogue	catalog	Katalog
caulk	calk	stemmen
chequer	checker	Gitterkammer
disk	disc	Scheibe
draught	draft	Entwurf; Zug
gauge	gage	Lehre; Spur
mould	mold	Gießform
tyre	tire	Reifen
vice	vise	Schraubstock

Richtige Silbenbetonung

achteckig a. octagonal
Alkali n alkali
Aluminium n *(US)* aluminum
Anode f anode
anzeigen v. t. record
Aussparung f. recess
Barometer n barometer
Benzol n benzol
diagonal a. diagonal
Diagramm n diagram
Durchmesser m diameter
Element n element
Feuchtigkeitsmesser m hygrometer
Gemenge n compound
Genauigkeit f accuracy
Intrument n instrument
Intervall n interval
Kontakt m contact
Maschinist m machinist
mengen v. t. compound

Metallograph m metallographer
Methode f method
Mikrometer n micrometer
Modul m module
Paragraph m paragraph
pendeln v. i. reciprocate
Peripherie f periphery
Profil n profile
Programm n. program
Prozeß m process
Pyrometer n pyrometer
sechseckig a. hexagonal
Stenograph m stenographer
Symbol n symbol
System n system
Temperatur f temperature
Toleranz f tolerance
Umfang m periphery
Verfahren n procedure
Vertrag m contract

Technische Kurzzeichen

a	(Beschleunigung, m/s^2), acceleration	E	(elektr. Feldstärke), electric field strength
A	(elektr. Stromstärke; Ampere), electric current, ampere	E	(Energie), energy
A	(Fläche, Oberfläche), area, surface	f	(Brennweite), focal length
		f	(Frequenz), frequency
Ah	(Amperestunde), ampere-hour	F	(elektr. Kapazität), capacitance
Al	(Aluminium), aluminium; *(US)* aluminum	F	(Kraft), force
		F_G	(Gewichtskraft), weight force
at	(techn. Atmosphäre), bar; 1 bar = 0,1 N/mm^2	ff.	(feuerfest), refractory
		g	(örtliche Fallbeschleunigung), acceleration of free fall
atm	(physikalische Atmosphäre), *cf.* bar	G	(elektr. Leitwert), conductance
atü	(Atmosphärenüberdruck), *cf.* bar; 1 atü = 0,980 665 bar	GG	(Grauguß), grey cast iron
		Gl.	(Gleichung), equation
b	(Breite), width	GS	(Stahlguß), cast steel
B	(magnetische Flußdichte), magnetic flux density, magnetic induction	grd	(Grad), degree
		G.T.	(Gewichtsteil), part by weight
		h	(Höhe, Tiefe), height, depth
bar	(Bar, Einheit von Kraft durch Fläche), bar = 0,1 N/mm^2	h	(relative Luftfeuchtigkeit), relative humidity
C	(elektr. Kapazität), capacitance	h	(Stunde), hour
cd	(Candela), candela, luminous intensity	H	(Brinellhärte), Brinell hardness
		H	(Induktivität), inductance
d	(Durchmesser), diameter	H	(magnetische Feldstärke), magnetic field strength
D	(elektr. Flußdichte), electric flux		
DIN	(Deutsche Industrienorm), German Standard	H	(magnetischer Leitwert), permeance
dm	(Dezimeter), decimeter	H_u	(spezifischer Heizwert; früher: unterer Heizwert), specific calorific value
DVM	(Deutscher Verband für die Materialprüfungen der Technik), German Association for Testing Materials		
		HF	(Hochfrequenz), radio frequency; high frequency
		hl	(Hektoliter), hectolitre
E	(Elastizitätsmodul), modul of elasticity	HM	(Hartmetall), carbide metal
		Hz	(Frequenz), frequency

I	(elektr. Stromstärke), electric current	Mol	(Stoffmenge), amount of substance
I	(Lichtstärke), luminous intensity	n	(Drehzahl), rotational speed
J	(Energie, Arbeit), energy	N	(Kraft), force
J	(Joule), joule	p	(Bewegungsgröße, Impuls), momentum
J	(Trägheitsmoment), moment of inertia	p	(Druck), pressure
J	(Wärme, Wärmemenge), heat, quantity of heat	p	(Leistung), power
		PS	(Pferdestärke), horsepower
J/kg	(spezifischer Brennwert; früher: oberer Heizwert), specific combustion value	q	(Querschnitt), cross-section, cross-sectional area
J/kg	(spezifischer Heizwert; früher: unterer Heizwert), specific calorific value	Q	(elektr. Leistung), electric charge, quantity of electricity
		Q	(Wärmemenge, Wärme), quantity of heat, heat
K	(Kelvin), kelvin	QS	(Quecksilber), mercury column
K	(Temperatur), temperature	r	(Halbmesser), radius
kg	(Kilogramm), kilogramme	R	(elektr. Widerstand, Wirkwiderstand), resistance
konz	(konzentriert), concentrated		
kp	(Kilogramm-Kraft), kilogramme-force	rad	(Austrahlungswinkel), angle of emission
kVA	(Kilovoltampere), kilovolt-ampere	rad	(Phasenwinkel, Verlustwinkel), phase angle, loss angle
kWh	(Kilowattstunde), kilowatt-hour		
l	(Länge), length	R.-T.	(Raumteil), part by volume
l.w.	(lichte Weite), internal width	s	(Sekunde), second
L	(Drall, Drallimpuls), angular momentum	s	(Weglänge), length of path
		S	(Fläche, Oberfläche), area, surface
L	(Induktivität), inductance		
lfd.	(laufend), running	S	(Scheinleitwert, modulus of admittance
lg	(Fertiglänge), finished length		
m	(Brennweite), focal length	S	*electr*. (Scheinleistung), apparent power
m	(Masse, Gewicht), mass, weight		
m	(Wellenlänge), wavelength	SW	(Schlüsselweite), width across flats
M	(Drehmoment), moment of force, torque		
		S.Z.	(Säurezahl), acid number
m	(Minute), minute	t	(Celsius-Temperatur), Celsius temperature
M_b	(Biegemoment), bending moment		
μm	(Mikrometer), 1/1000 mm, micrometer	t	(Zeit, Dauer), time, duration

T	(Schwingungsdauer), periodic time	V	(Volumen, Rauminhalt), volume, capacity
T	(Torsionsment), torque	VA	(Voltampere), volt-ampere
U	(elektrische Spannung) potential difference	v.H.	(vom Hundert), per cent
U/min	(Umdrehungen je Min.), cf. n (Drehzahl)	VOB	(Verdingungsordnungen für Bauleistungen), contract procedure for building works
v	(Geschwindigkeit), velocity, speed	W	(mechanische Leistung), power
v	(spezifisches Volumen), specific volume	W	(Strahlungsleistung), radiant flux, radiant power
v	(Stoffmenge), amount of substance	W	(Widerstandsmoment), section modulus
V	(elektrische Spannung), potential difference	W.E.	(Wärmeeinheit), thermal unit
		WS	(Wassersäule), water column
V	(Volt), volt	Z	(Impedanz), impedance
		zul	(zulässig), permissible

Technische Kurzwörter

CGS-Einheit (phys.) (Zentimeter-Gramm-Sekunde-System) metric system of (physical) units

E-Schweißen (Lichtbogenschweißen) arc welding

GI-Stahl (Gruben-I-Stahl) I-beams for mining purposes

GU-Stahl (Gruben-U-Stahl) mine channel steel

HD-Öl (Schmieröl für Dieselmotoren) heavy-duty oil

HF-Drossel (Hochfrequenzdrossel) radio frequency choke

I-Profil (Doppel T-Stahl) double T-beam section, I-beam

IP-Stahl (Parallelbreitflanschträger) parallel broad flanged double T-steel

LD-Stahl (Linz-Donawitz-Stahl, Sauerstoff-Konverterstahl) oxygen converter steel

MIG-Schweißen (Metall-Inertgas-Schweißen) inert gas metal arc welding

SG-Schweißen (Schutzgas-Lichtbogenschweißen) inert gas arc welding

SK-Verfahren (Sauerstoff-Konverterverfahren) oxygen steelmaking process

SM-Stahl (Siemens-Martin-Stahl) open-hearth steel

UP-Schweißen (Unterpulver-Schweißen) submerged arc welding

WIG-Schweißen (Wolfram-Inertgas-Schweißen) tungsten inert gas welding

Richtlinien für die Umrechnung

Bis zur endgültig vollzogenen Umstellung des britischen Maßsystems in der Praxis der angewandten Technik erscheint die folgende Liste mit Umrechnungsfaktoren der Maßeinheiten beider Systeme nicht überflüssig.
Zwar findet man die gesuchten Endwerte in vielen Handbüchern sowie in den Umrechnungstafeln von DIN 4890-4893 und B. S. 350, doch ist fraglich, ob diese Tabellen jeder stets griffbereit bei sich hat.
Der Techniker dürfte in der Regel einen Rechenschieber in der Tasche haben. Deshalb wird hier eine alphabetisch geordnete Liste aller Maßgrößen, wie sie in der Praxis des Berufsalltags vorkommen, völlig ausreichen.
Für die richtige Umrechnung erscheinen einige Anmerkungen zweckmäßig.

Längenmaße
Für industrielle Messungen gilt für 1 Zoll der gesetzlich festgelegte Wert von 25,4 mm.
Die Umrechnung metrischer Längenmaße in Zoll ist grundsätzlich in Bruchwerten anzugeben, also z. B. 4⁹⁄₆₄″. Zu diesem Zweck ist das Resultat erforderlichenfalls auf- oder abzurunden (z. B. 19 mm = ¾″). Nur bei Umrechnungen höchster Genauigkeit (z. B. Feinstablesungen) läßt sich der Dezimalzoll nicht vermeiden.
Dies gilt insbesondere für Maßgrößen der Physik und der Meßtechnik.
Im Bauwesen und in der Fördertechnik sind Zentimeter und Meter bzw. feet und yard die gängigen Längeneinheiten.
Für Maße in der Oberflächenverfeinerung gilt hier wie drüben das Mikron (O.001 mm).
Die Längenmaße im Werkzeugmaschinen- und Werkzeugbau sind Millimeter bzw. inch.
Die Längenmaße im Kraftfahrwesen sind Milimeter bzw. inch, zur Kennzeichnung von Geschwindigkeiten Kilometer bzw. mile.

Gewichtsmaße
Bei größeren Werten ist die Gewichtseinheit Kilogramm (kg) bzw. pound (lb.), bei kleineren Gramm (g) bzw. ounce (oz.). Läßt sich bei der Umrechnung von Gewichtsangaben in Gramm zunächst nur ein Dezimalwert ermitteln (z. B. 11.29 ozs.), ist dieser durch Auf- oder Abrunden in einen Bruch umzuwandeln, also 11¼ osz.
Dieser Hinweis gilt nur für den Gewichtsbereich von 15 ... 800 g. Unterhalb 15 g bzw. ½ oz. wird das Gewicht des Teils zweckmäßig pro Dutzend errechnet. Höhere Gewichte als 800 g sind in pounds anzugeben.
Ein gängiges Handelsgewicht ist das hundredweight mit dem etwa irreführenden Kurzzei-

chen cwt., fälschlich zuweilen auch centweight genannt. Es ist zwischen dem hundredweight in England und Amerika zu unterscheiden, da 1 Imp. cwt. = 50.8 kg, 1 US cwt. = 45.4 kg.

Raummaße
Bei Flüssigkeiten ist das Liter als Maßeinheit für industrielle Zwecke in gallons (gals.) umzurechnen, kleinere Litermengen in pints (pts.). Zu beachten ist der unterschiedliche Umrechnungswert für den britischen und amerikanischen Sprachbereich, denn 1 Imp. gallon entspricht 4,5 l, 1 US gallon nur 3.78 l. Dementsprechend ist 1 Imp. pint = 0.57 l, 1 US pint = 0.47 l.

Rohrdurchmesser und Blechdicken
Von Gas- und Wasserrohren, also gewöhnlichen Leitungsrohren (pipes), wird stets der Innendurchmesser, von hochwertigen Stahlrohren (tubes) der Außendurchmesser angegeben.
Der Durchmesser von Drahterzeugnissen sowie die Dickenabmessungen von Blech werden im allgemeinen nicht in Milimeter bzw. inches angegeben, sondern mit der entsprechenden Lehrennummer bezeichnet, z. B. Stahldraht BWG 10 (= 3.4 mm) bzw. Stahlblech U. S. S. G. 12 (= 2.78 mm).

Drücke oder Belastungen
Anstelle der im Prüfwesen früher üblichen Angabe für Drucke oder Belastungen in pounds per square inch (psi.) ist neuerdings diejenige in tons per square inch (t/sq. in.) getreten. Im metrischen Maßsystem wurde entsprechend der Normung die alte Einheit Kilogramm je Quadratmilimeter (kg/mm^2) durch Kilopound je Quadratmilimeter (kp/mm^2) ersetzt.

Umrechnungsfaktoren

Atmosphäre, technische (metrische) (at)
- 0.968 atmospheres
- 1 Kilogramm je Quadratzentimeter
- 10 Meter Wassersäule von + 4° C
- 735.5 Milimeter Quecksilbersäule von 0° C
- 14.2 pounds per square inch

Boiler horsepower (B. H. P.)
- 16.86 Kilogramm je Quadratzentimeter je Stunde
- 34.5 pounds per 10 square feet per hour
- 0.9 Quadratmeter Kesselheizfläche
- 10 square feet Kesselheizfläche

British thermal unit (B. t. u.)
- 778.0 foot-pounds
- 0.39 foot-ton
- 0.252 Kilokalorie
- 107.56 Meterkilogramm
- 1054.6 Wattsekunden

British thermal unit per second (B. t. u./sec.)
- 1.414 horsepower
- 778.0 foot-pounds per second
- 0.252 Kilokalorie je Sekunde
- 1.054 Kilowatt
- 1.433 Pferdestärken

Circular inch (cir. in.)
- 5.07 Quadratzentimeter
- 0.78 square inch

Cubic foot (cu. ft.)
- 1728 cubic inches
- 28.3 Liter
- 64.4 pounds

Cubic inch (cu. in.)
- 16.4 Kubikzentimeter
- 0.016 Liter

Cubic yard (cu. yd.)
- 27 cubic feet
- 46656 cubic inches
- 168.2 gallons (Imp.)
- 0.76 Kubikmeter
- 764.56 Liter

Foot (ft.)
- 12 inches
- 0.305 Meter
- 0.303 yard
- 30.48 Zentimeter

Foot per minute (ft./min.)
- 0.305 Meter je Minute
- 0.508 Zentimeter je Sekunde

Foot per second (ft./sec.)
- 60 feet per minute
- 1.1 Kilometer je Stunde
- 18.3 Meter je Minute
- 0.3 Meter je Sekunde
- 30.48 Zentimeter je Sekunde

Foot-pound (ft. lb.)
- 0.138 Meterkilogramm

Gallon (Imp.) (gal.)
- 4.546 Liter
- 8 pints

Gallon (US) (gal)
- 3.785 Liter
- 8 pints

Gramm (g)
- 15.43 grains
- 0.001 Kilogramm
- 0.035 ounce
- 0.002 pound
- 0.07 poundal

Gramm je Kubikzentimeter (g/cm^3)
- 62.43 pounds per cubic foot

Hektar (ha)
- 100 Ar
- 2.47 acres
- 10000 Quadratmeter

Horsepower (hp)
- 33000 foot-pounds per minute
- 550 foot-pounds per second
- 0.75 Kilowatt
- 76.0 Meterkilogramm je Sekunde
- 1.01 Pferdestärken
- 745.7 Watt

Horsepower-hour (hphr)
- 990 foot-tons
- 0.75 Kilowattstunde
- 1.01 Pferdestärkenstunde

Hundredweight (Imp.) (cwt.)
- 50.8 Kilogramm
- 112 pounds

Inch (in.)
- 0.08 foot
- 0.025 Meter
- 25.4 Millimeter
- 2.5 Zentimeter

Kilogramm (Gewicht) (kg) (s. a. Kilopond)
- 35.27 ounces
- 2.205 pounds
- 0.001 Tonne

Kilokalorie (kcal)
- 3.97 British thermal units
- 3086 foot-pounds
- 0.001 Kilowattstunde
- 426.9 Meterkilogramm

Kilometer (km)
- 3280.8 feet
- 0.62 mile
- 1093.6 yards

Kilometer je Stunde (km/h)
- 54.68 feet per minute
- 27.78 Zentimeter je Sekunde

Kilopond je Quadratmeter (kg/m^2)
- 1 Milimeter Wassersäule (+ 4° C)
- 0.001 pound per square inch
- 0.045 poundal per square inch

Kilopond je Quadratmilimeter (kg/cm^2)
- 0.635 ton per square inch

Kilopond je Quadratzentimeter (kg/mm^2)
- 1 Atmosphäre
- 0.97 atmosphere
- 32.84 feet of water
- 394 inches of water
- 10 Meter Wassersäule (+ 4° C)
- 735.58 Milimeter Quecksilbersäule
- 14.22 pounds per square inch

Kilowatt (kW)
- 737.4 foot-pounds per second
- 1.34 horsepower
- 102 Meterkilogramm je Sekunde
- 1.36 Pferdestärken
- 1000 Watt

Kilowattstunde (kWh)
- 1327 foot-tons
- 1.34 horsepower-hours
- 860.58 Kilokalorien
- 1.36 Pferdestärkenstunden
- 1000 Wattstunden

Kubikmeter (cbm, m^3)
- 35.31 cubic feet
- 1.31 cubic yards

Kubikzentimeter (ccm, cm^3)
- 0.061 cubic inch
- 0.001 Kilogramm
- 1000 Kubikmillimeter
- 0.035 ounce

Liter (l)
- 0.035 cubic foot
- 61.02 cubic inches
- 0.220 gallon (Imp.)
- 0.264 gallon (US)
- 1000 Kubikzentimeter
- 1.76 pints (Imp.)
- 2.205 pounds

Megapond (Mp)
- 1000 Kilopond

Meter (m)
- 3.28 feet
- 39.37 inches
- 1.09 yards

Meter je Minute (m/min)
- 0.05 feet per second

Meter je Sekunde (m/s)
- 196.85 feet per minute
- 3.28 feet per second
- 3.6 Kilometer je Stunde

Meterkilogramm (mkg)
- 7.23 foot-pounds

Mikron (µ)
- 0.001 Millimeter

Mil
- 0.001 inch
- 0.025 Millimeter

Mile (mi.)
- 5280 feet
- 1.609 Kilometer
- 1609 Meter
- 1760 yards

Mile per hour (mi./h)
- 1.609 Kilometer je Stunde
- 26.82 Meter je Minute

Milimeter (mm)
- 0.03937 inch
- 1000 Mikron

Milimeter Wassersäule (mm W. S.)
- 1 Kilogramm je Quadratmeter

Ounce (oz.)
- 437.5 grains
- 28.35 Gramm
- 0.062 pound

Pferdestärken (PS)
- 543.5 foot-pounds per second
- 0.986 horsepower
- 0.735 Kilowatt
- 75 Meterkilogramm
- 735.5 Watt

Pferdestärkenstunde (PSh)
- 0.986 horsepower-hour
- 632.9 Kilokalorien
- 0.735 Kilowattstunde

Pint (Imp.) (pt.)
- 0.473 Liter

Pound (lb.)
- 27.68 cubic inches
- 7000 grains
- 453.59 Gramm
- 0.453 Kilogramm
- 16 ounces

Pound per square inch (lb./sq. in.)
- 703.066 Kilogramm je Quadratmeter
- 0.070 Kilogramm je Quadratzentimeter
- 144 pounds per square foot

Poundal (pdl.)
- 14.10 Gramm
- 0.141 Kilogramm
- 0.031 pounds

Quadratkilometer (km^2)
- 247 acres

Quadratmeter (m^2)
- 10.76 square feet
- 1550 square inches

Quadratzentimeter (cm^2)
- 0.155 square inch

Slug
- 32.17 pounds

Square foot (sq. ft.)
- 0.093 Quadratmeter
- 144 square inches

Square inch (sq. in.)
- 645.159 Quadratmillimeter
- 6.45 Quadratzentimeter

Square mil (sq. mil)
- 128 circular mils

Ton (t) (long ton, 2240 lbs.)
- 1016 Kilogramm
- 2240 pounds
- 1.016 Tonnen (metrische)
- 1.12 tons (of 2000 lbs.)

Ton (t) (short ton, 2000 lbs.)
- 907 Kilogramm
- 2000 pounds
- 0.89 ton (of 2240 lbs.)
- 0.907 Tonne (metrische)

Ton per square inch (t/sq. in.)
- 1.57 Kilogramm je Quadratmillimeter
- 2240 pounds per square inch
- 0.157 Tonne je Quadratzentimeter

Tonne (metrische) (t)
- 0.984 long ton
- 1.102 short tons

Umdrehungen je Minute (U/min)
- 0.105 radian per second

Winkelminute
- 0.296 angular mil

Yard
- 3 feet
- 36 inches
- 0.91 Meter
- 91.44 Zentimeter

Zentimeter
- 0.3937 inch
- 10 Millimeter

Die Abkürzungen im Wörterverzeichnis

a.	adjective	met.	metallurgy
acoust.	acoustics	metallo.	metallography
adv.	adverb	metrol.	metrology
aerodyn.	aerodynamics	n	neuter noun
auto.	automotive engineering	nucl.	nucleonics
ball.	ballistics	opt.	optics
carp.	carpentry	p.a.	participial adjective
chem.	chemistry	photo.	photography
civ. eng.	civil engineering	phys.	physics
colloq.	colloquially	pl.	plural
com.	commercially	p.p.	past participle
cryst.	crystallography	radiogr.	radiography
electr.	electrical engineering	railw.	railways
electron.	electronics	s.a.	see also
engrav.	engraving	scaffold.	scaffolding
f	feminine noun	specif.	specifically
geol.	geology	techn.	technology
geom.	geometry	tel.	telephony
hydrodyn.	hydrodynamics	telec.	telecommunication
hydromech.	hydromechanics	telegr.	telegraphy
hydr. eng.	hydraulic engineering	telev.	television
incorr.	incorrectly	UK	British English
m	masculine noun	US	American English
mach.	machinery	v.i.	intransitive verb
magn.	magnetism	v.r.	reflexive verb
math.	mathematics	v.t.	transitive verb
mech.	mechanics		

Die Sachgebiete zur Abgrenzung der Begriffe im Wörterverzeichnis

acoustics	Akustik
aerodynamics	Aerodynamik
automotive engineering	Kraftfahrzeugtechnik
ballistics	Ballistik
bearing	Lagertechnik
bicycle	Fahrrad
blast furnace	Hochofen
blasting	Strahlen
boring	Aufbohren
broaching	Räumen
building	Bautechnik
centerless grinding	spitzenloses Schleifen
chemistry	Chemie
civil engineering	Tiefbau
cold work	Kaltarbeit
commerce	Handelswirtschaft
compressive molding	Formpressen
concrete	Betonbau
copying	Nachformverfahren
cost accounting	Kostenrechnung
crystallography	Kristallographie
drawing	Zeichnung
drilling	Vollbohren
electricity	Elektrotechnik
electroerosion	Elektroerosionsverfahren
electronics	Elektronik
engraving	Gravierfräsen
extruding	Strangpressen
forging	Schmiedetechnik
founding	Gießereitechnik
gaging	Lehrenkontrolle
gear cutting	Verzahnungstechnik
gearing	Getriebetechnik
geology	Geologie
geometry	Geometrie

grinding	Schleifen
heat treatment	Wärmebehandlung
hydraulic engineering	Wasserbau
hydraulics	Hydraulik
hydrodynamics	Hydrodynamik
hydromechanics	Hydromechanik
indexing	Teilen
injection molding	Spritzgießen
kinetics	Kinetik
ladle	Gießpfanne
lapping	Läppen
lathe	Drehmaschine
lubrication	Schmiertechnik
machinery	Maschinenbau
magnetism	Magnetismus
manipulation	Hantierung
material testing	Werkstoffprüfung
mathematics	Mathematik
mechanics	Mechanik
metal cutting	Zerspantechnik
metallography	Metallographie
metallurgy	Metallurgie
metrology	Meßtechnik
miller	Fräsmaschine
molding	Formerei
motorcycle	Motorrad
nucleonics	Kernphysik
operation	Arbeitsvorgang
optics	Optik
painting	Anstrichtechnik
photography	Fotografie
physics	Physik
planer	Hobelmaschine
planing	Hobeln
plastics	Kunststofftechnik
power press	Presse
profiler	Nachformfräsmaschine
programming	Programmsteuerung
radar	Funkmeß

radial	Auslegerbohrmaschine
radio	Funktechnik
radiography	Röntgenographie
railways	Eisenbahnbau
reactor	Reaktorenbau
result	Arbeitsergebnis
road building	Straßenbau
rolling mill	Walzwerk
sawing	Sägen
scaffolding	Gerüstbau
shaper	Waagerechtstoßmaschine
shell molding	Formmaskenverfahren
slotter	Stoßmaschine
steelmaking	Stahlwerksbetrieb
surface finish	Oberflächenbeschaffenheit
surveying	Vermessungskunde
technology	Technologie
telecommunication	Nachrichtenwesen
telegraphy	Telegrafie
telephony	Fernsprechntechnik
television	Fernsehtechnik
threading	Gewindeschneiden
time study	Zeitstudienwesen
tool	Werkzeug
traffic	Verkehrstechnik
turbine	Turbine
welding	Schweißtechnik
work study	Arbeitsstudium
woodworking	Holzbearbeitung

Von Wörterbüchern

1. Man verlange von ihnen nichts Unmögliches. Ihre sparsame und kritische Benutzung ist ein Zeichen von Berufserfahrung.

   Kein Wörterbuch ist *foolproof*. Niemand und nichts ist vollkommen. Irrtümer dürfen jedoch nur sporadisch auftauchen und müssen belanglos sein.

2. Sie erteilen nur demjenigen Rat, der damit umzugehen weiß.

   Man suche kein Wort darin, das nicht hineingehört. In einem elektrotechnischen Wörterbuch soll man nicht die Terminologie der Botanik, in einem für das Bauwesen nicht die der Stahlerzeugung suchen. Es erspart Zeit und Ärger.

3. Nicht jedes Stichwort kann sämtliche Deutungsvarianten erfassen.

   Der Lexikograph eines Fachwörterbuches ist kein Enzyklopädist. Außerdem würde keiner die Arbeit bezahlen wollen.

4. Auslassungen sind nicht entscheidend.

   Der Rezensent freut sich, wenn er eine entdeckt. Der kundige Benutzer weiß sich zu helfen. Er findet den Begriff oft unter einer anderen Benennung.

5. Endlose Reihen von Wortverbindungen (Komposita) sind unnötiger Ballast.

   Der Benutzer des Buches kann sie sich leicht selber bilden, indem er die Stammwörter zusammensetzt. Komposita stehlen den Platz für wichtige Begriffe.

6. Man suche im Buch keine veralteten Benennungen, die lexikographisch Museumswert haben. Der jetzt gültige Ausdruck ist maßgebend.

   Ungezählte Begriffe haben in den letzten zwei Jahrzehnten ihr Gesicht geändert. Die früheren Benennungen ebenfalls aufzunehmen, wäre Platzverschwendung.

7. Jedes technische Wörterbuch sollte die Möglichkeit bieten, metrische Werte und Einheiten in das britische Maßsystem umzurechnen.

   Eine Forderung, die gern übersehen wird. Unter 2000 sfpm kann sich mancher nichts vorstellen. Ohne klare Vorstellung aber arbeitet der Mensch im dunkeln.

8. Fachwörterbücher mit über 30% sachgebietsfremdem Wortbestand gehören in den Papierkorb.

   Die beliebte Methode des *padding* verkennt die Intelligenz des Benutzers. Prüfe das Buch vor dem Ankauf auf seinen Substanzwert.

9. Wörterbuch mit mehr als 10% unstreitigen Fehlübersetzungen, Irrtümern, Ungenauigkeiten und Druckfehlern sollte schon der Verleger makulieren.

Eine einzige Fehlübersetzung genügt oft, den Sinn eines Satzes zu verdrehen und Schaden zu stiften. Dieser übersteigt meistens den Wert des Buches.

10. Sie sind wie Uhren, eine etwas ungenaue ist besser als keine.

Das beste: kein Wörterbuch zu schreiben, keines herauszugeben und vor allem, keines benutzen zu müssen.

A

abändern v.f. modify; alter; (Texte:) amend

abbauen v.t. (≈ abbrechen:) dismantle; (chem.) decompose

abbeizen v.t. pickle; (Leder:) dress

abbinden v.t. & v.i. (Zement:) set

Abblaseventil n blow-off valve

abblenden v.t. (auto.) dim the headlights

Abblendfußschalter m (auto.) dimmer pedal

Abblendlicht n (auto.) dimmed light

Abbrand m (met.) melting loss; (electr.)oxidation

abbrausen v.t. rinse (off), spray

abbrechen v.t. break [off]; (≈ abmontieren:) dismantle; (≈ unterbrechen:) discontinue

abbremsen v.t. brake

abbrennen v.t. & v.i. burn off (down); (welding) flash

Abbrennschweißung f constant temperature pressure welding

Abbrennstumpfschweißung f flash welding f flash butt welding

abbröckeln v.i. crumble away; chip off

Abbruch m demolition; wrecking

abdämmen v.f. (building) dam

Abdampf m exhaust steam; (chem.) evaporation

ABDAMPF ~, exhaust steam ~: (Labor-Geräte:) evaporating ~

abdecken v.t. cover; guard

Abdeckung f cover, covering; cowl

abdichten v.f. seat; make tight; pack; lute

Abdichtung v.f. seal; packing; gasket; (manipulation) sealing

abdrehen v.t. turn off; (mach.) turn; (Schleifscheiben:) true; (Schrauben:) unscrew

ABDREH ~, turning ~; truing ~

Abdrehgerät n trueing device

abdrosseln v.t. throttle; (≈ verzögern:) slow down

Abfall m waste, refuse; decrease, drop; (≈ Schräge:) slope

ABFALL~, (~beseitigung, ~erzeugnis, ~fett, ~gut, ~kohle, ~säure, ~stoff, waste~)

abfasen v.t. chamfer, bevel

abfedern v.t. spring-mount

abflachen v.t. flatten

abflächen v.t. surface, face

Abflachung f (operation:) flattening; (result:) flat

abflämmen v.t. (met.) flame-scarf

abfluchten v.t. align, make flush

Abfluchtung f alignment

ABFLUSS ~, (~hahn, ~öffnung, rohr, ~rinne, ~ventil) discharge ~, drain ~

Abfuhr f disposal, removal

Abgabeleistung f (electr.) power output

Abgas n exhaust gas, waste gas

ABGAS ~, (~heizung, ~kanal, ~leitung, ~schalldämpfer, ~vorwärmer) exhaust gas ~

abgenutzt p.a. worn [out]

abgesetzt p.a. stepped, shouldered

abgesetzter Meißel, offset cutting tool

abgestumpft p.a. blunt, dull

abgießen v.t. pour; cast; teem

Abgleich m balance; adjustment; (radio) tracking

abgleichen v.t. (electr.) balance; (radio) gang, track

abgraten v.t. deburr; (forging) trim

ABGRAT ~, (~fräser, ~maschine, ~messer, ~presse, ~schere, ~stempel, ~werkzeug) trimming ~

abgreifen v.t. pick off, pick up; (Maße:) take a measurement; caliper, calliper; (electr.) tap

Abguß m cast, pouring; (≈ Gußstück:) casting

abheben v.t. lift, take off, remove; (founding) strip; (e. Drehmeißel:) clear [from], relieve

Abhebung f (Fräser:) lift

Abhilfe f remedy

Abhitze f waste heat

Abhub m lifting, removal; stripping; clearing, relief; retraction; s.a. abheben

abkanten v.t. (≈ abschrägen:) chamfer; (≈ umlegen:) fold; (Fugen:) edge

Abkantpresse f (UK) folding press; (US) press brake

abklopfen v.t. (Zunder:) descale; (molding) rap

abkneifen v.t. pinch off

abkröpfen v.t. offset

abkürzen v.t. shorten; cut off

ablagern v.t. (Holz:) season; – v.r. settle

Ablagerung f deposition; deposit

ablängen v.t. cut to length

ablassen v.t. drain, discharge

ABLASS ~, (~hahn, ~schraube, ~ventil) drain ~

Ablauf m outlet, discharge; outflow; (mach.) cycle; expiration

ABLAUF ~, (~rinne, ~rohr, ~rutsche) discharge ~

Ablaufhaspel f (rolling mill) pay-off reel

Ablaufrollgang m (rolling mill) delivery table

ablehren v.t. (US) gage; (UK) gauge

ableiten v.t. (≈ fortleiten:) pass off; (Flüssigkeiten:) discharge; (≈ herleiten:) derive (from)

ablenken v.t. deflect

Ablenkrolle f guide pulley

Ablenkung f (electr.) deflection

Ablesbarkeit f readability

ABLESE ~, (~genauigkeit, ~lupe, ~mikroskop) reading ~

ablesen v.t. read

Ableseskala f direct reading dial

Ablesung f reading

Abluft f exhaust air

Abluftstutzen m air vent

Abmaß n allowance, variation in size

abmessen v.t. measure; gage; caliper

Abmessung f dimension; size; measurement

abmontieren v.t. disassemble, demount

Abnahme f acceptance; inspection; (≈ Beseitigung:) removal; (≈ Rückgang:) decrease; reduction; (rolling) draft

ABNAHME ~, (~prüfung, ~toleranz, ~versuch) acceptance ~

Abnahmebeamter m quality inspector

Abnahmeinspektor m works inspector

Abnahmelehre f inspection gage

Abnahmeverweigerung f rejection

Abnahmevorschrift f quality specification

abnehmbar a. removable; detachable

abnehmen v.t. accept; remove; detach; – v.i. decrease

Abnehmer m (com.) customer; receiver; buyer

abnutzen v.t. & v.i. wear

Abnutzung f wear

abplatten — Absorption

abplatten v.t. (woodworking) raise
Abplattung f (woodworking) panel raising cut
abpressen v.t. (Rohre:) run a hydraulic pressure test
abreißen v.t. tear off, pull off; – v.i. (Lichtbogen:) break
ABRICHT ~, (in Verbindung mit Schleifarbeiten:) dressing ~
abrichten v.t. (woodworking) hand plane; (starke Hölzer:) size; (Schleifscheiben:) dress
Abrichtgerät n dressing device
Abrichtlineal n straightedge
Abrichtplatte f surface plate
ABRIEB ~, abrasion ~
Abriß m draft; sketch; (surveying) layout
Abrißpunkt m (auto.) firing point; (surveying) bench mark
Abrißzündung f (auto.) make-and-break ignition
ABROLL ~, rolling ~
Abrollhaspel f uncoiler
Abrollwalze f pull-off roll
abrunden v.t. round; radius
Abrundung f rounding; radius
Absackwaage f sack filling balance
Absatz m (mach.) shoulder, step; (chem.) sediment; (com.) paragraph
Absatzwelle f shouldered shaft
ABSAUG ~, (~flasche, ~kolben, ~öffnung, ~pumpe) suction ~; (~gebläse, ~raum) exhaust ~
absaugen v.t. (Flüssigkeiten:) suck off, draw off; (Gase, Staub:) exhaust
abschalten v.t. disconnect, disengage; (electr.) switch off
Abschaltleistung f (electr.) breaking capacity; interrupting rating

Abschaltung f disconnection, disengagement; (electr.) switching off
Abscheider m separator; (Öl:) trap
ABSCHER ~, (~bolzen, ~festigkeit, ~stift, ~wirkung) shear ~, shearing ~
abscheren v.t. shear [off]
Abscherkupplung f shear-pin clutch
abschirmen v.t. screen; shield
abschleifen v.t. grind off; (mit Schmirgel:) sand
abschleppen v.t. (auto.) tow off
ABSCHLEPP ~, (~dienst, ~haken, ~kran, ~seil) towing ~
Abschleppwagen m (auto.) wrecking car
abschmelzen v.t. melt off (away), fuse
Abschmelzsicherung f fusible cut-out
abschmieren v.t. lubricate, grease
Abschmierfett n (auto.) chassis grease
Abschnitt m section; (math.) sector
abschopfen v.t. (rolling mill) crop, top
abschrägen v.t. bevel, chamfer; (welding) scarf
abschrauben v.t. unscrew, screw off
ABSCHRECK ~, (~mittel, ~riß, ~spannung, ~temperatur, ~tiefe, ~wirkung) quenching ~
Abschreckalterung f quench-age hardening, quench aging
Abschreckbiegeversuch m quench bending test
abschrecken v.t. (Stahl:) quench; (Gußeisen:) chill
Abschreckhärte f quench hardness
Abschreckung f (Stahl:) quenching
absetzen v.t. deposit; (metal cutting) shoulder, step; – v.r. settle out
absichern v.t. (electr.) protect by fuses
absorbieren v.t. absorb
Absorption f absorption

ABSORPTIONS — Abwärtsschweißung

ABSORPTIONS ~, (~apparat, ~fähigkeit, ~farbe, ~flüssigkeit, ~gefäß, ~vermögen, ~wärme) absorption ~

Absorptionsvermögen n absorption capacity

ABSPAN ~, cutting ~

abspannen v.t. unclamp; (Werkstücke:) unload; (Masten:) stay, rig

ABSPERR ~, (~schieber, ~ventil) shutoff ~, stop ~

Absperrhahn m shutoff valve

absperren v.t. shut off, stop

Absperrmittel n (painting) sealing agent

abspritzen v.t. spray off; (Drahtringe:) sull-coat

Abstand m distance; spacing, space; clearance

ABSTECH ~, (~automat, ~drehmaschine, ~maschine, ~meißel, ~support, ~vorgang) cutting-off ~, cutoff ~

abstechen v.t. (metal cutting) cut-off, part-off; (met.) tap

abstecken v.t. (e. Verteilerfeld:) set out

absteifen v.t. stiffen, reinforce; brace; (building) shore; (Schlacke:) scotch

Abstich m (met.) tapping, tap

ABSTICH ~, (~bühne, ~schlacke) tapping ~; (~loch, ~öffnung, ~pfanne, ~rinne) tap ~

ABSTIMM ~, (~bereich, ~kondensator, ~schärfe, ~skala) tuning ~

abstimmen v.t. time, synchronize; (radio) tune

abstreichen v.t. scrape off; (Schlacke:) skim; (molding) strike off

abstreifen v.t. (Kokillen:) strip

Abstreifer m (Schlacke:) skimmer; (Blockformen:) stripper; (Öl:) scraper

Abstreifformmaschine f stripping plate molding machine, stripper

Abstreifring m (Öl:) scraper ring

Abstrich m (met.) scum, dross

abstufen v.t. grade, vary; step; (painting) shade

Abstufung f (speeds) grading, variation; (Getrieberäder:) cone; (Welle:) shoulder; (painting) shade

abstumpfen v.t. blunt, dull; – v.i. get dull

abstützen v.t. support, back up; brace; (Gebäude:) shore

ABTAST ~, (~stift, ~verfahren) tracer ~; (~folge, ~geschwindigkeit, ~strahl) scanning ~

abtasten v.t. (mach.) trace; (telev.) scan

abteufen v.t. (civ.eng.) sink

Abtrag m, **Abtragung** f (metal cutting) removal

abtragen v.t. (metal cutting) remove; (drawing) plot

abtrennen v.t. cut-off, part-off; separate

Abtrieb m (gearing) output drive

ABTRIEBS ~, (~drehzahl, ~welle) output ~

ABTROPF ~, (~gestell, ~platte, ~schale) draining ~

Abtropfblech n drip pan

abtropfen v.i. drip off, drain

Abtropftisch m drain table, drip table

Abwärme f waste heat

ABWÄRTS ~, (~bewegen, ~drücken, ~gleiten, ~schwenken) ~ downward, ~ down

ABWÄRTS ~, (~bewegung, ~hub, ~schwenkung) downward ~

Abwärtsschweißung f vertical downwelding

Abwärtstransformator *m* step-down transformer

Abwasser *n* waste water; (Abwässer:) sewage

ABWASSER ~, (~beseitigung, ~chlorung, ~klärung, ~reinigung) sewage ~

Abwasserkanal *m* sewer

abweichen *v.i.* deviate; differ; vary

Abweichung *f* deviation; difference; variation

abwracken *v.t.* salvage

abwürgen *v.t.* (Gewinde:) strip; (Motor:) stall

abzapfen *v.t.* draw off; tap

abziehen *v.t.* draw off, withdraw; (Flüssigkeiten:) siphon off; (Gußblockformen:) strip; (Schlacke:) tap; (Schleifscheiben:) true; (≈ herausziehen:) extract; – *v.i.* (Gase:) escape

Abziehformkasten *m* snap flask

Abziehhülse *f* withdrawal sleeve

Abziehmutter *f* withdrawal nut

Abziehstein *m* bench stone, whetstone

Abziehvorrichtung *f* withdrawing tool

Abzug *m* hood (for gases); iussue (of gases); *(photo.)* contact-print

Abzugshaube *f* hood

Abzugsrohr *n* off-take

ABZWEIG ~, (~dose, ~klemme, ~leitung, ~stutzen) branch ~

Abzweigmuffe *f* cable jointing sleeve

Aceton *n* acetone

acetonlöslich *a.* acetone-soluble

Acetylen *n* acetylene

Acetylensauerstoffbrenner *m* oxyacetylene torch

Acetylensauerstoffschweißen *n* oxyacetylene welding

ACHS ~, *(auto.)* (~antrieb, ~aufhängung, ~druck, ~gehäuse, ~lager, ~last, ~stummel, ~sturz, ~stütze, ~welle) axle ~

Achsabstand *m* center (centre) distance; *(auto.)* wheel base

Achsholzen *m (auto.)* king pin, pivot pin

Achse *f* (≈ Tragachse:) axle; *(math.)* axis, center line

Achsenkreuz *n* coordinate system of axes, axes of coordinates

Achsenschnitt *m* axial section

achsensenkrecht *a.* (≈ lotrecht:) perpendicular

achsgerecht *a.* endwise

Achsial ~, cf. Axial ~

Achslage *f* axial position

achsparallel *a.* axially parallel

Achsschenkel *m* axle journal; *(auto.)* steering swivel

ACHSSCHENKEL ~, *(mach.)* journal ~; *(auto.)* steering swivel ~

Achsstand *m* wheel base

achteckig *a.* octagonal

Achtpolröhre *f* hectode

achtseitig *a.* octahedral

Achtspindelautomat *m* eight-spindle automatic machine

achtstufiges Getriebe, eight-speed gear drive

Ackerschlepper *m* farm tractor

Ader *f (electr.)* core; conductor; *(geol.)* vein, lode

Adjustageabteilung *f* finishing department

Adjustagemaschine *f* dressing and straightening machine

adjustieren *v.t.* adjust, set

agglomerieren *v.t.* agglomerate

Aggregat *n* aggregate, unit

Akkordarbeit f piece work
Akkordberechnung f rate fixing
Akkordsatz m piece rate
Akkordsystem n piece rate plan
Akkumulator m storage battery, accumulator
AKKUMULATOREN ~, (~batterie, ~ladung, ~zelle) accumulator ~
Alarmeinrichtung f warning system
ALAUN ~, (~hütte, ~lauge, ~lösung, ~pulver, ~stein, ~wasser) alum ~
Alaunerde R alumina
alitieren v.t. aluminize
Alkali n alkali
ALKALI ~, (~beständigkeit, ~gehalt, ~lauge, ~lösung, ~rückstand) alkali ~
Alkydharz n alkyd resin
Alligatorhaube f alligator hood
Allradantrieb m (auto.) all-wheel drive
Allstromempfänger m all-mains receiver
Allwetterverdeck n all-weather hood
Allzweckmaschine f all-purpose machine
altern v.t. ~v.i. (Stahl:) age; (Leichtmetall:) age-harden; (Holz, Lehrwerkzeuge:) season
alterungsbeständig a. (met.) insusceptible to aging, non-aging
Alterungshärtung f age-hardening
Alterungsneigung f age-hardening susceptibility
Alterungssprödigkeit f precipitation brittleness, embrittlement by aging
Altwagen m (auto.) second-hand car
alumetieren v.t. alumetize
aluminieren v.t. aluminum coat
Aluminium n (US) aluminium; (UK) aluminium
ALUMINIUM ~, (~blech, ~bronze, ~draht, ~farbe, ~guß, ~legierung, ~leiter, ~überzug) aluminum ~
Aluminiumdruckguß m die-cast aluminum
Aluminiumknetlegierung f wrought aluminum alloy
aluminiumhaltig a. (Legierung:) aluminum-bearing
aluminothermische Schmelzschweißung, thermit fusion welding
aluminothermische Schweißung, thermit welding
aluminothermisches Preßschweißverfahren, thermit pressure welding
AMATEUR ~, (~empfänger, ~sender) amateur ~
Amboß m anvil
AMBOSS ~, (~bahn, ~gesenk, ~horn, ~stöckel) anvil ~
Aminoplast-Preßmasse f aminoplastic compression molding material
Ammoniak n ammonia
ammoniakfrei a. free from ammonia
Amperemeter n ammeter, amperemeter
Amperestundenzähler m amperehour meter
Amphibienfahrzeug n amphibian vehicle
Amplitude f (electr., magn.) amplitude
AMPLITUDEN ~, (~modulation, ~sieb, ~verzerrung) amplitude ~
Amt n (tel., telegr.) office, telephone exchange, central office
AMTS ~, (~anruf, ~batterie, ~wähler) exchange ~
Amtsfreizeichen n dialling tone
Amtswähler m junction selector
Amtszeichen n dialling tone
Amylalkohol m amyl alcohol

ANALOG ~, (~rechner, ~technik) analogue ~
Analyse *f* analysis
Analysenwaage *f* analytical balance
anbauen *v.t. (techn.)* attach, mount (on), flange-mount
anblasen *v.t. (blast furnace)* blow on (in)
Anbaugerät *n* attachment
Anbaumotor *m* flange-mounted motor
anbringen *v.t.* attach, mount, fit, fix; install
Anbruch *m* incipient fracture (*or* crack)
Anbruchsicherheit *f* resistance to incipient cracking
ändern *v.t.* change; vary; modify; correct; amend
Änderung *f* change; variation; modification, correction; amendment
andrehen *v.t.* (Schrauben:) tighten; (Motor:) start
andrücken *v.t.* press against, spin; – *v.i.* bear on
ANEINANDER ~, (~drücken, ~fügen) ~together
ANFAHR ~, (~beschleunigung, ~eigenschaft, ~moment) starting ~
anfahren *v.t.* (Maschine, Motor:) start; – *v.i.* run against
ANFANGS ~, (~bahn, ~bereich, ~druck, ~festigkeit, ~lagerluft, ~moment, ~spiel, ~stadium, ~stellung) initial ~
ANFAS ~, (~kurve, ~meißel, ~werkzeug) chamfering ~
anfasen *v.t.* chamfer, bevel
anfassen *v.t.* touch, grip
anfertigen *v.t.* make, manufacture
anfeuchten *v.t.* moisten, wet; (Formsand:) temper
anflanschen *v.t.* flange-mount

anfressen *v.t.* corrode; (Ofenfutter:) erode
Angel *f* (e. Werkzeuges:) tang
angenähert *p.a.* approximate
angreifen *v.t. (chem.)* corrode; attack
angreifend *p.a. (chem.)* corroding, aggressive
angrenzend *p.a.* adjacent, adjoining
Anguß *m* lug
anhaften *v.i.* adhere to
anhalten *v.t.* stop, bring to rest; *(auto.)* hitchhike
Anhänger *m (auto.)* trailer
ANHÄNGER ~, (~achse, ~bremse, ~fahrgestellt, ~kupplung) trailer ~
Anhängeschild *n* tie-on tag
Anilinfarbstoff *m* aniline dye
Anker *m* anchor; *(electr., magn.)* armature; stator; rotor; *(mach.)* tie rod; *(tel., telegr.)* stay; *(blast furnace)* belt
ANKER ~, (~blech, ~eisen, ~kern, ~klemme, ~leiter, ~nut, ~rückwirkung, ~spannung, ~spule, ~stab, ~strom, ~welle, ~wicklung, ~widerstand) armature ~; (~bolzen, ~mast, ~platte, ~schraube) anchor ~
Ankersäule *f* (e. S. M.-Ofens:) buckstay
ankörnen *v.t.* center
Anlage *f* installation; plant; equipment; *(gearing)* contact, bearing
ANLAGE ~, (~fläche, ~flansch) contact ~
ANLASS ~, (~leistung, ~magnet, ~moment, ~motor, ~regler, ~schalter, ~strom, ~stufe, ~ventil ~wicklung), starting ~
Anlaßbeständigkeit *f (heat treatment)* retentivity of hardness
anlassen *v.t. (auto.)* start; (Stahl:) temper
Anlasser *m* (Motor:) starter

ANLASSER ~, (~kabel, ~motor, ~ritzel, ~schalter, ~zahnkranz) starter ~
Anlaßfarbe f temper color
Anlaßhebel m *(motorcycle)* kick-starter
Anlaßofen m tempering furnace
Anlaßöl n tempering oil
Anlaßprödigkeit f temper brittleness
Anlaßventil n *(auto.)* starting valve
ANLAUF ~, (~bedingungen, ~drehzahl, ~kondensator, ~moment, ~strom) starting ~
anlaufen v.i. start; (gegen:) strike (against); (≈oxidieren:) become dull, tarnish
Anlauffarbe f tempering color, temper color
Anleitung f instruction; guide
anlenken v.t. couple
anmontieren v.t. mount on (to)
annageln v.t. tack
ANNÄHERUNGS ~, (~formel, ~wert) approximate ~
ANODEN ~, (~batterie, ~dichte, ~gleichrichter, ~kupfer, ~modulation, ~rückkopplung, ~schlamm, ~spannung, ~stecker, ~strom, ~transformator) anode ~
anodenmechanisch a. electrolytic
anodisch a. anodic
anpassen v.f. & v.r. adapt (to), adjust, suit
Anpassung f adaptation, matching
ANPASSUNGS ~, (~dämpfung, ~kreis, ~transformator) matching ~
Anpassungsfähigkeit f adaptability
Anpreßdruck m contact pressure
anpressen v.t. press against
Anregung f *(nucl.)* excitation
anreichern v.t. enrich, concentrate
anreißen v.t. (mittels Reißnadel:) scribe, mark off; *(drawing)* trace, draw, plot
Anreißlinie f layout line
Anreißplatte f surface plate
Anreißprisma n V-block
Anreißspitze f scriber point
Anreißzeug n layout tool
Anreißzirkel m beam trammels
Anriß m incipient crack
Anruf m *(tel.)* call; ringing
ANRUF ~, (~klappe, ~klinke, ~lampe) calling ~
anrufen v.t. *(tel.)* call, ring, phone
Ansatz m (Welle:) shoulder; (Schraube:) neck; (Zapfen:) lug
ANSATZ ~, (~drehen, ~fräsen, ~schleifen, ~schmieden) shoulder ~
Ansätze mpl. (am Konverter:) kidneys; (im Schmelzofen:) accretions
ansaugen v.t. suck (in, up)
Ansaughub m suction stroke
Ansaugkrümmer m *(auto.)* air intake
Ansaugluft f induction air
Ansaugrohr n *(auto.)* inlet manifold
Ansaugstutzen m (e. Pumpe:) intake
Ansaugventil n inlet valve
Anschaffungskosten pl. first cost
Anschlag m impact; *(mach.)* stop dog
ANSCHLAG ~, (~buchse, ~finger, ~hebel, ~knagge, ~nase, ~nocken, ~schiene, ~stift, ~walze) stop ~
anschlagdrehen v.t. turn multiple diameters
Anschlagleiste f *(sawing)* fence
Anschlaglineal n T-square
Anschlagschleifen n shoulder grinding
Anschlagwinkel m try square, back square
Anschluß m *(electr.)* connection

anschlußfertig *a. (electr.)* wired ready for connection

Anschlußgleis *n* spur track

Anschlußkabel *n (electr.)* power supply cable; *(tel.)* subscriber's cable

Anschlußklemme *f (electr.)* terminal

Anschlußleistung *f (electr.)* connected load

Anschlußleitung *f (electr.)* lead wire; *(tel.)* subscriber's line

Anschlußmaße *npl.* dimension diagram

Anschlußnummer *f (tel.)* subscriber's number

Anschlußschnur *f* flexible cord

Anschlußstecker *m* attachment plug

Anschlußstutzen *m* connecting piece, connecting branch

Anschlußwert *m (electr.)* connected load

Anschnallgurt *m (auto.)* safety strap

anschneiden *v.t. (metal cutting)* start cutting; *(founding)* gate

Anschnitt *m (metal cutting)* first cut; *(threading)* start; *(founding)* gate

Anschnittwinkel *m (threading)* lead angle

anschrägen *v.t.* chamfer, bevel

anschrauben *v.t.* screw on; bolt to

anspitzen *v.t.* point; *(cold work)* cold-swage

ansprechen *v.i.* respond (to)

anstählen *v.t.* steel-face

Anstalt *f (techn.)* shop

anstauchen *v.t.* (Bolzenköpfe:) head

anstellen *v.f.* (Werkzeug:) set, position; (Motor:) start; (Lager:) adjust

Anstellvorrichtung *f (rolling mill)* housing screw

Anstellwinkel *m (metal cutting)* setting angle

anstirnen *v.t.* spot face

anstoßen *v.t. & v.i.* strike (against); (Fugen:) abut

anstreichen *v.t.* paint

Anstreicher *m* painter

Anstrich *m* paint coat; paintwork; coat; coating

Anstricherneuerung *f* renewal of paintwork

Antenne *f(US)* antenna; *(UK)* aerial

ANTENNEN ~, (~isolation, ~kapazität, ~kopplung, ~leistung, ~stab, strom, ~verlust) antenna ~ aerial ~

Antennenverstärkung *f* directive gain

Antidröhnpappe *f (auto.)* anti-drumming sheet

antreiben *v.t.* drive; gear; power

Antrieb *m* drive; actuation

ANTRIEBS ~, (~element, ~kraft, ~kupplung, ~leistung, ~moment, ~motor, ~rad, ~riemen, ~ritzel, ~scheibe, ~seite, ~spindel, ~zahnrad) driving ~

Antriebsdrehzahl *f* input speed

Antriebskasten *m* (e. Werkzeugmaschine:) gearbox

Antriebskegelrad *n* bevel pinion

Antriebswelle *f* drive shaft

anvisieren *v.t.* sight

Anvisiermikroskop *n* sighting microscope

anwärmen *v.t.* warm up, heat up (slightly)

Anwärmofen *m* heating furnace

Anweisung *f* instruction, direction

anwerfen *v.t. (auto.)* crank

Anwurfmotor *m* crank start motor

Anwurfwand *f (building)* roughcast wall

ANZAPF ~, *(electr.)* (~spannung, ~umschalter) tap ~

Anzapfdampfmaschine *f* bleeder-type steam engine

anzapfen v.t. (electr.) tap; (Dampf:) bleed, extract
Anzapfstelle f (Turbine:) extraction point
Anzapfstutzen m steam extraction branch
Anzapftransformator m tapped voltage transformer
Anzapfturbine f back pressure turbine
Anzapfturbogenerator m bleeding turbogenerator
Anzapfung f (electr.) tap, tapping; (turbine) bleeding
Anzapfventil n bleeder valve
anzeichnen v.t. mark
Anzeige f reading; indication; record
ANZEIGE ~, (~bereich, ~genauigkeit, ~instrument, ~gerät, ~vorrichtung) indicating ~
Anzeigeleuchte f (auto.) indicator lamp
anzeigen v.t. indicate; record; register
Anzeiger m indicator; recorder; meter; registering instrument
anziehen v.t. pull; draw; (Schrauben:) tighten; (magn.) attract; – v.i. (Bremse:) grip
Anziehungskraft f (magn.) attractive force
Anziehvermögen n (auto.) starting power
Anzug m (Keil:) taper; (Motor:) start; (Spindel:) draw-in; (Walzkaliber:) taper
Anzugmutter f tightening nut
Anzugsvermögen n. (auto.) pulling power
Anzünder m lighter
Apparat m apparatus; instrument; device
Apparateschnur f flexible cord
Apparatur f apparatus; outfit; equipment
arbeiten v.i. work; operate; (Gerät:) function
Arbeitgeber m employer
Arbeitnehmer m employee

ARBEITS ~, (~beispiel, ~bewegung, ~ebene, ~fläche, ~flur, ~genauigkeit, ~höhe, ~kolonne, ~lehre, ~leistung, ~plan, ~spannung, ~stellung, ~verfahren, ~zeit) working ~, operating ~
ARBEITS ~, (metal cutting) (~belastung, ~druck, ~geschwindigkeit, ~hub, ~kurve, ~länge, ~meißel, ~stellung) cutting ~
Arbeitsablauf m (cost accounting) operating sequence; (mach.) work cycle
Arbeitsanalyse f (work study) operation analysis
Arbeitsauftragsanalyse f job analysis
Arbeitsbereich m (e. Maschine:) capacity
Arbeitsbewertung f job evaluation
Arbeitsflüssigkeit f (electroerosion) electrolyte
Arbeitsgang m operation; work; cycle
Arbeitsgrube f (auto.) repair pit
Arbeitskraft f (cost accounting) manpower
Arbeitsmangel m lack of work
Arbeitsplatzleuchte f machine lamp
Arbeitsseite f (e. Walzgerüstes:) off-side
arbeitssparend a. labor-saving
Arbeitsspindel f (lathe) workspindle, main spindle
Arbeitsstück n workpiece, workpart, work
Arbeitstag m workday
Arbeitstisch m worktable
Arbeitstür f (e. Schmelzofens:) charging door; (e. Roheisenmischers:) padding door
Arbeitsvermögen n working capacity; (phys.) kinetic energy; (Werkstoffprüfung) energy of deformation
Arbeitsvorgang m operation

Arbeitsweise f method of operation (or working); (≈ Praxis:) practice; (≈ Hergang:) procedure

Arcatom-Schweißung f atomic hydrogen welding

Argonschweißung f argonarc welding

Armaturen fpl. fittings

Armaturenbrett n instrument board

Armaturentafel f instrument panel

Armco-Eisen n ingot iron

armieren v.t. (US) armor; (UK) armour; (Beton:) reinforce

Armstütze f arm rest

arretieren v.t. lock, arrest

Arretierung f arrest

Arretiervorrichtung f locking device

Asbest m asbestos

ASBEST ~, (~dichtung, ~faser, ~gewebe, ~kitt, ~mehl, ~packung, ~pappe, ~platte, ~schnur, ~schürze, ~wolle, ~zement) asbestos ~

Asche f ash; slag; cinder

Aschengrube f ashpit

Asphalt m asphalt; (incorr.) bitumen

ASPHALT ~, (~auskleidung, ~beton, ~bitumen, ~decke, ~lack, ~kitt, ~pflaster) asphalt ~

asphaltieren v.t. asphalt

Asphaltmischmakadam m bitumen macadam

Asphaltsplitt m bituminized chips, asphalt-coated chips

Asphaltstraße f bituminous road

Astloch n knot hole

Asynchrongenerator m asynchronous alternator

Asynchronmotor m induction motor

ATOM ~, (~gewicht, ~gitter, ~kern, ~meiler, ~physik, ~technik, ~teilchen, ~wärme, ~zahl) atomic ~

atomisieren v.t. atomize

Atomzertrümmerung f nuclear fission

Attest n certificate

atü abbr. (= Atmosphärenüberdruck) number of (metric) atmospheres above atmospheric pressure

ÄTZ ~, (~kali, ~kalk, ~lösung, ~natron) caustic ~

ätzen v.t. (metallo.) etch; (chem.) corrode

Ätzwirkung f corrosive action

Audion n (radio) audion, grid detector

aufarbeiten v.t. (Werkzeuge:) recondition, dress

Aufbau m (auto.) car body; (building) construction; (chem.) composition; (≈ Montage:) erection

Aufbaueinheit f (mach.) basic [machine] unit

aufbauen v.t. construct, build up; install, erect; (chem.) compose

Aufbauschneide f (metal cutting) build-up edge

aufbereiten v.t. (Erz:) dress; (Formsand, Kohlen:) prepare; (Werkzeuge:) recondition; redress

Aufbereitung f dressing; preparation; s.a. aufbereiten

Aufbereitungsanlage f dressing plant; preparation plant

aufbewahren v.t. store

Aufblähung f bulging; (Koks:) swelling

Aufblaseverfahren n (steelmaking) top blowing process

aufbocken v.t. (auto.) jack up

aufbohren v.t. bore

Aufbohrung f borehole, bore

Aufeinanderfolge f sequence; succession

aufeinanderfolgend *a.* successive
Aufenthaltsraum *m* (e. Betriebes:) messroom
auffalzen *v.t.* seam on
aufflanschen *v.t.* flange-mount
auffüllen *v.t.* refill; (Öl:) top up
Aufgabe *f* function; job; task; (von Werkstücken:) loading, feeding
Aufgabegefäß *n* (shell molding) dump box
Aufgabeschnecke *f* screw feeder
Aufgabetrichter *m* feed hopper; (Hochofen:) distributor
Aufgabezeit *f* (shell molding) coating time
aufgeben *v.t.* charge; load, feed
aufgenommene Leistung, input
aufhängen *v.t.* suspend (from)
Aufhängung *f* suspension
aufklappbar *v.t.* carburize
Aufkohlung *f* carburization
Aufkohlungsmittel *n* recarburizer, carburizer
aufladen *v.t.* (electr.) charge (a battery); (Material:) load
Aufladeventil *n* (auto.) charging valve
Auflage *f* base; bearing; rest; support; (Hartmetall:) tip
Auflagedruck *m* bearing pressure
Auflagemetall *n* (= Überzugsmetall) cladding material
Auflager *n* support
Auflagerfläche *f* bearing surface
Auflicht *n* (opt.) incident light
auflösen *v.t.* dissolve
auflöten *v.t.* solder (on); braze (on)
Aufmaß *n* allowance
aufmauern *v.t.* brick up
Aufnahme *f* reception; (chem.) absorption; (electr.) input; (mach.) holding fixture; (met.) pick-up; (radio) recording; (tool) adaptor
Aufnahmefähigkeit *f* (chem.) absorption power; (mach.) holding capacity
Aufnahmeflansch *m* holding flange
Aufnahmegerät *n* (Schallaufzeichnung) tape recorder
Aufnahmeleistung *f* (electr.) input power
aufnehmen *v.t.* receive, accommodate; take up; absorb
Aufputzverlegung *f* on plaster laying
aufreiben *v.t.* (mittels Reibahle:) ream
Aufreißen *n* (road building) scarification, scarification, scarifying
Aufriß *m* (drawing) elevation
aufschrauben *v.t.* screw on, bolt on (to)
aufschrumpfen *v.t.* shrink on
Aufschüttung *f* (road building) filling
aufschweißen *v.t.* weld on; deposit, build up
Aufschweißlegierung *f* building-up alloy, hard-facing alloy
Aufschweißplättchen *n* carbide tip
Aufseher *m* (civ.eng.) supervisor
aufspannen *v.t.* clamp; bolt; mount
Aufspannfutter *n* chuck
Aufspannplatte *f* (planer) platen
Aufspannschlitten *m* (lathe) compound slide rest
Aufspanntisch *m* (e. Werkzeugmaschine:) worktable
Aufspannut *f* T-slot
Aufspannwinkel *m* (tool) angle plate
Aufspannzeit *f* (für Werkstücke:) loading time
aufspeichern *v.t.* store, accumulate
Aufstampfboden *m* (Formerei) bottom board

aufstampfen v.t. stamp on; (Herdfutter:) tamp down; (Formsand:) ram up
Aufsteckdorn m (US) arbor; (UK) arbour
Aufsteckfräser m shell end mill
Aufsteckrad n (Wechselradgetriebe:) change gear, loose gear
Aufsteckreibahle f shell reamer
Aufsteckschlüssel m (US) socket wrench; (UK) box spanner
aufstellen v.t. erect; install
Aufstellung f erection, installation
Aufstellungsmöglichkeit f layout facility
Aufstickung f (met.) nitrogen pick-up
Aufstromvergaser m up-draught carburettor
auftanken v.t. (auto.) top up, refuel
Aufteilung f (von Kosten:) allocation; division
Auftragegeschwindigkeit f (welding) rate of deposition
auftragen v.t. (painting) coat; apply a coat; (welding) deposit, build up; (Meßwerte:) plot
Auftraggeber m contractor
auftragschweißen v.t. hard face, hard surface
Auftraglöten n coat soldering
Auftragnehmer m customer
Auftragschweißen f deposition welding, surface-fusion welding
Auftragung f deposition; deposit
Auftreffwinkel m (electr., opt.) angle of incidence
Auftrieb m (aerodyn.) lift; (hydrodyn.) buoyancy
Auf und Abbewegung f raising and lowering
Aufwand m (cost accounting) expenditure

Aufwandschlüssel m (cost accounting) expense code
AUFWÄRTS ~, (~bewegung, ~hub, ~schwenkung) upward ~
aufwärtsschweißen v.t. weld up-hand
Aufwärtsschweißen f vertical up-welding
aufweiten v.t. expand; (Rohre:) flare out
Aufweiteprüfung f drift test, flaring test
Aufweitewalzwerk n becking mill
Aufwickelhaspel f recoiler
aufzeichnen v.t. plot; draft; register, record
Aufzug m lift, elevator; (Lasten:) hoist
Aufzugseil n hoisting rope, elevator rope
Aufzugswinde f hoist winch
Augenbolzen m eye bolt
ausbaggern v.t. dredge, excavate
ausbalancieren v.t. counterbalance
Ausbau m removal, withdrawal; (Demontage:) disassembly
ausbauchen v.t. bulge out, belly out
ausbessern v.t. repair; (road building) patch
ausbeulen v.t. flatten, planish
Ausbeulwerkzeug in planishing tool, bumping tool
ausbeuten v.t. exploit
Ausbeutung f exploitation
ausbilden v.t. design
Ausbildung f (techn.) design
ausblasen v.t. (e. Schmelzofen:) blow out; (Dampf:) exhaust
ausbleichen v.t. (painting) fade
ausbohren v.t. drill
Ausbohrung f (operation:) drilling; (result:) drill hole
ausbreiten v.t. flatten, hammer; – v.r. spread [out], (phys.) propagate
Ausbreiteprobe f flattening test
Ausbringen n yield, output

ausbuchsen — ausholen

ausbuchsen v.t. (e. Lager:) line (out); (mit Weißmetall:) babbitt

ausdehnen v.t. (längs:) extend; (allseitig:) expand

Ausdehnung f extension; expansion; extent

Ausdehnungskoeffizient m coefficient of expansion

AUSDREH ~, (~futter, ~meißel, ~spindel, ~werkzeug) boring ~

ausdrehen v.t. bore

Ausdrückmaschine f (Kokerei) [mechanical] pusher

auseinandernehmen v.t. disassemble

Ausfahrt f (Autobahn:) exit point

Ausfall m (mach.) breakdown; failure

ausfällen v.t. precipitate

ausflicken v.t. (Ofenfutter:) reline, repair; (road building, welding) patch

ausfluchten v.t. align, level out

Ausfluchtung f alignment

AUSFLUSS ~, (~geschwindigkeit, ~öffnung, ~rohr, ~ventil) discharge ~

Ausfressung f (Ofenfutter:) scouring, erosion

ausfugen v.t. (building) joint

ausführen v.t. carry out, perform, do

Ausführung f ($\approx$ Durchführung:) execution; ($\approx$ Bauart:) design; ($\approx$ handwerkliche Ausführung:) workmanship; (e. Oberfläche:) finish

ausfüttern v.t. (Lager:) line; (Herd:) fettle

Ausgang m exit

Ausgangskabel n (tel., telegr.) outgoing cable

Ausgangskapazität f (electron.) output capacitance

Ausgangsleistung f (radio) power output

Ausgangsmaß n basic size

Ausgangsspannung f output voltage

Ausgarzeit f (steelmaking) killing period

ausgelaufener Block, (met.) bled ingot

ausgeleiert p.p. worn away, completely worn

ausgießen v.t. (Lager:) babbitt; (Fundament:) grout

Ausgleich m balance; compensation, correction

AUSGLEICH ~, (= DIFFERENTIAL ~) (~achse, ~gehäuse, ~getriebe, ~getrieberad, ~kegelrad, ~ritzel) differential ~

ausgleichen v.t. compensate; counterbalance; adjust; balance out; equalize; (met.) soak

Ausgleichgrube f (met.) soaking pit

AUSGLEICHS ~, (electr.) (~spule, ~wicklung) compensating ~

Ausgleichschicht f (road building) levelling course

Ausgleichsglühen n (met.) soaking

Ausgleichszylinder m (power press) dash pot

ausglühen v.t. soft-anneal, anneal; ($\approx$ normalglühen:) normalize

Auguß m (ladle) lip, nozzle

Aushalsemeißel m necking tool

aushalsen v.t. (mach.) neck

Aushärtegeschwindigkeit f (shell molding) cure rate

aushärten v.t. (shell molding) cure

Aushärteofen m (shell molding) curing oven

Aushärtung f (met.) precipitation hardening; age-hardening

Aushaustempel m punching die

ausholen v.t. (woodworking) route

ausixen *v.t. (welding)* double vee out
auskehlen *v.f.* groove; channel
auskesseln *v.t.* (e. Bohrung:) trepan
auskitten *v.t.* putty
auskleiden *v.t.* (Schmelzofen:) line
auskolken *v.i. (building)* underwash; *(metal cutting)* crater
Auskolkung *f (building)* scouring, underwashing, subsurface erosion; *(metal cutting)* crater
auskragen *v.i.* cantilever, project, overhang
auskreuzen *v.t. (welding)* vee out
Auskreuzwinkel *m (welding)* angle of bevel
auskuppeln *v.t.* declutch, uncouple; disconnect
ausladen *v.i.* (≈ auskragen) overhang
Ausladung *f* (baulich:) reach; (radial:) radius; *(press, shaper)* throat; *(tool)* overhang
Auslaß *m* outlet, discharge opening; exhaust
AUSLASS ~, (~hub, ~kanal, ~krümmer, ~leitung, ~schlitz, ~ventil) exhaust ~
Auslastung *f* utilization
Auslauf *m* (e. Gewindes:) run-out
auslaufen *v.i.* (Flüssigkeiten:) leak out; (Drehzahlen:) slow down
Auslaufknie *n (building)* outlet elbow
Auslaufrollgang *m (rolling mill)* runout table
Auslaufseite *f (rolling mill)* delivery side
Auslaufzeit *f* slow-down time
auslegen *v.t.* lay out; design
Ausleger *m (planer)* rail; *(radial)* arm; (e. Kranes:) boom, jib
AUSLEGER ~, (~bogen, ~brücke, ~kran, ~träger) cantilever ~

Auslegerbohrmaschine *f* radial drilling machine, radial
Auslegung *f* layout; design
auslenken *v.t.* deflect
Auslenkung *f* deflection
Auslesepaarung *f* sorting
auslöschen *v.t. (welding)* extinguish
Auslöschung *f (welding)* extinction
AUSLÖSE ~, (in Verbindung mit Schaltbewegungen:) (~bock, ~druck, ~genauigkeit, ~mechanik, ~nocken, ~stange, ~vorgang) tripping ~, release ~
auslösen *v.t.* release; uncouple; trip
Auslösung *f* release; tripping; disengagement; *(electr.)* circuit opening
ausmauern *v.t.* wall up, brick up, lay bricks
Ausmauerung *f* brick lining
ausmessen *v.t.* measure
Auspuff *m (auto.)* exhaust
AUSPUFF ~, *(auto.)* (~blende, ~dichtung, ~gase, ~geräusch, ~hub, ~kanal, ~krümmer, ~leitung, ~rohr, ~schalldämpfer, ~topf) exhaust ~
Auspufftopf *m* (motorcycle) silencer
ausrasten *v.t.* disengage
ausreiben *v.t. (mach.)* ream
ausrichten *v.t.* straighten; level; adjust; align
ausrücken *v.t.* disconnect, disengage; (e. Kupplung:) unclutch; (Getriebe:) throw out of mesh
ausrüsten *v.t.* equip; *(building)* remove the form
Ausrüstung *f* equipment; outfit
ausschachten *v.t.* excavate
Ausschachtung *f* excavating
ausschalen *v.t. (building)* strip
ausschalten *v.t. (electr.)* open the circuit; (Schaltgeräte: switch off; (Hebel:)

Ausschalter 43 **Ausstrahlung**

disengage; (Kupplung:) unclutch; (Motor:) stop
Ausschalter *m* circuit-breaker, cutout; trip
ausscheiden *v.t.* separate; settle out
Ausscheidungshärtung *f (met.)* precipitation hardening
Ausschlag *m* (e. Zeigers:) deflection
Ausschmelzverfahren *n (founding)* lost wax molding, investment molding
ausschneiden *v.t.* cut out; (≈ ausstanzen:) blank; (Holz:) recess
Ausschnitt *m (math.)* sector
Ausschuß *m* scrap; waste
Ausschußlehrdorn *m* 'not go' screw plug gage
Ausschußlehre *f* not go' gage
ausschütten *v.t. (concrete)* dump
Ausschüttwaage *f* sack-shaking balance
ausschwenken *v.t.* swing out (sideways)
ausseigern *v.i. (met.)* segregate
Ausseigerung *f (met.)* segregation
außen *adv.* outside, outward, external
AUSSEN ~, (als Bestimmungswort von Verben) (~drehen, ~einstechen, ~kopieren, ~läppen, ~schleifen) ~ externally
AUSSEN ~, (als Bestimmungswort von Zerspanungsvorgängen und -maschinen) external ~; (als Bestimmungswort meßtechnischer Begriffe) outside ~
Außenantenne *f (US)* outdoor antenna; *(UK)* outdoor aerial
Außenarbeit *f* outdoor work
aussenden *v.t. (radio)* transmit
Außengewinde *n* male thread, external thread
Außengewindeschneiden *n* screwcutting
aussenken *v.t.* countersink; counterbore
Außenlack *m* exterior varnish

AUSSENRÄUM ~, (~maschine, ~nadel, ~vorrichtung, ~werkzeug) surface broaching ~
außenräumen *v.t.* surface-broach
Außenrundläppmaschine *f* external cylindrical lapping machine
Außenrundräummaschine *f* circular surface broaching machine
Außentaster *m* outside caliper
außermittig *a.* eccentric
Außermittigkeit *f* eccentricity
aussetzen *v.t.* stop; interrupt; expose; – *v.i.* fail
aussetzend *p.a.* intermittent
aussieben *v.t. (radio)* filter [out]
ausspannen *v.t.* unclamp; (Werkstücke:) unload
aussparen *v.t.* recess
Aussparung *f* recess
ausspülen *v.t.* rinse, wash out, flush; scavenge
ausstatten *v.t.* equip
Ausstattung *f* (e. Maschine:) equipment, attachment
ausstechen *v.t. (mach.)* recess
Ausstechmeißel *m* recessing tool
aussteifen *v.t.* (e. Schalung:) stiffen
Aussteuerung *f (radio)* level control
Ausstoß *m* output
ausstoßen *v.t. (mech.)* eject; *(opt., phys.)* emit; *(coking)* push out
Ausstoßer *m (power press)* ejector
Ausstoßmaschine *f (Kokerei)* coke pusher ram
Ausstoßseite *f* (e. Koksofen:) discharge end, ram side
ausstrahlen *v.t.* radiate; emit
Ausstrahlung *f* radiation

Aussuchpaarung f selective assembly
austauschbar a. interchangeable, replaceable
Austauschbarkeit f interchangeability; replacebility
Austauschbau m interchangeable assembly
austauschen v.t. interchange; exchange; replace
Austauschmotor m exchange engine
Austauschwerkstoff m substitute material
austenitischer Stahl, austenitic steel
Austragband n discharge conveyor
Austragungsschurre f (concrete) pouring chute
Austritt m (electron.) exit
Austrittseite f delivery end; (Walzgerüst:) catcher's side
Austrittsgeschwindigkeit f discharge velocity; (Walzgut:) delivery speed
Austrittswinkel m (opt.) angle of emergence
austrocknen v.t. dry; (Holz:) season
ausvauen v.t. (welding) vee out
auswaschen v.t. (building) scour, underwash
auswechseln v.t. replace, exchange; renew
auswerfen v.t. (Werkstücke:) eject
Auswerfer m ejector
auswuchten v.t. balance
Auswurf m discharge, ejection; (e. Konverters:) slopping
auszeichnen v.t. mark out
ausziehbar a. telescopic
ausziehen v.t. pull out, extract, withdraw; (Kurven:) trace
Ausziehleiter f extending ladder

Auto n automobile, motorcar, passenger car
Autobagger m tired mechanical shovel
Autobahn f highway, turnpike
Autobus m bus, motor-coach
Autobusbahnhof m bus terminal
Autoempfänger m auto radio
Autofahrer m motorist
Autogengerät n (welding) gas welding equipment
Autogenschweißen f autogenous welding, gas welding, oxyacetylene welding
Autogen-Technik f gas welding technology
Autoheber m car jack
Autokarte f road map
Automat m (mach.) automatic machine, automatic lathe, automatic; (electr.) automatic cut-out
Automatendreher m automatic lathe operator
Automatendreherei f automatic turning shop
Automatenlegierung f free-cutting alloy
Automatenmessing n free-cutting brass
Automatenstahl m automatic screw steel, free-cutting steel
Automatik f automatic mechanism; automaticity
automatisieren v.t. automate, automatize
Automatisierung f automation
Automobil n automobile, motor-car
Automobilgetriebe n automobile transmission
Automobilindustrie f automotive industry
Automobilreparaturwerkstatt f motorcar repair shop
Automobilreparaturwerkzeug n motor tool

Automobilschraubenschlüssel *m* automotive wrench
Automobilwinde *f* motorcoach jack
Autoöl *n* motor oil
Autoradio *n* automobile radio equipment
Autoreifen *m (US)* tire; *(UK)* tyre
Autoschlosser *m* garageman
Autostraße *f* motor road
Autoverkehr *m* motor traffic
Autowinde *f* motorcar jack
AXIAL ~, (~kolbenmotor, ~kolbenpumpe, ~kraft, ~lager, ~last, ~luft, ~schnitt, ~schub, ~verschiebung, ~winkel) axial ~

Axialbelastung *f* thrust
Axialkegelrollenlager *n* taper roller thrust bearing
Axialpendelrollenlager *n* self-aligning roller thrust bearing
Axialschrägkugellager *n* angular contact thrust ball bearing
Axialzylinderrollenlager *n* cylinder roller thrust bearing
Azeton *n cf.* Aceton
Azetylen *n cf.* Acetylen
Azetylenlampe *f* acetylene torch

B

Backe f (≈ Schneidbacke:) die; (≈ Spannbacke:) jaw; (e. Bremse, e. Setzstockes:) shoe
Backenbrecher m jaw crusher
Backenfutter n jaw chuck
Backkohle f caking coal
Backstein m solid brick
Badnitrieren n obs. cf. Salzbadnitrieren
Bagger m (naß:) dredger; (trocken:) excavator
Baggerkran m shovel crane
Baggerlöffel m bucket, shovel
baggern v.t. dredge; excavate
Baggerung f (road building) earth excavating
Bahn f track, path; (Amboß, Hammer:) face (= Führungsbahn) way; (ball.) orbit; railway, railroad; (plastics) length; (Papier:) strip
BAHN ~, (~damm, ~fracht, ~gelände, ~netz, ~versand) railway ~, railroad ~
Bahnkörper m track bed
Bahnlinie f (e. Gesenkes:) flow line
Bahnwärter m lineman
BAJONETT ~, (~fassung, ~scheibe, ~verschluß) bayonet ~
Bake f (radio) beacon
Balken m beam, girder
BALKEN ~, (building) (~biegepresse, ~lehrgerüst, ~rüttler, ~trägerbrücke, ~waage) beam ~
Balkenherdofen m walking beam furnace
Ballenpresse f baling press
ballig a. crowned, convex
BALLIG ~, (~dreheinrichtung, ~fräseinrichtung, ~schleifvorrichtung, ~verzahnen) convex ~
Balligkeit f convexity, crowning; (Walze:) camber
Ballonreifen m (auto.) balloon tire (or tyre)
Bananenstecker m (radio) banana pin
Band n 1. (acous., magn., plastics) tape; 2. (conveying) belt; 3. (grinding) abrasive belt; 4. (met.) strip; 5. (radio, auto.) band
BAND ~, 1. (~ableser, ~antenne, ~aufnahmegerät, ~speicher, ~steuerung, ~system) tape; 2. (~förderer) conveyor; 3. (~schleifen, ~schleifmaschine, ~spannung) belt; 4. (~stahl, ~walzwerk) strip; 5. (~breite, ~filter) band~; s. a. 1–5
Bandage f (auto.) tire; (UK) tyre
BANDAGEN ~, (~bohr- und drehbank, ~presse, ~walzwerk) tire ~; (UK) tyre ~
Bandbremse f band brake
Bandeisen n hoop iron; band iron
Bandkupplung f rim clutch
Bandmaß n tape rule
Bandmikrofon n ribbon microphone
Bandstahl m strip steel
Bandwalzen n continuous rolling
Banjoachse f banjo-axle
Bank f (≈ Drehbank:) lathe; (≈ Hobelbank:) planer; (≈ Werkbank:) bench
BANK ~, (~amboß, ~arbeit, ~hammer, ~schere, ~schraubstock) bench ~
Bankett n (auto.) bench, berm
Bär m (techn.) (civ.eng.) tup; (forging) ram; (met.) skull
Barren m bar, ingot
Basalt m basalt

basisch *a.* basic
basische Zustellung, basic lining
Batterie *f (auto.)* battery
BATTERIE ~, (~empfänger, ~gehäuse, ~gerät, ~gestell, ~kasten, ~klemme, ~ladegerät, ~schlamm, ~spannung, ~zündschalter) battery ~
Batteriesäureprüfer *m (auto.)* battery tester
Bau *m* design; construction; manufacture; building, building work
BAU ~, (~arbeit, ~aufseher, ~behörde, ~geräte, ~gewerbe, ~grube, ~material, ~mörtel, ~platz, ~polizei, ~rundholz, ~sand, ~stein, ~stelle, ~techniker, ~terrain, ~unternehmer, ~wesen) building ~
Bauart *f* design; type; pattern
Baubeschlag *m* ironmongery
Baublech *n* structural steel sheet
Baubreite *f* overall width
Baubüro *n* site office
Baudrehkran *m* slewing construction crane
Baueinheit *f* basic construction unit
Bauelement *n* construction element
bauen *v.t.* build; design; construct
Bauer *m (techn.)* builder
baufällig *a.* dilapidated, decaying
Baugerüst *n* scaffold
Baugewerbewerkzeug *n* contractors' tool
Bauglas *n* glass for building purposes
Baugröße *f* size
Baugrube *f* building pit
Baugrubensohle *f* foundation pit base
Baugrund *m* foundation soil, subsoil
Baugruppe *f (mach.)* basic unit assembly group

Bauhandwerker *m* building trade worker
Bauherr *m* proprietor
Bauhof *m* contractors' yard
Bauhöhe *f* headroom; height overall
Bauholz *n* structural timber
Bauindustrie *f* contracting industry
Bauingenieur *m* civil engineer
Baukasteneinheit *f (e. Maschine:)* basic machine unit
Baukastensystem *n* modular design
Baukunde *f* architecture
Baulänge *f* length overall
Bauleiter *m* site engineer
Baumschinen *fpl.* construction machinery, contractor's machinery
Baumaschinenindustrie *f* construction machinery industry
Baumaschineningenieur *m* construction machinery engineer
Baumaß *n* constructional dimension
Baumeister *m* master builder
Baumuster *n* model, type
BAUMWOLL ~, (~dichtung, ~faser, ~garn, ~riemen, ~seil) cotton ~
Baumwollgewebe *n* cotton duck, canvas
Baumwollhartgewebe *n* laminated fabric of cotton
Bauplanung *f* project planning
Baupumpe *f* contractors' pump
Bauschnittholz *n* sawn building timber
Baustahl *m* engineering steel; machinery steel; structural steel
Baustahlgewebe *n* building steel lathing; steel fabric
Baustellenbüro *n* site office
Baustoff *m* building material; *(mach.)* (≈ Werkstoff:) engineering material
Bautechnik *f* construction engineering
bautechnisch *a.* architectonic

Bauteil n (civ. eng.) building component; (mach.) structural part
Bauweise f design; construction
Bauwerk n structure, construction; building
Bauwerksbeton m structural concrete
Bauwerkzeug n construction tool
Bauwesen n civil engineering
Bauwinde f hand power winch
Bauwirtschaft f contracting industry
Bauzaun m building fence
Bauzeichnung f construction drawing
beanspruchen v.t. stress; strain; load
Beanspruchung f stress; strain; load
beanstanden v.t. complain; object
Beanstandung f complaint; rejection; objection
Bearbeitbarkeit f (spanlos:) workability; (zerspanend:) machinability
bearbeiten v.t. treat; process; work; machine
Bearbeitung f (1. (spanlos:) working; 2. zerspanend:) machining; 3. (Oberflächen:) treatment
BEARBEITUNGS ~, (~bedingung, ~beispiel, ~eigenschaft, ~genauigkeit, ~güte, ~kosten, ~länge, ~plan, ~technik, ~verfahren, ~verhalten, ~vorgang, werkzeug, ~zeit, ~zugabe) 1. working ~; 2. machining ~
Bearbeitungsriefe f tool mark
beaufsichtigen v.t. supervise, control, inspect
Becherwerk n bucket elevator
Bedachung f roofing
Bedachungsmaterial n roofing material
bedienen v.t. (e. Maschine:) operate; (Schaltgeräte:) control
Bedienung f (von Maschinen:) operation; (von Schaltorganen:) control

BEDIENUNGS ~, (~anleitung, ~anweisung, ~bühne, ~griff, ~hebel, ~organ, ~vorschrift) operating ~
Bedienungsschild n instruction plate
Bedienungsschlüssel m (US) wrench; (UK) spanner
Bedienungtafel f (e. Werkzeugmaschine:) control panel; (als Hängetafel:) control pendant
Bedienungszeit f (e. Maschine:) machine-handling time
Bedienungszentrale f (e. Werkzeugmaschine:) central control panel
beeinträchtigen v.t. impair, impede
befestigen v.t. fasten; clamp; fix; bolt; secure; mount
Befestigung f fastening; clamping; mounting
Befestigungsschraube f clamping bolt; fastening bolt (or screw)
begehbar a. traversable
begichten v.t. (met.) burden
Begichtung f (met.) charging, burdening
BEGICHTUNGS ~, (~anlage, ~bühne, ~kübel, ~wagen) charging ~
Begleitelement n accompanying element
begradigen v.t. straighten
begrenzen v.t. limit; bound
Begrenzungsleuchte f (auto.) width indicator lamp
Begrenzungslinie f boundary line
Behälter m container, receptacle; tank
behandeln v.t. treat
Behandlung f treatment
Beharrungsvermögen n inertia
Beheizung f firing, heating
Behelfsausrüstung f improvised equipment
Behelfsbau m temporary structure

Behelfskonstruktion *f* makeshift construction
behelfsmäßig *a.* provisional
behobeln *v.t.* plane
beidseitiges Schweißen, all-position welding
Beifahrer *m (auto.)* assistant driver; *(motorcycle)* sidecar rider
Beil *n* hatchet; (zweihändiges:) axe
Beilegering *m (milling)* spacing collar
beistellen *v.t. (building construction)* supply; *(mach.)* set, adjust
Beistellung *f* adjustment; *(grinding)* infeed
Beitel *m* chisel
Beiwagen *m (motorcycle)* sidecar
BEIZ ~, (~bad, ~fehler, ~korb, ~lösung, ~riß) pickling ~
Beizbrüchigkeit *f* acid brittleness, hydrogen embrittleness
Beize *f* pickling solution
beizen *v.t. (met.)* pickle; (Farbe, Holz:) stain
Beiznarbe *f (surface finish)* pit
Beizzusatz *m* inhibitor
bekleiden *v.t. (civ. eng.)* panel
Bekleidungsblech *n* panelling sheet
Belag *m* coat; lining; *(road building)* covering
belasten *v.t. (mech.)* load
Belastung *f* loading; *(electr., mech.)* load; *(phys.)* strain
BELASTUNGS ~, (~änderung, ~ausgleich, ~diagramm, ~fähigkeit, ~kurve, ~prüfung, ~schwankung, ~spitze, ~verteilung) load ~
belegen *v.t.* (Oberflächen) coat; face; (≈ bestücken:) tip
beleuchten *v.t.* illuminate
Beleuchtung *f* lighting set

BELEUCHTUNGS ~, (~anlage, ~einrichtung, ~stärke, ~technik) lighting ~
Beleuchtungskörper *m* illuminator
Beleuchtungsmesser *m* lux-meter, photometer
belichten *v.t.* expose
Belichtung *f (photo.)* exposure
Belichtungsmesser *m* exposure meter
Belichtungszeit *f* exposure time
belüften *v.t.* ventilate
belüfteter Beton, air-entrained concrete
Belüftung *f* ventilation; aeration
Belüftungsanlage *f (auto.)* ventilating equipment
Belüftungsmittel *n (concrete)* air entraining compound
Belüftungsschlauch *m (auto.)* ventilating hose
bemaßen *v.t.* dimension
bemessen *v.i.* dimension; proportion; (Leistungen:) rate
benetzen *v.t. (concrete)* wet, moisten
Benzin *n* benzine; *(US)* gasoline; *(UK)* petrol
BENZIN ~, (~behälter, ~einspritzmotor, ~einspritzpumpe, ~filter, ~förderpumpe, ~kanister, ~leitung, ~lötlampe, ~motor, ~pumpe, ~standanzeiger, ~tank, ~zapfstelle) gasoline ~, petrol ~
Benzol *n* benzene; (als Handelserzeugnis:) benzol
BENZOL ~, (~gewinnung, ~wäsche, ~wäscher) benzol ~
beratender Ingenieur, consulting engineer
Beratungsdienst *m* advisory service
Berechnungsfestigkeit *f* calculated strength
Bereich *m* range; region; reach; sphere

Bereifung f *(auto.)* tires; *(UK)* tyres
Bereitung f *(concrete)* preparation
Bergbau m mining
bergen v.t. salvage
Bergkiefer f mountain pine
Bergwerk n mine
Bericht m report
berieseln v.t. spray, wash, sprinkle; *(met.)* scrub
Berieselungsturm m (= Wäscher) scrubber
Bernsteinlack m amber varnish
Berufsausbildung f vocational training
Berufsgenossenschaft f employers' liability insurance company
beruhigen v.t. (Schmelzbad:) kill, quiet
beruhigter Stahl, killed steel
Beruhigungszuschlag m *(met.)* addition for quieting steel
Berührung f contact; touch
BERÜHRUNGS ~, (~elektrizität, ~fläche, ~ebene, ~korrosion, ~linie, ~spannung) contact
besäumen v.t. trim, edge; (Holz:) square; (Walzgut:) shear
Besäumschere f trimming shear
beschädigen v.t. damage; spoil (Oberflächen:) mar
Beschaffenheit f (äußere:) finish, appearance; (innere:) condition, quality
beschaufeln v.t. *(turbine)* blade
beschicken v.t. charge; feed; *(blast furnace)* burden
Beschickung f charging; charge
BESCHICKUNGS ~, (~bühne, ~mulde, ~seite, ~tür, ~wagen) charging ~
Beschlag m mounting, fitting
beschleunigen v.t. accelarate
Beschleuniger m accelerator

Beschleunigungsfußhebel m *(auto.)* accelerator pedal
Beschleunigung f acceleration
Beschleunigungsmesser m accelerometer
Beschleunigungsvermögen n *(auto.)* accelerating power
Beschneidemaschine f trimming machine; squaring machine
beschneiden v.t. trim; shear; square
beseitigen v.t. eliminate; remove; remedy
Beseitigung f removal; elimination
BESETZT ~, (~lampe, ~zeichen) busy ~
BESSEMER ~, (~betrieb, ~birne, ~roheisen, ~schmelze, ~stahl, ~verfahren) Bessemer ~
Bestand m *(cost accounting)* inventory; (≈ Vorrat:) stock
beständig a. resistant, immune (to), insusceptible; *(phys.)* stable
Beständigkeit f stability; immunity (to), resistance (to)
Bestandsaufnahme f stock-taking
Bestandsplan m inventorial plan
Bestandteil m ingredient; substance; *(metallo.)* constituent
bestimmen v.t. *(chem.)* determine; (e. Lage:) locate
bestoßen v.t. trim; edge; square
bestrahlen v.t. *(radiogr.)* expose to rays
Bestrahlung f irradiation; (Sand:) blasting
bestücken v.t. (Maschine mit Werkzeugen:) tool; (Schneidwerkzeuge:) tip; hard-face
Bestückung f *(Hartmetall:)* tipping
betätigen v.t. manipulate; operate; actuate; *(electr.)* energize
Betätigungsschalter m *actuating switch*
Beton m concrete

BETON ~, (~balken, ~bau, ~bauwerk, ~brecher, ~decke, ~einpreßmaschine, ~fundament, ~mischer, ~mischung, ~mörtel, ~pfeiler, ~platte, ~presse, ~rüttelgerät, ~rüttelstampfer, ~spritzmaschine, ~stahlmatte, ~stampfer, ~straße, ~straßenbau, ~verdichtung, ~werkstein) concrete ~

Betoneisenschneider *m* reinforcement bar cutter

Betonformstahl *m* deformed reinforcing steel

Betonieranlage *f* concreting plant

betonieren *v.t.* cast concrete

Betonstahl *m* reinforcing steel bars

Betrieb *m* plant, factory; workshop; shop; operation; practice; working

BETRIEBS ~, (~anweisungen, ~bedingungen, ~daten, ~drehzahl, ~einrichtung, ~handbuch, ~kosten, ~leistung, ~sicherheit, ~störung, ~verhältnisse, ~vorschrift) operating ~

Betriebsanlage *f* plant, installation

betriebsbereit *a.* ready for operation (*or* service)

betriebsfertig *a* ready for operation

Betriebsingenieur *m* production engineer

Betriebskostenstelle *f* non-administrative cost center

Betriebsleiter *m* works superintendent

Betriebsmann *m* production man

betriebsmäßig *a.* under normal service conditions

Betriebsmittel *n* production facility

betriebssicher *a.* reliable in service

Betriebsspannung *f* working voltage, line voltage

Betriebsstoff *m* (*auto.*) fuel

Betriebsstromkreis *m* working circuit, service circuit

Betriebstechnik *f* production engineering

Betriebsunfall *m* shop accident

Betriebsunterbrechung *f* service interruption

Betriebswerkstatt *f* engineering workshop

Betriebswirt *m* applied economics engineer

Betriebswirtschaft *f* applied economics

Betriebswirtschaftlichkeit *f* production economy, operational economics

Bett *n* (e. Maschine:) bed

BETT ~, (~kröpfung, ~mulde, ~neigung, ~schmierung, ~verlängerung, ~verwindung, ~wanne) bed ~

Bettführung *f* bedways, guideways

Bettkoks *m* (*founding*) bed coke

Bettprisma *n* (*lathe, planer*) V-way

Bettschlitten *m* (*lathe*) carriage; (*miller*) saddle

Bettschlittendrehmaschine *f* (*US*) saddle-type turret lathe; (*UK*) combination turret lathe

Bettung *f* (*building*) bedding

beugen *v.t.* (*opt.*) diffract

Beugung *f* (*opt.*) diffraction

bewährt *p.a.* approved

Bewässerung *f* irrigation

beweglich *a.* (waagerecht:) movable; (senkrecht:) floating; (allseitig:) flexible

Beweglichkeit *f* movability

Bewegung *f* movement, motion; (Tisch:) traverse

Bewegungsanalyse *f* motion analysis

Bewegungsenergie *f* kinetic energy

Bewegungslehre *f* kinematics

Bewegungssitz *m* running fit

Bewegungsstudie f motion study
Bewegungsumkehr f reversal of travel
bewehren v.t. reinforce, armor
bewehrtes Kabel, armored cable, sheathed cable
Bewehrung f reinforcement, armoring; *(UK)* armouring
BEWEHRUNGS ~, (~matte, ~stab, ~stahl) reinforcing ~
Bewertung f assessment; rating
Bewitterungsversuch m weathering test
Bezirksknotenamt n district tandem office
Bezirksnetz n (tel.) tandem toll circuit
bezogene Formänderung, *(forging)* relative deformation
bezogene Größe, referenced quantity
Bezug m reference
BEZUGS ~, (~ebene, ~frequenz, ~kante, ~profil, ~punkt, ~spannung, ~temperatur) reference ~
Bezugsfeile f flexible file
Bezugsformstück n *(copying)* master component
BIEGE ~, (~beanspruchung, ~fähigkeit, ~festigkeit, ~maschine, ~probe, ~spannung, ~versuch, ~zange) bending ~
Biegeermüdungsversuch m transverse fatigue test
Biegegesenk n snaker
Biegemaschine f (Feinblech:) folding machine; (Grobblech:) bending machine
biegen v.f. bend; deflect
Biegepresse f bending press; bulldozer; *(US)* press brake
Biegeschlagversuch m bending impact test

Biegeschwellfestigkeit f pulsating bending fatigue strength
Biegesteife f resistance to deflection
Biegewechselfestigkeit f bending fatigue strength, endurance limit in reversed bending
Biegsamkeit f ductility, pliability, flexibility
Biegung f bend, bending; deflection; (Querbiegung:) flexure
Bieter m tendering firm
Bilanz f balance
Bild n photo; illustration; *(opt., radio, radar)* image, picture, map; *(surface finish)* pattern
BILD ~, (~ebene, ~feld, ~fläche, ~schärfe) image ~
Bildabtastung f scanning
Bildfrequenz f picture frequency
Bildfunk m photoradio
bildlich a. graphic(al)
Bildröhre f picture tube
bildsam a. plastic; ductile
Bildschirm m (telev.) viewing screen, picture screen
Bildsignal n picture signal
Bildtelegraphie f telephotography
Bildwerfer m picture projector
Bimetallthermometer n bimetallic thermometer
Bims m pumice
Bimsbeton m pumice concrete
Bimsstein m pumice
binäre Legierung, binary alloy
BINDE ~, (~draht, ~energie, ~kette, ~mittel, ~vermögen) binding ~
Bindekraft f cohesion
binden v.f. bind, bond; tie up; unite
Binder m *(painting)* binder
Binderfarbe f emulsion paint

Binderschicht f *(road building)* base course; binder course
Binderspachtel m binder filler
Bindezeit f (Zement:) setting time
bindiger Boden, cohesive soil
Bindung f bond, binding
Birne f *(electr.)* bulb; (met.) converter
Birnenbetrieb m converter practice
Birnenprozeß m *(met.)* converting process, air refining process
Bitterkalk m magnesian limestone
Bitumenanstrich m bitumen coat
Bitumenauskleidung f bitumen lining
Bitumenfarbe f bituminous paint
Bitumenlack m bituminous varnish
Bitumen-Straßenbau m asphalt paving
Bitumensplitt m asphalt-coated chips
bituminös a. bituminous
blank a. (Oberfläche:) bright; (Draht:) bare
Blankdraht m bare wire
blankgeglüht p.p. bright annealed
blankgewalzt p.p. bright cold rolled
blankgezogen p.p. bright drawn
blankglühen v.t. bright anneal
Blankglühung f bright anneal
Blanklack m clear varnish
blankschleifen v. grind bright
Blasdüse f *(converter)* blowing nozzle, injector nozzle; (Sand:) blast nozzle
Blase f bubble; (im Guß:) blowhole; (Stahloberfläche:) blister
Blasebalg m bellows
Blasedauer f *(met.)* time of blowing
Blasemeister m *(steelmaking)* blower
blasen v.t. *(steelmaking)* blow; (Preßluft:) blast
blasen mit Heißwind, blow hot
blasen mit kaltem Wind, blow cold

Blasenbildung f *(painting)* pimpling
Blasenstahl m blister steel
Blaslanze f *(steelmaking)* jetting lance
Blasstrahl m *(steelmaking)* jet
Blatt n (Axt:) body; (Bandsäge:) band; (Kreissäge:) web; (= Klinge:) blade; (Feder:) leaf, plate
Blattfeder f leaf spring, laminated spring
Blattlehre f screw pitch gage
Blaubruch m blue-brittleness
blaubrüchig a. blue-brittle
Blaubrüchigkeit f blue-brittleness
bläuen v.t. blue
Blauglühen n blue annealing
Blaupause f blueprint
Blausprödigkeit f cf. Blaubrüchigkeit
Blech n 1. (als Werkstoff, ohne Bezug auf die Blechstärke:) sheet metal; 2. (als Feinblecherzeugnis:) sheet; 3. (als Grobblecherzeugnis:) plate
Blech ~, 1. (~bearbeitung, ~bearbeitungsmaschine, ~kasten, ~käfig, ~lehre, ~packung, ~preßteil, ~rohr, ~schutz, ~teile, ~trommel, ~verarbeitende Industrie, ~walzwerk, ~waren, ~wärmofen, ~wanne) sheet metal ~; 2. (~bördelmaschine, ~doppler, ~gerüst, ~messing, ~paket, ~richtmaschine, ~straße, ~streifen, ~sturz, ~sturzenwärmofen, ~walze, ~walzwerk) sheet~; 3. (~bördelmaschine, ~biegemaschine, ~kröpfmaschine, ~richtmaschine, ~straße, ~träger, ~walze, ~walzwerk) plate ~
Blechausschnitt m blank
Blechduo n two-high plate mill
Blecheinlage f shim
Blechfalzmaschine f seaming machine
Blechgerüst n *(rolling)* plate mill

Blechhalter *m* blankholder
Blechhalterrahmen *m (power press)* die shoe
Blechhalterstößel *m (power press)* blankholder slide
Blechkantenhobelmaschine *f* plate-edge planer
Blechschere *f* tinners' snip
Blechschraube *f* self-tapping screw
Blech-, Stab- und Formeisenschere *f* combination slitting shears and bar cutter
Blechstärke *f* gage of a sheet (*or* plate)
Blechstreifen *m (Rohrherstellung)* skelp
Blechtafel *f* plate; sheet; panel
Blechtrio *n* three-high plate mill
Blechverkleidung *f* sheeting
Blei *n* lead
BLEI ~, (~akkumulator, ~auskleidung, ~bad, ~batterie, ~farbe, ~härtung, ~kabel, ~legierung, ~mantel, ~plombe, ~raffinerie, ~sammler, ~vergiftung, ~verglasung) lead ~
bleibende Dehnung, (= Bruchdehnung) elongation
bleibende Durchbiegung, permanent set
Bleiblech *n* sheet lead
Bleiglätte *f* litharge
bleihärten *v.t.* (Draht:) patent
Bleimennige *f* red lead
Bleischachtofen *m* lead blast-furnace
Bleiweiß *n* white lead
Blende *f (opt.)* diaphragm, aperture; *(building)* screen; (e. Auspuffes:) deflector
blenden *v.t.* dazzle, glare
Blendschutzscheibe *f (auto.)* anti-dazzle screen
Blindboden *m (building)* counterfloor
Blindflansch *m* blind flange, blank flange

Blindleistung *f (electr.)* wattless power
Blindleistungszähler *m* idle current wattmeter
Blindleitwert *m* susceptance
Blindstich *m (rolling mill)* blind pass, dummy pass
Blindstrom *m* reactive current, wattless current
Blindverbrauchszähler *m* reactive volt-ampere-hour meter; var-hour meter
Blindversuch *m* blank test
Blindwalze *f (rolling mill)* dummy roll, idle roll
Blindwiderstand *m (electr.)* reactance
Blinker *m (auto.)* flasher, flashing trafficator
Blinkerschalter *m (auto.)* flasher switch, trafficator switch
Blinkgeber *m (auto.)* flasher unit
Blinkleuchte *f (auto.)* flasher lamp
Blinklicht *n* flashing light
Blinklichtglühlampe *f* flasher bulb
Blinkmotor *m (auto.)* flasher motor
Blinkschlußleuchte *f (auto.)* flasher tail lamp
Blinksignal *n* flashlight signal
Blitzableiter *m* lightning conductor, lightning rod
blitzen *v.i.* flash
Blitzgespräch *n* lightning call
Blitzlicht *n (photo.)* flashlight, flash-bulb
Blitz-Rohrzange *f* grip wrench
Blitzschutz *m* lightning protection
Blitzschutzanlage *f* lightning protective system, lightning arrester equipment
Blitzschutzautomat *m* lightning arrester
Block *m (building)* block; *(met.)* ingot; *(gearing)* cluster
BLOCK ~, *(met.)* (~abstreifer, ~dreh-

bank, ~drücker, ~einsetzkran, ~einsetzmaschine, ~erzeugung, ~hebetisch, ~hobelmaschine, ~kran, ~kupfer, ~wagen, ~wärmofen, ~zange) ingot ~

Blockaufsatz *m* hot top, feeder head

Blockbauweise *f* block building system

blocken *v.t. (met.) (US)* bloom; *(UK)* cog

Blockflämmen *n* deseaming

Blockform *f (met.)* ingot mold

Blockgerüst *n (rolling mill)* blooming mill stand; *(UK)* cogging stand

blockieren *v.t. & v.i.* block, obstruct; lock; interlock

Blockkaliber *n* blooming pass; *(UK)* cogging pass

Blockkopf *m (met.)* top end of an ingot **verlorener** ~, crop end

Blocknickel *n* pig nickel

Blockrad *n (gearing)* cluster gear

Blockreversierduo *n* two-high reversing blooming mill

Blockschaltbild *n* block diagram

Blockschaltung *f (electr.)* single-unit circuit

Blockschere *f* bloom shears

Blockseigerung *f* ingotism

Blockstraße *f* blooming mill train; *(UK)* cogging mill train

Blocktrio *n* three-high blooming mill

Blockwalze *(US)* blooming roll; *(UK)* cogging roll

Blockwalzen *n (US)* blooming; *(UK)* cogging

Blockwalzgerüst *n (UK)* blooming mill stand; *(UK)* cogging mill stand

Blockwalzwerk *n (US)* blooming mill; *(UK)* cogging mill

Blockwender *m (rolling mill)* manipulator, ingot tilter

Bluten *n (painting)* bleeding

Bock *m* (≈ Stützbock:) pedestal, stand, trestle; (≈ Hebebock:) jack; (≈ Auflagebock:) rest, support

Bockschere *f* bench shear

Bockwinde *f* hoisting crab

Boden *m* ground; floor; soil; bottom; base; (e. Kessels:) end

Bodenabstand *m (auto.)* road clearance

Bodenbelag *m* floor covering

Bodenkorrosion *f* soil corrosion

Bodenmechanik *f* soil mechanics

Bodenplatte *f* baseplate, bottom plate

Bodenrad *n (lathe)* main drive gear

Bodensatz *m* sediment

Bodenschwelle *f (building)* sill

Bodenwindkonverter *m* bottom blown converter

Bogen *m (electr., math.)* arc; *(building)* arch; (≈ Kurve:) curve; (Rohre:) bend

BOGEN ~, (~entladung, ~lampe, ~licht) arc ~

Bogenbrücke *f* arch(ed) bridge

bogenförmig *a.* arched, arc-shaped

Bogenmaß *n* circular measure

Bogensäge *f* coping saw

Bogenverband *m (building)* arch bond

Bogenverzahnung *f* spiral tooth system

Bogenzirkel *m* wing divider

Bohle *f* plank

BOHR ~, 1. (Vollbohrarbeit:) (~arbeit, ~bereich, ~bild, ~buchse, ~druck, ~einheit, ~einrichtung, ~elektrode, ~futter, ~kopf, ~leistung, ~loch, ~maschine, ~motor, ~schlitten, ~spindel, ~stellung, ~support, ~tisch, ~vorrichtung, ~werkzeug) drilling ~, drill~; 2. (Aufbohrarbeit:) (~arbeit, ~bild, ~einheit, ~einrichtung, ~futter, ~hülse,

~kopf, ~meißel, ~meißelhalter, ~pinole, ~schlitten, ~spindel, ~stange, ~technik, ~tisch, ~vorrichtung, ~vorschub, ~werkzeug) boring ~
Bohrapparat *m* portable drill
Bohrautomat *m* automatic boring machine; automatic drilling machine
Bohrbank *f* boring mill
Bohremulsion *f* cutting solution
bohren *v.t.* 1. (≈ vollbohren:) drill; 2. (≈ aufbohren:) bore; 3. (e. Gewinde:) tap; 4. *(woodworking)* bore
Bohrer *m* (≈ Spiralbohrer:) twist drill; (≈ Gewindebohrer:) tap; (für Holz:) auger bit
Bohrerfutterkegel *m* drill socket
Bohrerhalter *m* drill chuck
Bohrerzapfen *m* drill tang
Bohrgrat *m* burr
Bohrknarre *f* ratchet brace
Bohrleier *f* brace
Bohrloch *n* drill hole; borehole, bore
Bohröl *n* soluble oil
Bohrspäne *mpl.* drillings; borings
Bohrtiefe *f* 1. depth of drill hole; 2. depth of borehole
Bohr- und Drehmaschine *f* vertical boring and turning mill
Bohrung *f* drill hole; borehole; hole
Bohrungsläppmaschine *f* internal lapping machine
Bohrungslehre *f* internal caliper gage
Bohrungstiefe *f* depth of hole
Bohrwasser *n* diluted soluble oil
Bohrwerk *n* boring mill
Bohrwinde *f* bit brace
Bolzen *m* bolt; pin; stud
Bolzenabschneider *m* bolt clipper
Bolzenautomat *m* automatic stub lathe

Bolzengewinde *n* screw thread, male thread
Bolzenschießgerät *n* bolt driving gun
Bolzenschweißverfahren *n* stud welding process
bombieren *v.t.* camber
Bombierung *f* camber
bondern *v.t.* bonderize
Bootslack *m* spar varnish
Bootsmotor *m* motorboat engine
Bord *n* (e. Lagers:) shoulder, lip
Bördelautomat *m* automatic flanging machine (*or* press)
Bördelhöhe *f* flange depth
Bördelmaschine *f* (für Grobbleche:) flanging machine; (für Feinbleche:) beading machine
bördeln *v.t.* (warm:) flange; (kalt:) bead; (mit Drahteinlage:) wire
Bördelnaht *f (welding)* double flanged seam
Bördelnietung *f* flanged seam riveting
Bördelpresse *f* flanging press
Bördelrohr *n* flanged pipe
Bördelstoß *m* double-flanged butt joint
Bördel-Stoßkante *f* abutting edge of flange
Borke *f* bast
Böschung *f* slope
Bowdenzugkabel *n (auto.)* Bowden control cable
Boxermotor *m (auto.)* engine with horizontally opposed cylinders
Brachzeit *f* downtime
Bramme *f (met.)* cast slab, slab ingot
BRAMMEN ~, (~schere, ~straße, ~tiefofen, ~walze, ~walzwerk) slabbing ~, slab ~
Brandmauer *f* fire wall

Brandriß *m* thermal crack, hot crack, heat crack, fire-crack
Brecheisen *n* crowbar
brechen *v.t. (mech.)* break; fracture; rupture; *(opt.)* refract; – *v.i.* break
Brecher *m* crusher, breaker
Brechsand *m* crushed stone sand
Brechungsmesser *m (opt.)* refractometer
Brechungsvermögen *n (opt.)* refractive power
Brechungszahl *f (opt.)* refractive index
Breitband *n* wide strip
Breitbandwalzwerk *n* wide strip mill
Breite *f* width
breiten *v.t. (forging)* flatten
Breitflachstahl *m* wide flat steel, universal steel
Breitflanschprofil *n* wide flange section
Breitflanschträger *m* H-beam, wide flanged beam
Breitfußschiene *f* flat-bottomed rail
Breitstrahlscheinwerfer *m (auto.)* broad beam headlamp
BREMS ~, (~, (~anlage, ~backe, ~band, ~belag, ~belastung, ~druckmesser, ~dynamometer, ~flüssigkeit, ~fußhebel, ~gestänge, ~hebel, ~klotz, ~kolben, ~kupplung, ~leitung, ~manschette, ~nabe, ~nachstellung, ~pferdestärke, ~regler, ~schuh, ~trommel, ~wächter) brake ~, braking ~
Bremsdrehmoment *n* braking torque
Bremse *f* brake
bremsen *v.t.* brake
Bremsfeld *n (radio)* retarding field
Bremsgitter *n (electron.)* suppressor grid
Bremsgitterspannung *f* suppressor grid voltage
Bremskegel *m* friction cone

Bremsleistung *f* brake horsepower
Bremsleuchte *f (auto.)* stop lamp
Bremslicht *n* stop light
Bremslichtschalter *m (auto.)* stoplamp switch
Bremslüfter *m* brake-lifting magnet
Bremslüftmagnet *m* magnetic brake
Bremsluftmanometer *n (auto.)* air brake pressure gage
Bremsmagnet *m* magnetic brake, braking magnet
Bremsmoment *n* braking torque
Bremsmotor *m* self-braking motor
Bremsöl *n (auto.)* brake fluid
Bremsring *m (auto.)* brake collar
Bremsscheibe *f* friction disc
Bremsschlauch *m (auto.)* flexible brake tubing
Bremsschlüssel *m (auto.)* brake expander mechanism
Bremsschluß-Kennzeichenleuchte *f* stop/tail/number plate lamp
Bremsschlußleuchte *f (auto.)* stop/tail lamp
Bremsspannung *f* retarding-field potential
Bremsspur *f (auto.)* skid track
Bremsstrahl *m* retardation jet
Bremsstrecke *f (auto.)* braking distance
Bremssubstanz *f (nucl.)* moderator
Bremsung *f (nucl.)* retardation
Bremsvermögen *n* braking power
Bremsvorrichtung *f* braking mechanism
Bremswelle *f (auto.)* brake cross shaft
Bremswirkung *f* braking effect
Bremszaum *m* Prony brake
brennen *v.t.* burn; calcine, roast; sinter; bake
Brenner *m* burner; *(welding)* torch, blowpipe

Brennerdüse f burner nozzle
Brennerzange f burner plier
Brennfläche f *(opt.)* focal surface
Brennfleck m focal spot
Brenngas n fuel gas
brennhärten v.t. flame-harden
brennhobeln v.t. gouge
Brennkammer f combustion chamber
Brennkraftmaschine f internal combustion engine
Brennlinie f focal line, focal curve
Brennmaterial n fuel
Brennofen m (Kalk:) burning kiln
Brennpunkt m (Öl:) burning point, fire point; *(opt.)* focus
Brennputzen n hot deseaming
Brennschneiden n flame-cutting, torch cutting, gas cutting
brennschweißen v.t. flash-weld
Brennstoff m fuel
BRENNSTOFF ~, (~behälter, ~leitung, ~pumpe, ~technik, ~verbrauch, ~zufuhr) fuel ~
Brett n board
Brettfallhammer m board drop hammer
brikettieren v.t. briquet
Brille f *(mach.)* (≈ Setzstock:) steadyrest; (e. Stopfbüchse:) gland; *(welding)* goggles
Brinell-Härtemeßverfahren n Brinell hardness test
Brinell-Härtezahl f Brinell hardness number
brinellieren v.t. brinell
Brinell-Kugeldruckprobe f Brinell ball hardness test
bröckeln v.i. crumble
brodeln v.i. *(radio)* bole, hum
Bronze f bronze

BRONZE ~, (~gießerei, ~guß, ~pulver) bronze ~
Bruch m breakage; rupture; (≈ Schrott:) scrap; *(metallo.)* fracture; *(math.)* fraction
BRUCH ~, (~beanspruchung, ~belastung, ~festigkeit, ~probe, ~spannung, ~versuch, ~zähigkeit) fracture ~
Bruchaussehen n fracture appearance
Bruchdehnung f elongation
bruchfest a. fracture-proof
Bruchfläche f fractured surface, fracture
Bruchgefahr f risk of breakage
Bruchgefüge n grain structure
Bruchgestein n quarry rock
Bruchgrenze f ultimate stress limit
brüchig a. brittle, short; fragile
Brüchigkeit f shortness, brittleness; fragility
Bruchlast f ultimate load, load at rupture
Bruchmodul m modulus of rupture
bruchsicher a. break-proof
Bruchsicherheit f safety against fracture, resistance to failure
Bruchspan m segmental chip
Bruchstelle f point of fracture
Bruchstrich m *(math.)* fraction line
Bruchstück n fragment
Bruchteil m *(math.)* fraction
Bruchteilen n *(mach.)* incremental indexing
Brücke f *(building, electr., mach., tel.)* bridge; (Kohlebürste:) rocker
BRÜCKEN ~, (~bau, ~bauer, ~geländer, ~pfeiler, ~rampe, ~schaltung, ~zufahrt) bridge ~
Brückenbaufirma f bridge-building firm
Brückendrehkran m rotary bridge crane
Brückenkran m bridge crane, gantry crane

Brückenlaufkran *m* travelling bridge crane
Brückenwaage *f* scale platform, weighbridge
Brüden *m* water vapor
Brumm *m (radio)* hum
brummen *v.i. (radio)* hum, buzz
Brummfaktor *m* ripple ratio
Brummfrequenz *f* hum frequency
Brummspannung *f* hum voltage, ripple voltage
Brummton *m* humming noise
Brummzeichen *n* buzzer signal
brünieren *v.t.* brown, brown-finish
Brunnenbau *m* well construction
Brust *f (mach.)* (e. Werkzeugschneide:) cutting face, face, rake
Brustfreiwinkel *m* (e. Drehmeißels:) relief angle
Brustleier *f* breast drill
Brustmikrofon *n* breast transmitter
Brustwinkel *m* cutting angle
Brüreaktor *m* breeder reactor
Bruttolohn *m* gross income
Buch *n (tel.)* directory
Buchenholz *n* beech-wood
Buchse *f* sleeve; (Kette:) barrel; (Lager:) bushing; (Zylinder:) liner
Büchse *f* (= Dose) tin, box; *(lubrication)* cup
Buckelblech *n* buckled plate
Buckelnaht *f (welding)* projection weld
Buckelschweißen *n* projection welding
Buckelschweißung *f* projection weld
Bügel *m* stirrup; frame; (Kopfhörer:) harness; (Säge:) frame; (Stromabnehmer:) bow; (Vorhangschloß:) shackle
Bügeleisen *n* smoothing iron, flat iron
Bügelhaken *m* shackle-eye
Bügelsäge *f* hacksaw; (Maschine:) power hacksaw
Bügelschraublehre *f* micrometer caliper
Bühne *f* platform; floor; stage
Bund *m* collar; (Welle:) shoulder
BUND ~, (~bolzen, ~mutter, ~schraube, ~welle) collar ~
Bundaxt *f* carpenters' axe
Bündel *n* bundle, bunch; *(opt.)* beam; *(nucl.)* stream
bündeln *v.t.* bundle
Bundesstraße *f* Federal road
bündig *a.* flush
Bundschleifen *n* shoulder grinding
Bunker *m* bunker, bin
Bunsenbrenner *m* Bunsen gas burner
Buntmetall *n* brass and bronze
Büro *n* office
Bürste *f (electr.)* [carbon] brush
BÜRSTEN ~, (~brücke, ~feuer, ~halter, ~joch, ~verschiebung, ~verstellung) brush ~
Bus *m* omnibus, bus; motor-coach
Butylacetat *n* butyl acetate

C

c, C, *s. a.* K and Z
Cabriolet *n* convertible, cabriolet
Camlock-Spindelkopf *m* camlock spindle nose
C-förmige Presse, gap frame press
Charge *f* charge; heat
Chargierbühne *f* charging platform, charging floor
chargieren *v.t.* charge; load; feed
Chargierkran *m* charging crane
Chargiermulde *f* charging box
Chargierschwengel *m* peel
Chassis *n (auto.)* chassis
Chauffeur *m* driver
Chilesalpeter *n* Chili saltpeter, nitrate of soda
Chlor *n* chlorine
Chlorbenzol *n* chlorobenzene
chloren *v.t.* chlorinate
chlorieren *v.t.* chlorinate
Chlorkali *n* potassium chloride
Chlorkalk *m* bleaching powder
Chlorung *f* chlorination
Chrom *n* chromium, chrome
Chromgelb *n* chrome yellow
Chrom-Molybdänstahl *m* chrome-moly steel
Chromnickelstahl *m* chrome-nickel steel
Chromstahl *m* chrome steel, chromium steel
Codewandler *m (electron.)* decoder
Cowper *m* Cowper stove
Coupé *n* coupé
Crackbenzin *n* cracked gasoline

D

DACH ~, (~antenne, ~rinne, ~schindel) roof ~
DACH ~, (~filz, ~filznagel, ~nagel, ~schiefer, ~ziegel) roofing ~
Dachbedeckungsblech *n* roofing sheet
Dachbedeckungsmaterial *n* roofing material
Dachdecker *m* roof tiler
Dachfirst *m* ridge
dachförmig *a.* V-shaped
Dachneigung *f* pitch of roof
Dachpappe *f* tar-board
Dachschindelnagel *m* shingle nail
Dachsparren *m* rafter
Dachspriegel *m (auto.)* tilt
Damm *m* embankment; dam, dike
Dammar *n (painting)* dammar
Dammbau *m* embanking
Dämmplatte *f* insulating slab
dämmen *v.t. (hydr.eng.)* dam up; insulate
Dämmschicht *f* insulating layer
Dämmstoff *m* insulating material
Dampf *m* vapor; *(UK)* vapour; (Wasserdampf:) steam; (≈ Rauch:) smoke; – pl. fumes
DAMPF ~, (~absperrhahn, ~absperrschieber, ~antrieb, ~bad, ~druck, ~druckmesser, ~entnahme, ~entwicklung, ~erzeuger, ~hammer, ~heizung, ~kessel, ~kolben, ~kraft, ~kraftwerk, ~leitung, ~leitungsrohr, ~schieber, ~schlange, ~spannung, ~strahl, ~turbine, ~überdruck, ~überhitzer, ~ventil, ~zylinder) steam ~
dampfartig *a.* vaporous
dampfdicht *a.* steam-tight

dampfen *v.i.* steam; fume, smoke
dämpfen *v.t. (acous.)* damp; *(mech.)* cushion; absorb; *(electr.)* attenuate; (Licht:) dim
dampfförmig *a.* vaporous
Dampfkesselblech *n* boiler plate
Dampfkesselkohle *f* steam-coal
Dampfkraftmaschine *f* steam power engine
Dampfmaschine *f* steam engine
Dampfmesser *m* steam gage
Dämpfung *f (acous.)* muting; *(electr.)* attenuation; *(mech.)* cushioning; (Licht:) dimming; (Schwingungen:) damping; absorption
Dampfverteiler *m* steam header
Dämpfungsausgleich *m* attenuation compensation
Dämpfungsfähigkeit *f* damping power
Dämpfungsfaktor *m* attenuation coefficient
Dämpfungsfeder *f* shock absorber spring
Dämpfungsmaterial *n (acous.)* muffling material
Dämpfungsvermögen *n* damping power
Dämpfungswicklung *f* damping winding
Dämpfungswiderstand *m* buffing resistance
Dauer *f* duration; period; interval
Dauerbeanspruchung *f* endurance stress
Dauerbelastung *f* constant load
Dauerbetrieb *m* continuous operation
Dauerbiegefestigkeit *f* bending fatigue strength
Dauerbiegemaschine *f* fatigue bending machine

Dauerbiegeversuch m fatigue bending test
Dauerbruch m fatigue failure
Dauerdehngrenze f creep limit
Dauererdschluß m continuous earth
Dauerfestigkeit f fatigue limit, fatigue strength
Dauerform f permanent mold
Dauerformguß m permanent mold casting
Dauerformmaschine f permanent mold machine
Dauergeschwindigkeit f *(auto.)* sustained speed, cruising speed
Dauerkerbschlagversuch m notched bar impact endurance test
Dauerkurzschluß m sustained shortcircuit
Dauerladegerät n *(electr.)* trickle charger
Dauerlauf m continuous running; continuous cycle
Dauerleistung f (Motor:) constant power
Dauermagnet m permanent magnet
Dauermagnetstahl m permanent magnet steel
Dauerparker m *(auto.)* day parker
Dauerprüfung f endurance test, fatigue test
Dauerruf m *(tel.)* permanent ringing
Dauerschlagbiegefestigkeit f repeated impact bending strength
Dauerschlagfestigkeit f resistance to repeated impact
Dauerschlagversuch m repeated impact test
Dauerschlagzugversuch m repeated impact tension test
Dauerschwingbeanspruchung f repetition of dynamic stress
Dauerschwingfestigkeit f fatigue strength, endurance limit of stress

Dauerspannung f *(electr.)* constant voltage
Dauerstandfestigkeit f creep limit, creep resistance
Dauerstandstreckgrenze f time-yield
Dauerstandversuch m creep test, endurance test
Dauerstrom m constant current
Dauertauchversuch m continuous immersion test
Dauerverdrehfestigkeit f endurance limit in torsion
Dauerverdrehversuch m endurance torsion test
Dauerversuch m endurance test, fatigue test
Dauerversuchsmaschine f endurance testing machine, fatigue testing machine
Dauerwarmfestigkeit f creep strength at elevated temperatures
Dauerwechselfestigkeit f limiting fatigue stress, rotary beam endurance limit
Dauerzugfestigkeit f endurance tensile strength
Dauerzugversuch m endurance tension test
Dauerzustand m steady state, permanence; perpetuity
Deckanstrich m finishing coat, top coat
Decke f cover; coat; *(building)* ceiling; floor
Deckel m cover; lid; (e. Lagers:) cap
deckeln v.t. cover
Deckenbalken m overhead beam
Deckenbeleuchtung f ceiling light
Deckenbeton m pavement concrete
Deckenlampe f ceiling lamp
Deckenleuchte f roof lamp, ceiling lamp
Deckenschalter m ceiling switch
Deckentransmission f overhead lineshaft

Deckenvorlegewelle f overhead countershaft
Deckfähigkeit f (painting) covering power
Deckfurnier n surface veneer
Deckkraft f (painting) covering power
Decklage f (welding) top run
Deckraupe f (welding) top layer (of a weld)
Deckschicht f top layer, top coat
dehnbar a. elastic; ductile
Dehnbarkeitsprüfung f ductility test
dehnen v.t. extend; stretch; elongate
Dehnfuge f expansion joint
Dehngrenze f creep limit
Dehnung f (≈ Bruchdehnung:) elongation; (≈ elastische Dehnung:) stress; stretch; (≈ Ausdehnung:) expansion; (≈ Längung:) extension
Dehnungsfuge f expansion joint
Dehnungsmesser m extensometer, strain gage
Dehnungsriß m expansion crack
Dehnverhalten n creep characteristics
Dehnverlauf m creep curve characteristics
dekapieren v.t. pickle
Dellennaht f (welding) projection weld
Dellenschweißung f projection welding
Deltaschaltung f delta connection
Demontage f disassembly
demontieren v.t. disassemble, dismantle
Desinfektionsmittel n disinfection solution
Desoxidation f deoxidation
Desoxidationsmittel n deoxidizer
desoxidieren v.t. deoxidize
Destillat n distillate
Destillationsgas n by-product gas
Destillationskokerei f by-product coke-oven plant
Destillationsrückstand m distillation residue

destillieren v.t. distil
Detektorempfänger m crystal set
Detektorkreis m detector circuit
Dezimalbruch m decimal fraction
Dezimalwaage f decimal balance
diagonalverrippt a. diagonally braced
Diagrammpapier n graph paper
Diamant m diamond
DIAMANT ~, (~abdrehgerät, ~abrichtgerät, ~fassung, ~halter, ~läppscheibe, ~pulver, ~säge, ~spitze, ~splitter, ~trennscheibe, ~werkzeug) diamond ~
diamantbestückt a. diamond-tipped
dicht a. dense; tight; leak-proof
Dichte f density; tightness; (Guß:) compactness; (Sand:) closeness; (≈ spez. Gew.:) specific gravity
Dichtemesser m densimeter
dichten v.t. (techn.) seal, make tight, pack; lute
DICHT ~, (~gewinde, ~mittel, ~naht, ~spalt, ~wirkung) sealing ~
Dichtring m packing ring
Dichtstoff m jointing material
Dichtung f (mach.) packing; seal; sealing joint
Dichtungsfilz m packing felt
Dichtungskitt m sealing cement
Dichtungsmanschette f gasket
Dichtungsmasse f sealing compound
Dichtungsnutring m U-shaped sealing ring of composition material
Dichtungsring m joint washer; packing ring; [oil] retainer ring
Dichtungsscheibe f gasket
Dichtungsschraube f sealing screw
Dichtungsschweißung f caulk weld
dick a. thick; heavy; (Flüssigkeiten:) viscous

Dicke f thickness
Dickenmesser m thickness gage
dickflüssig a. viscous; (Öl:) thick-bodied
Dickflüssigkeit f consistency
Dickglas n plate glass
Dickmantelelektrode f (welding) heavily coated electrode
Dickten- und Abrichthobelmaschine f timber sizer
Diele f board
Dielektrikum n dielectric
dielektrischer Verlustfaktor, power factor
Dielektrizitätskonstante f dielectric constant; permittivity
Dienstgespräch n service call
Dienstleistung f service
Dienstleistungskosten pl. service cost
Dienstleitung f (tel.) service line, order line
DIESEL ~, (~einspritzleitung, ~elektrisch, ~feldbahnlokomotive, ~kraftstoff, ~motor, ~öl, ~triebwagen) diesel ~
Dieselaggregat n diesel driven generating set
Dieselkraftstoffilter m (auto.) diesel fuel oil filter
Dieselpumpenaggregat n diesel-engined pumping set
Dieselschlepper m diesel-engined tractor
Dieselzugmaschine f diesel-engined motive unit
Dieselzylinderkopf m diesel engine cylinder head
Differential n (gearing) differential drive; (math.) differential
DIFFERENTIAL ~, (~bewegung, ~bremse, ~flaschenzug, ~gehäuse, ~getriebe, ~gewinde, ~gleichung, ~kegelrad, ~kondensator, ~teilen, ~teilgerät,

~transformator, ~welle, ~wicklung, ~widerstand) differential ~
diffundieren v.i. diffuse
Digital-Analog-Umsetzer m digital-to-analogue converter
Digitalrechenmaschine f digital computer
Dimension f dimension, size
dimensionieren v.t. dimension
dimensionslos a. dimensionless, non-dimensional
Diode f diode; two-electrode valve
Diodenbegrenzer m diode limiter
Diodengleichrichter m diode detector
Diodenröhre f diode; two-electrode valve
Dioptriefernrohr n diopter telescope
DIPOL ~, (~antenne, ~feld, ~moment) dipole ~
DIREKT ~, (~ablesung, ~anzeige, ~teilen, ~teilverfahren, ~wahl) direct ~
Distanzscheibe f spacer washer, shim
Dochtöler m wick oiler
Dochtschmierung f wick oiling
Dolomitbrennofen m dolomite calcining kiln
Dolomitkalk m magnesian limestone
Dolomitzustellung f dolomite lining
DOPPEL ~, (~brechung, ~dreieckschaltung, ~duowalzwerk, ~hub, ~kehlnaht) double ~
Doppelakkumulator m two-cell accumulator
Doppelbetrieb m. (tel.) duplex operation
doppelbördeln v.t. double-seam
Doppeldiode f duodiode
Doppelerdschluß m polyphase earth
doppelfalzen v.t. (cold work) double-sema
Doppelfräsmaschine f duplex milling machine

doppelgleisige Eisenbahnbrücke, double-track railway bridge
Doppelhebel *m* push and pull lever, twin lever
Doppel-I-Naht *f* double-I butt weld
Doppelkondensator *m* dual capacitor
Doppelkopfschiene *f* bullhead rail
Doppelkurbel *f* dual crank
Doppellager-Exzenterpresse *f* double-eccentric press
Doppellamellenkupplung *f* double multiple disc clutch
Doppellangfräsmaschine *f* duplex planomiller
Doppelleiter *m (electr.)* twin conductor
Doppelmaulschlüssel *m* double head wrench; *(UK)* double ended spanner
doppeln *v.t.* (Bleche:) double
Doppelnutmotor *m* double squirrel cage motor
Doppelplanfräsmaschine *f* duplex fixed-bed miller
doppelpolig *a.* bipolar
Doppelprismaführung *f (lathe)* double V-guide
Doppelreihenmotor *m* double-tandem engine
Doppelringschlüssel *m* double-ended box wrench
Doppelrollgabelschlüssel *m* adjustable double end wrench
Doppelschaltung *f (automatic)* double indexing
Doppelschlittenräummaschine *f* dual-ram broaching machine
Doppelschlußmotor *m* compound-wound motor
doppelseitig *a.* double-ended

Doppelspindelfräsmaschine *f* two-spindle fixed-bed-type miller
Doppelständer-Exzenterpresse *f* straight-sided eccentric press
Doppelständerfräsmaschine *f* planer-type milling machine; double column miller
Doppelständerhobelmaschine *f* double-standard planer
Doppelständerkurbelpresse *f* straight-sided crank press
Doppelständerstanzautomat *m* straight-sided power punching press
doppelständrige Abgratpresse, *f* straight-sided trimming press
Doppelstecker *m* two-pin plug, twin plug
Doppelsupport *m (lathe)* connected rests
Doppeltarifzähler *m (electr.)* two-rate meter
Doppeltaster *m* inside and outside caliper
Doppel-T-Träger *m* I-beam, I-girder
Doppel-Tulpen-Naht *f* double-U butt weld
Doppelung *f (surface finish)* lamination
Doppel-U-Stoß *m (welding)* double-U-butt joint
Doppelweggleichrichtung *f* two-way rectification
doppelwirkend *a.* double-acting
doppelwirkende Presse, double-action press
Doppelzapfenschlüssel *m* face spanner
Döpper *m* header: *(UK)* snap die
Doppler *m* (Blech:) doubling machine, doubler
Dorn *m (tube rolling)* mandrel, piercer; *(seam welding)* contact bar; *(milling)* arbor
Dornelektrode *f (welding)* contact bar
dornen *v.t. (rolling mill)* pierce

Dornpresse f piercing press
Dornring m (milling) collar
Dornstange f (rolling mill) piercer rod
Dose f (electr.) socket; (≈ Büchse:) tin; (Holz:) box
Dosenlibelle f levelling indicator
Dosenschalter m branch switch
Dosiereinrichtung f (plastics) metering device
dosieren v.t. (chem.) dose; (concrete) batch
Dosiermaschine f dosing machine
Dosierung f (concrete) batching; (chem.) dosing
Dosierungspumpe f dosing pump
Dosis f (nucl.) dose
Draht m wire
DRAHT ~, (~abschneider, ~ader, ~bürste, ~bewehrung, ~funk, ~gaze, ~geflecht, ~gewebe, ~glas, ~haspel, ~klemme, ~lehre, ~netz, ~ring, ~schneider, ~seil, ~seilrolle, ~seiltrieb, ~sieb, ~speichenrad, ~stärke, ~stift, ~straße, ~umspinnung, ~walze, ~zange, ~ziehbank, ~zieherei, ~zug) wire ~
Drahteinlegemaschine f wiring machine
drahteinlegen v.t. wire
Drahtkaliber n (rolling mill) wire rod pass
drahtlos a. (telegr.) wireless
drahtlose Telefonie, wireless telephony
Drahtrundfunk m wired radio
Drahtseilbahn f cableway
Drahtseilbrücke f cable suspension bridge
Drahtseilklemme f wire rope clip
Drahtwalzen n wire rod milling, rod milling, wire milling
Drahtwalzerei f [wire] rod milling
Drahtwalzwerk n wire-rod mill
Drahtzange f cutting plier

Draisine f trackmotor car
Drall m twist, spiral; (phys.) angular momentum, spin; (von Drahtlitzen:) lay; (e. Schnecke:) helix
Drallausgleich m (ball.) twist balance
drallförmig a. spiral; helical
Drallführung f (rolling mill) twist guide
Drallstahl m (concrete) twisted steel
Drängraben m drainage trench
dränieren v.t. drain
Dränrohr n drain pipe
Dränung f drainage
Draufsicht f (drawing) plan view
Drehachse f axis of rotation; (≈ Drehzapfen:) fulcrum pin, pivot pin
Dreharbeit f (metal cutting) lathe work
Drehautomat m automatic [lathe]
Drehbank f. cf. Drehmaschine
drehbar a. rotary; (in der gleichen Ebene:) swivelling; (allseitig:) pivoting
Drehbewegung f rotary motion
Drehbild n turned surface pattern
Drehbohrmeißel m rotary boring tool
Dreh-, Bohr- und Abstechmaschine f turning, drilling, boring, and cutting-off lathe
Drehbolzen m fulcrum pin, pivot
Drehbrücke f swing bridge
Drehdorn m lathe mandrel
Drehdurchmesser m (US) swing; (UK) turning diameter
Dreheinrichtung f (lathe) turning attachment
Dreheiseninstrument n moving-iron instrument
Dreheisenoszillograph m soft iron oscillograph
drehen v.t. (mach.) turn; (mech.) rotate, revolve; (≈ schwenken:) swing, swivel

Dreher *m (mach.)* lathe operator
Dreherei *f* lathe shop; (≈ Arbeit:) lathe work
Drehfeder *f (auto.)* torsion spring
Drehfederstab *m (auto.)* torsion bar
Drehfederung *f* torsion bar suspension
Drehfeld *n (electr.)* rotary field
Drehfeldinstrument *n* rotating field instrument
Drehfeldmesser *m (electr.)* induction wattmeter
Drehfeldmotor *m* induction motor
Drehfeldrichtungsanzeiger *m* phase-sequence indicator
Drehfenster *n* pivoting window
Drehfensterriegel *m* latch for pivoting windows
Drehfestigkeit *f* torsional strength
Drehflügelfenster *n (auto.)* vent wing
Drehfutter *n* lathe chuck
Drehgelenk *n* swivel-joint
Drehgeschwindigkeit *f* speed of rotation; *(lathe)* turning speed
Drehgestell *n* bogie
Drehgriff *m* star handle; *(motorcycle)* twist-grip
Drehhebel *m* pivoted lever
Drehherdofen *m* rotary hearth furnace
Drehherz *n* lathe dog
Drehimpuls *m (phys.)* angular momentum
Drehkeilkupplung *f* rolling key clutch
Drehknopfausschalter *m* rotary switch
Drehkolbengetriebe *n* rotary piston mechanism
Drehkolbenverdichter *m* rotary piston compressor
Drehkondensator *m* variable capacitor
Drehkran *m* rotary crane, swing crane, slewing crane

Drehkranz *m (auto.)* turntable
Drehkreuz *n* capstan handle
Drehlänge *f (lathe)* distance between centers
Drehleiter *f* turntable ladder
Drehling *m* cf. Drehzahn
Drehmagnet *m* moving magnet
Drehmagnetgalvanometer *n* moving-magnet galvanometer
Drehmagnetinstrument *n* moving-magnet instrument
Drehmantel *m (radial)* sleeve
Drehmaschine *f* lathe, turning machine
Drehmeißel *m* turning tool, lathe tool
Drehmeißelhalter *m* lathe tool holder
Drehmitte *f* center of rotation; *(lathe)* center line
Drehmoment *n* torque
Drehmomentenwandler *m* torque converter
Drehmomentregler *m* torque regulator
Drehmomentschlüssel *m* torque wrench
Drehofen *m* rotary furnace; (Zement:) rotary kiln
Drehpunkt *m* fulcrum
Drehregler *m* rotary type regulator; (≈ Regeltransformator:) induction regulator
Drehriefe *f* tool mark
Drehrohrofen *m* cylindrical kiln
Drehrost *m* revolving grate
Drehschalter *m* rotary switch
Drehscheibe *f* turntable
Drehschieber *m (hydraulics)* cylindrical rotary valve
Drehschiebermotor *m* rotary valve engine
Drehschlitten *m (lathe)* turning slide rest
Drehschraubstock *m* swivel vise
Drehschwingung *f* torsional vibration

Drehschwingungsdämpfer *m* torsional damper

Drehschwingungsfestigkeit *f* torsional fatigue limit

Drehschwingungsversuch *m* repeated torsion test

Drehsinn *m* sense of rotation

Drehspäne *mpl.* turnings

Drehspannung *f (mat. test.)* torsional stress

Drehspindel *f* workspindle, main spindle

Drehspindelkopf *m (lathe)* spindle head, spindle nose

Drehspitze *f* lathe center

Drehspule *f* moving coil

Drehspulgalvanometer *m* moving-coil galvanometer

Drehspulinstrument *n* moving-coil instrument

Drehspulmeßwerk *n* moving-coil measuring mechanism

Drehstabfederachse *f (auto.)* torsion bar axle assembly

Drehstahl *m cf.* Drehmeißel

Drehstrom *m* three-phase alternating current

DREHSTROM ~, (~anlasser, ~generator, ~kommutatormotor, ~kreis, ~motor, ~netz, ~transformator, ~umspanner, ~wicklung, ~zähler) three-phase ~

Drehsupport *m* slide rest

Drehteil *n* (als Bauteil:) swivel, pivoted member; ($\approx$ gefertigtes Teil:) turned part; ($\approx$ zu fertigendes Teil:) lathe work; (e. Konsolfräsmaschine:) table-base; (e. Supports:) swivel slide

Drehtiefe *f (lathe)* depth of cut

Drehtisch *m* rotary table

Drehtransformator *m* induction regulator

Drehtrommel *f* rotary drum

Drehtrommelfräsmaschine *f* drum miller

Drehtrommelröstofen *m* rotary calcining kiln

Drehumformer *m* rotary converter

Dreh- und Bohrmaschine *f* turning, drilling and boring lathe

Dreh- und Bohrwerk *n* vertical turning and boring mill

Drehung *f* rotation, revolution; turn; *(nucleon.)* spin

Drehungsfeder *f* torsion spring

Drehvorrichtung *f (lathe)* turning attachment

Drehwähler *m* rotary selector

Drehwechselfestigkeit *f* torsional fatigue strength

Drehwerk *n (mach.)* face-plate lathe

Drehwerkzeug *n* lathe tool

Drehwinkel *m* angle of rotation

Drehzahl *f* speed; ($\approx$ Umdrehungszahl:) number of revolutions

DREHZAHL ~, (~änderung, ~bereich, ~einstellung, ~geber, ~hebel, ~minderung, ~regelung, ~reihe, ~schalter, ~schaltung, ~schild, ~steuerung, ~stufe, ~stufung, ~tabelle, ~verstellung, ~vorwahl, ~verhalten, ~vorwähler, ~wahl, ~wechsel) speed ~

Drehzahlautomatik *f* automatic speed change

DREHZAHLEN ~, (~schild, ~tabelle, ~verhältnis, ~wechsel) speed ~

Drehzahlmesser *m (auto.)* revolution counter, speedometer, tachometer; *(mach.)* speed indicator

Drehzahlregler *m* speed regulator, speed governor

Drehzahlwächter *m* automatic speed selector
Drehzahlwähler *m* speed counter
Drehzahn *m* cutter bit
Drehzapfen *m* fulcrum pin, pivot
Dreiachshinterkipper *m* six-wheeler end tipper
Dreiachslastkraftwagen *m* six-wheeler truck
Dreiachssattelzugmaschine *f* six-wheeler tractor
Dreibackenfutter *n* three-jaw chuck
dreidimensional *a.* three-dimensional
Dreieck *n* (geom.) triangle
Dreieckschaltung *f* delta connection
Dreieckspannung *f* delta voltage
Dreieckwarnzeichen *n* (auto.) triangular warning sign
dreifach *a.* triple, threefold
Dreifachkondensator *m* three-gang capacitor
Dreifachleitung *f* (electr.) three-core cable
Dreifachschalter *m* three-point switch
Dreifachstecker *m* three-pin plug
Dreiganggetriebe *n* three-speed gear
Dreigangkondensator *m* three-gang capacitor
Dreikantfeile *f* three-square file
Dreilagenstahlblech *n* soft center steel sheet
Dreileiterkabel *n* three-core cable
DREIPHASEN ~, (~anlasser, ~gleichrichter, ~motor, ~strom, ~transformator) three-phase ~
dreipolig *a.* three-pole, three-pin
Dreipolröhre *f* triode
Dreirädergetriebe *n* three-gear drive
Dreirad-Lieferkraftwagen *m* three-wheel delivery van
Dreiradwagen *m* three-wheeler
Dreiseitenkipper *m* three-way tipper
Dreistiftstecker *m* three-pin plug
Dreistufenmotor *m* three-speed motor
DREIWEGE ~, (~bohrmaschine, ~feinstbohrwerk, ~gewindebohrmaschine, ~hahn, ~schalter) three-way ~
dressieren *v.t.* (Bleche:) cold reduce, cold finish, level, dress
Dressierwalzwerk *n* skin pass mill
Drillbohrer *m* spiral ratchet drill
Drillwulststahl *m* twisted steel bars of deformed squares
Drossel *f* (≈ Drosselklappe:) throttle valve; (≈ Drosselspule:) choke coil; (radio) reactor
Drosselkopplung *f* impedance coupling
Drosselmodulation *f* choke modulation
drosseln *v.t.* throttle; (speeds) slow down
Drosselregler *m* throttling-type governor
Drosselventil *n* throttle valve
Drosselwiderstand *m* choke impedance
Drosselwirkung *f* (electr.) choking effect
Druck *m* pressure; load; thrust; compression; push
DRUCK ~, (~abfall, ~anzeiger, ~ausgleich, ~behälter, ~entlastung, ~gas, ~gefälle, ~höhe, ~kontakt, ~leitung, ~messer, ~ölung, ~regelung, ~regelventil, ~schalter, ~schlauch, ~schmierapparat, ~schwankung, ~stufe, ~ventil, ~verlust, ~wasser) pressure ~
Druckaufgabe *f* (shell molding) pressure investment
Drückbank *f* spinning lathe
Druckbeanspruchung *f* compressive stress
Druckbelastung *f* compressive load application

Druckempfänger m *(telegr.)* printing-receiving apparatus
drücken v.t. press, push; squeeze; (Hohlkörper:) spin; (Gewinde:) roll
Drücker m *(rolling mill)* pusher
Drückerschaltung f trigger switch
Druckfarbe f printing ink
Druckfeder f compression spring
Druckfestigkeit f compression strength
Druckfirnis m litho varnish
Druckgefäß n (e. Druckgießmaschine:) gooseneck
druckgießen v.t. pressure die-cast
Druckgießmaschine f die-casting machine
Druckguß m pressure die-casting, pressure casting
DRUCKGUSS ~, (~form, ~legierung, ~verfahren) die-casting ~
Druckgußteil n pressure die-casting
Druckknopf m *(electr.)* push button, press button
Druckknopfschalter m push button switch
Druckknopfsteuerung f push-button control
Druckkugellager n ball thrust bearing
Druckluft f (≈ Preßluft) compressed air
DRUCKLUFT ~, (~anlaßventil, ~auflade-ventil, ~bremse, ~futter, ~hebebock, ~horn, ~kippvorrichtung, ~pendelwischer, ~preßformmaschine, ~ramme, ~scheibenwischer, ~schlauch, ~servobremse, ~signalhorn, ~spannung, ~speicher, ~zylinder) air ~, air-operated ~, pneumatic ~
Druckluftnietung f pneumatic riveting
Druckminderventil n pressure reducing valve
Drucköl n hydraulic oil

Drucköler m force feed oiler
Druckölgetriebe n hydraulic transmission
Druckrollenlager n roller thrust bearing
Druckschalter m push-button switch
Druckschmierpresse f grease gun
Druckschmierung f force-feed lubrication
Druckschraube f thrust bolt
Druckspannung f compression stress
Druckstange f (s. Presse:) pitman
Druckstelle f *(surface finish)* drag mark
Drucktaste f press key, push key
Druckstahlläppen n liquid honing
Druckumlaufschmierung f force-feed circulation oiling
Druckventil n delivery valve
Druckwelle f *(explosive metal-forming)* shock wave
Druckzugschalter m push-pull switch
Dübel m dowel [pin]
dübeln v.t. dowel
Dübelschweißung f stud welding
Dunkelkammer f *(photo.)* dark-room
Dünnblechschweißung f light-gage sheet steel welding
dünnflüssig a. highly fluid, of low viscosity
Dünnflüssigkeit f fluidity
Dünnmantelelektrode f *(welding)* light coated electrode
dünnwandig a. thin-walled
Dunst m vapor; *(UK)* vapour; (Schleier:) haze
Duo n two-high rolling mill
Duoblechwalzwerk n two-high plate mill
Duoblockwalzwerk n two-high blooming (or cogging) mill
Duo-Feinblechwalzwerk n two-high sheet rolling mill
Duo-Fertiggerüst n two-high finishing stand

Duogerüst n two-high stand
Duomaßwalzwerk n two-high sizing mill
Duoreversierblechwalzwerk n two-high reversing plate mill
Duoreversierblockstraße f two-high reversing blooming train
Duoreversiergerät n two-high reversing stand
Duoreversierwalzwerk n two-high reversing mill
Duostopfenwalzwerk n two-high piercing mill
Duostreckwalze f two-high rougher
Duouniversalwalzwerk n two-high universal mill
Duowalzgerüst n two-high rolling mill stand
Duowalzwerk n two-high rolling mill
Duplexverfahren n (met.) duplexing process
durchbiegen v.t. bend, deflect; – v.r. sag
Durchbiegeversuch m deflection test
Durchbiegung f bend; deflection; (≈ Durchhang:) sag
Durchbiegungsfestigkeit f transverse bending strength
Durchbluten n (painting) bleeding
durchbrennen v.i. (Sicherung:) blow
Durchbruchöffnung f (power press) opening in back
durchdringen v.t. & v.i. penetrate
Durchdringung f penetration
durchfallendes Licht, transmitted light
durchfließen v.t. & v.i. flow through, pass through
Durchfluß m flow; (electr.) passage
Durchflußanzeiger m flow indicator
Durchflußmesser m flow meter
Durchflußregelung f flow control

durchführen v.t. carry through, perform; conduct; (Leitungen:) pass through
Durchführung f performance; (electr.) wall entrance
Durchführungsisolator m bushing insulator
Durchführungswandler m bushing current transformer
Durchgang m passage; pass, travel; (power press) throat; (electr.) passage
Durchgangsgespräch n through call, transit call
Durchgangshahn m straightway cock
Durchgangsloch n throughhole
Durchgangsschleifen n through-feed grinding
Durchgangsschraube f bolt
Durchgangsventil n straightway valve
Durchgangsverkehr m (tel.) through traffic
Durchgangswiderstand m (electr.) volume resistance
durchgegangene Sicherung, blown fuse
durchgehende Bohrung, through-hole
Durchgriff m (electron.) reciprocal of amplication
Durchhang m (e. Riementrums:) sag, slack
Durchhärtung f through hardening, full hardening
durchlässig a. permeable, pervious
Durchlässigkeit f (magn.) permeability; (opt.) transmission factor
Durchlauf m passage, travel, traverse
Durchlauffräsmaschine f continuous milling machine
Durchlaufglühofen m continuous annealing furnace
Durchlaufkopieren n continuous copying

Durchlaufmischer *m (concrete)* continuous mixer

Durchlaufofen *m* continuous heating furnace

Durchlaufplan *m (work study)* process chart

Durchlaufschleifen *n* through-feed grinding

Durchlaufschmierung *f* non-circulatory lubrication

Durchlaufstoßofen *m* pusher-type furnace

Durchlaufverzinnungsanlage *f* continuous tinning line

Durchlicht *n (opt.)* transmitted light

durchlüften *v.t.* aerate

Durchlüfung *f* aeration

Durchmesser *m* diameter

Durchmesserteilung *f (gearing)* diametral pitch

Durchsatz *m* (e. Ofens:) throughput

durchschalten *v.t. (tel.) (UK)* connect through; *(US)* put through

Durchschaltung *f (tel.)* through-connection

durchscheinend *p.a.* translucent

Durchschlag *m (ball.)* penetration; *(electr.)* breakdown, puncture; *(tool)* drift punch

durchschlagen *v.t.* pierce; – *v.i. (electr.)* puncture; (Sicherungen:) blow out

Durchschlagfestigkeit *f (electr.)* disruptive strength, dielectric strength

Durchschlagsicherung *f* blow-out fuse

Durchschlagspannung *f (electr.)* breakdown voltage

Durchschlagspannungsmeßgerät *n* disruptive voltage measuring apparatus

Durchschleifen *n* through-feed grinding

durchschnittlich *a.* average

durchsichtig *a.* transparent

durchsickern *v.i.* percolate, trickle through

Durchsickerung *f* percolation

durchsieben *v.t.* screen, sift

durchspülen *v.t.* flush out

Durchsteckwandler *m* bushing transformer

Durchstoßofen *m* pusher type furnace

Durchstrahlung *f (electron.)* irradiation

durchtränken *v.t.* soak; impregnate

durchwärmen *v.t.* soak

durchweichen *v.t.* soak

Durchweichungsgrube *f* soaking pit

Durchzugformmaschine *f* stripping plate machine

Durchzugskraft *f* (Riemen, Kette, seil:) pulling power

Düse *f* nozzle; (Brenner:) tip; (Windform:) tuyere; *(rocket)* nozzle

Düsenboden *m (met.)* tuyere bottom

Düsenbrenner *m* open-flame burner

Düseneinstellung *f* injection timing

Düsenkopf *m (met.)* tuyere nozzle

Düsenprüfgerät *n (auto.)* nozzle tester

Düsenstock *m (met.)* tuyere stock; pen stock

Düsenvergaser *m* spray carburetor

Düsenverschmutzung *f (auto.)* nozzle fouling

Dynamik *f* dynamics; *(radio)* volume range

Dynamikbegrenzer *m (radio)* volume compressor

Dynamikdehner *m (radar)* dynamic expander, volume expander

Dynamikentzerrer *m* volume compensator

Dynamikentzerrung *f (radar)* gamma compensation

Dynamikregelung *f (radar)* contrast control
Dynamikregler *m (radio)* compandor
Dynamit *n* dynamite

Dynamo *m* dynamo, direct current generator, *(motorcycle)* magneto
Dynamoanker *m (electr.)* dynamo armature
Dynamoblech *n* dynamo sheet

E

eben *a.* plane, flat, level
Ebene *f* plane, plane surface
Ebenheit *f* planeness
ebnen *v.t.* level, flatten, smooth
Echolot *n* echo depth sounder
Echolotung *f* echo sounding
Echounterdrückung *f* echo suppression
Eckaussteifung *f (building)* corner truss
Eckblech *n* gusset plate
Eckenmaß *n* width across corners
Eckenwinkel *m* corner radius; (e. Meißels:) included plan angle
eckig *a.* angular, cornered
Ecknaht *f (welding)* corner weld
Eckstoß *m* corner joint
Eckventil *n* angle valve
Edelgas *n* inert gas, rare gas
Edelgas-Lichtbogenschweißen *n* inert-gas arc welding
Edelgleitsitz *m* wringing fit
Edelhaftsitz *m* medium force fit
Edelmetall *n* precious metal, noble metal
Edelpassung *f* force fit
Edelstahl *m* fine steel, high-grade steel, high-quality steel
Edeltreibsitz *m* shrink fit
Edisonsockel *m (electr.)* Edison screw cap
Effektbeleuchtung *f* decorative lighting
egalisieren *v.t.* level, flatten
eichen *v.t.* calibrate
eichenlogarer Riemen, oak-tanned belt
Eichkondensator *m* calibration capacitor
Eichleitung *f* reference circuit
Eichmaß *n* standard measure, calibration standard

Eichnormale *f* calibrating standard
Eichstrich *m* calibration mark
Eigenantrieb *m* separate drive, individual drive
Eigendämpfung *f (electr.)* self-modulation
Eigenerregung *f* self-excitation
Eigenfrequenz *f* natural frequency
Eigengewicht *n* deadweight
Eigenheit *f* characteristic feature
Eigenlüftung *f* (Motor:) fan ventilation
Eigenschaft *f* property, characteristic; (≈ Güte:) quality
Eigenschwingung *f* natural vibration; resonant frequency
Eigenspannung *f (electr.)* natural voltage
Eigenwärme *f* sensible heat
Eigenzündung *f (auto.)* compression ignition
Eilgang *m (auto.)* overdrive; *(mach.)* rapid traverse
Eilgangvorlauf *m* (Schlitten:) rapid advance
Eilrücklauf *m* (Schlitten:) quick return motion
Eilselbstgang *m* (Schlitten:) power rapid traverse
Eilvorlauf *m* rapid advance
Eilvorschub *m (lathe)* rapid traverse rate of feed
Eimer *m* bucket, pail
Eimerleiter *f (hydr. eng.)* dredging ladder
Eimertrockenbagger *m* dredger excavator
Einachsanhänger *m* semi-trailer
Einachsschlepper *m* two-wheel tractor
einadrig *a. (electr.)* single-core

Einankerumformer *m* single-armature converter
einatomig *a.* monoatomic
Ein-Aus-Schalter *m* push button On-Off
Einbahnstraße *f* one-way road
Einbahnverkehr *m* one-way traffic
Einbau *m* installation, fitting, assembly; mounting; *(road building)* placement
einbauen *v.t.* install, incorporate; fit; build into; mount; locate
einbaufertig *a.* ready for assembly
Einbauinstrument *n* flush-type instrument
Einbaumaß *n* fitting dimension, mounting dimension, installation dimension
Einbaumotor *m* skeleton frame-type motor; *(auto.)* built-in engine
Einbauschalter *m (electr.)* flush mounting switch, recessed switch
Einbauscheinwerfer *m (auto.)* recessed headlamp
Einbautafel *f* built-in panel
Einbauvorschrift *f* mounting instruction
Einbauwagenheber *m* built-in jack
Einbauwinkler *m (auto.)* flush-type direction indicator
Einbauzeichnung *f* installation plan
Einbereichsuper *m* superhet
einbetonieren *v.t.* concrete, encase in concrete
einbetten *v.t.* embed, imbed
einbeulen *v.t.* bulge (inward)
Einbeulversuch *m* bulging test
einblasen *v.t.* blast in, blow in (into), inject
einblenden *v.t.* fade in; *(opt.)* blend; *(electron.)* gate
Einbrand *m (welding)* penetration
Einbrandtiefe *f (welding)* depth of penetration
Einbrennemaille *f* baking enamel
einbrennen *v.t. (welding)* penetrate; *(painting)* bake
Einbrenn-Grundemaille *f* stoving black
Einbrennsilber *n* fired-on silver
einbringen *v.t.* charge, load; *(concrete)* pour
eindimensional *a.* unidimensional
eindrehen *v.t.* (Schrauben:) screw tight; *(metal cutting)* recess; neck
Eindrehung *f* recess; neck
eindringen *v.i.* penetrate; ingress, enter; infiltrate
Eindringtiefe *f (electr., welding)* depth of penetration
Eindruck *m* impression, indent, indentation; *(fig.)* impression
eindrücken *v.t.* indent, impress
Eindruckschmierung *f* oil-shot system
Eindrucktiefe *f* depth of indentation
Eindruckzentralschmierung *f* centralized oil shot system
einebnen *v.t. (road building)* spread and level
einfach *a.* simple, plain; single
Einfacherdschluß *m* single earth
einfacher Schalter, one-way switch
Einfachfräsmaschine *f* plain milling machine
Einfachmaschine *f* single-purpose machine
Einfachschlüssel *m* single head engineers' wrench; *(UK)* single ended spanner
Einfachteilen *n* single indexing
Einfachtelegrafie *f* simplex telegraphy
Einfach-V-Naht *f (welding)* single-bevel butt weld
einfachwirkend *a.* single-acting
Einfahrt *f (auto.)* entry
Einfall *m (opt.)* incidence

einfallendes Licht, incident light
Einfallwinkel *m* angle of incidence
Einfang *m (nucl.)* capture
einfangen *v.t. (nucl.)* capture
einfetten *v.t.* grease; lubricate; (≈ einölen:) oil; (Leder:) stuff, dress
Einflammbrenner *m* single-jet blowpipe
Einflankenwerkzeug *n* single-flank tool
Einflugschneise *f* lane of approach
Einflußgröße *f (metrol.)* limiting quantity
einfügen *v.t.* insert
einführen *v.t.* feed into, lead in; enter; introduce
Einführungsrinne *f (rolling mill)* entering trough
einfüllen *v.t.* fill up; charge; *(shell molding)* invest
Einfüllöffnung *f* inlet; *(auto.)* filler hole
Einfüllschraube *f* filler plug
Einfüllsieb *n (auto.)* tank strainer
Einfüllstopfen *m* filler plug
Einfüllstutzen *m (auto.)* filler pipe
Einfülltrichter *m* hopper; *(auto.)* service funnel
Eingang *m (electr.)* input
eingängiges Gewinde, single thread
EINGANGS ~, (~drehzahl, ~frequenz, ~größe, ~kondensator, ~leitwert, ~transformator, ~wert, ~widerstand) input ~
Eingangsspannung *f* signal voltage
eingerüstiges Walzwerk, single-stand mill
Einglasung *f* glazing
eingleisig *a.* single-track
eingravieren *v.t.* engrave
eingreifen *v.t. & v.i.* engage; *(gears)* mesh, mate
Eingriff *m (gears)* engagement, mesh

Eingriffsbereich *m (gearing)* zone of contact
Eingriffsdauer *f (gearing)* period of engagement
Eingriffsflankenspiel *n (gearing)* contact backlash
Eingriffsteilung *f (gears)* contact ratio
Eingriffswinkel *m (gearing)* angle of action, pressure angle
Einguß *m (founding)* sprue, feeder
Eingußtrichter *m (founding)* pouring gate
einhalsen *v.t. (mach.)* neck
einhärten *v.t.* depth-harden
EINHEBEL ~, (~bedienung, ~gangschaltung, ~klemmung, ~schaltung, ~steuerung, ~wahlschalter) single-lever ~
Einheit *f* (Maß:) unit; (Zahl:) unity
einheitlich *a.* uniform; standardized
Einheitlichkeit *f* uniformity
EINHEITS ~, (~gewinde, ~motor) standard ~
Einheitsbohrung *f* basic hole
Einheitswelle *f* basic shaft
einhüllen *v.t.* envelop
einkapseln *v.t.* encase, enclose
einkerben *v.t.* notch, groove; nick
Einkerbung *f* notch, groove; indent; nick
einklinken *v.t.* latch
Einknopfabstimmung *f (radio)* ganged tuning; single dial control
Einkreisempfänger *m* single-circuit receiver
Einkurvenautomat *m* single-cam operated automatic turret lathe
Einlage *f* shim; (Hartmetall:) insert
Einlaß *m (auto.)* inlet, intake
Einlaßkrümmer *m (auto.)* inlet manifold
Einlaßschlitz *m (auto.)* admission port
Einlaßventil *n (auto.)* intake valve

Einlauf *m* inlet
Einlaufbogen *m (building)* inlet bend
Einlauftrichter *m* feed hopper
einlegen *v.t.* insert; position, place (into)
Einleiterkabel *n* single-core cable
Einmannbedienung *f* one-man control
einmitten *v.t.* center
einölen *v.t.* oil
einpassen *v.t.* fit, adjust
Einpaßschleifen *n* match grinding
Einpaßzugabe *f* fitting allowance
EINPHASEN ~, (~generator, ~gleichrichter, ~induktionsmotor, ~lokomotive, ~motor, ~strom, ~transformator, ~wechselstrom, ~wicklung, ~zähler) single-phase ~
einphasig *a.* single-phase
einplanieren *v.t. (road building)* spread and level
einpolig *a.* unipolar; (Stecker:) one-pin
einpoliger Schalter, single-contact switch
Einpreßmaschine *f (concrete)* pressure grouting machine
Einpreßmörtel *m* intrusion mortar
Einpreßverfahren *n (building)* pressure grouting process
Einprofilschleifscheibe *f* single-rib grinding wheel
einrammen *v.t. (civ.eng.)* drive
Einrammen *n (civ.eng.)* pile driving
einrasten *v.t. & v.i.* engage (a notch)
einreihiges Kugellager, single-row ball bearing
Einrichtearbeit *f (mach.)* setting-up work
einrichten *v.t.* equip; install; set up; arrange
Einrichter *m* [tool] setter
Einrichtezeit *f (mach.)* setup time; *(welding)* preparation time

Einrichtung *f* (≈ Ausrüstung:) equipment; (≈ Zurüstung:) attachment; (zum Spannen:) fixture
Einrillenscheibe *f (rope drive)* single-groove sheave
einrollen *v.t. (grinding)* crush-dress; *(sheet metal)* curl; *(gears)* burnish; *(tubes)* roll up
einrücken *v.t. (gears)* engage; *(lever)* shift; *(motor)* start
Einrückhebel *m* engaging lever
Einrückung *f* engagement; shifting; starting
einsacken *v.t.* bag, sack
Einsatz *m* insertion; insert; (e. Arbeitskolonne:) employment; (≈ Einsatzhärtung:) case-hardening; (≈ Einsatzschicht:) case; (≈ Ofeneinsatz:) charge; (≈ Paßstück:) adaptor; *(cost accounting)* input; (≈ Verwendung:) use, application; (Hartmetall:) bit; (e. Sicherung:) link
EINSATZ ~, *(met.)* (~kasten, ~mittel, ~ofen, ~pulver, ~stahl, ~topf) case-hardening ~
einsatzbereit *a.* ready for duty
Einsatzbrücke *f (lathe)* gap bridge
einsatzhärten *v.t.* case-harden, carburize
Einsatzmeißel *m* (= Drehzahn) cutter bit
Einsatzschicht *f (met.)* case
Einsatzstoff *m (accounting)* input material
Einsatzstück *n* (e. Bettkröpfung:) *(lathe)* gap bridge
einsaugen *v.t.* suck in
Einsaugluft *f (auto.)* induction air
einschalen *v.t. (building)* shutter
Einschaler *m (building)* form setter
EINSCHALT ~, (~hebel, ~moment, ~motor, ~strom) starting ~
Einschaltdruckknopf *m* start button

einschalten *v.t. (electr.)* switch in (on); (Hebel:) shift into position; (Kupplung:) throw into engagement; (Getriebe:) engage; (Motor, Maschine:) start

Einschaltknopf *m* start button

Einschaltstoß *m (electr.)* make impulse

Einschaltzustand *m (electr.)* on-position

Einscheibe *f* (Riementrieb:) single-cone pulley

Einscheibenantrieb *m* single-pulley drive

Einscheibenkupplung *f* single-plate [friction] clutch

einschieben *v.t.* slip into; insert

Einschiebermotor *m (auto.)* single-sleeve valve engine

Einschienenhängebahn *f* monorail hoist

Einschienenlaufkatze *f* single beam trolley

einschlägig *a.* relevant

einschlagen *v.t.* drive; smash; – *v.i.* strike, hit

einschleifen *v.t.* grind-in; (Ventile:) seat

einschließen *v.t.* include, enclose; (Gase:) occlude; (Verunreinigungen:) entrap

Einschluß *m (met.)* inclusion

einschmieren *v.t.* lubricate, grease; oil

einschnappen *v.t.* snap into place

einschneidiger Meißel, single-point cutting tool

einschnüren *v.t.* (zylindrische Körper:) neck

Einschnürung *f* reduction in area; (von Zerreißstäben:) necking

einschrauben *v.t.* screw in, screw home

Einschraublänge *f* length of thread engagement

einschreiben *v.t. (geom.)* inscribe

einschütten *v.t. (shell molding)* invest

einschwalben *v.t.* dovetail

einschwimmen *v.t.* (Brückensegmente:) float into position

EINSCHWING ~, (~störung, ~strom) transient ~

einschwingen *v.i. (acous.)* build up, resonate

EINSEITENBAND ~, (~betrieb, ~empfänger, ~modulation, ~sender, ~telefonie, ~übertragung) single-side band ~

einseltig *a.* single-ended, single-end, single-sided, one-sided

Einsenkbarkeit *f* hobability

einsenken *v.t.* (≈ versenken:) countersink; counterbore; (≈ Vertiefungen kalt eindrücken:) die-sink, hob

Einsenkstempel *m* die-hob, hub

Einsenkung *f* cavity

EINSETZ ~, (~kran, ~maschine, ~mulde, ~seite, ~wagen) charging ~

einsetzen *v.t.* insert; enter; use; *(heat treatment)* carburize; *(met.)* charge; (Kerne:) secure

einspannen *v.t.* clamp [in position], secure; (Werkstücke:) load; (Werkzeuge:) mount; (in e. Futter:) chuck

Einspannvorrichtung *f* clamping device, gripping device; *(mach.)* work-holding fixture

einspeisen *v.t. (radio)* feed

EINSPINDEL ~, (~automat, ~bohrmaschine, ~fräsmaschine, ~halbautomat, ~planfräsmaschine, ~vollautomat) single-spindle ~

einspindlige Maschine, single-spindle machine

EINSPRITZ ~, (~leitung, ~motor, ~pumpe, ~ventil) injection ~

Einspritzdauer f *(auto.)* injection period
Einspritzdüse f *(auto.)* injection nozzle
einspritzen v.t. inject
Einspritzgeschwindigkeit f *(Druckgießverfahren:)* rate of die casting
Einspritzung f *(auto.)* injection
Einspritzventil n *(auto.)* injector valve
Einspritzvergaser m atomizing carburetter
einspülen v.t. *(hydr.eng.)* sluice; *(civ.eng.)* water-jet
Einspülen n *(civ.eng.)* water-jet driving
EINSTÄNDER ~, 1. *(drill press, boring mill, slotter, vertical turret lathe)* single-column ~; 2. *(jig borer, planer, miller, grinder)* openside ~
Einständerbauart f 1. single-column construction; 2. open-side type construction
Einständer-Blechkantenhobelmaschine f openside plate planer
Einständer-Exzenterpresse f open-front press
Einständer-Exzenterpresse mit großer Ausladung, gap frame press
Einständer-Fräsmaschine f open-side milling machine
Einständer-Hobelmaschine f open-side planer
Einständer-Karusseldrehmaschine f single-column vertical turret lathe
Einständer-Langfräsmaschine f open-side planer-miller
Einständer-Räderziehpresse f gap frame reducing press
Einständer-Schleifmaschine f openside grinder
Einständer-Stoßmaschine f single-column slotter
EINSTECH ~, 1. (~hebel, ~hub, ~programmsteuerung, ~schleifen, ~schleifmaschine, ~schleifscheibe, ~schlitten, ~ventil, ~vorschub) plunge-cut ~; cf. einstechen
2. (~getriebe, ~gewindewalzmaschine, ~hebel, ~schleifen, ~schleifvorrichtung, ~tiefe, ~verfahren) infeed ~; cf. einstechen
3. (~kopieren, ~kurve, ~meißel, ~schieber, ~support, ~verfahren, ~vorschub, ~werkzeug) recessing ~; *cf.* einstechen
einstechen v.t. 1. *(plain grinding)* plunge-cut; 2. *(centerless grinding)* feed in; 3. *(lathe)* recess, neck
Einstechzustellung f 1. plunge-cut feed; 2. infeed
Einsteckriegel m mortise latch
Einsteckschloß n mortise dead lock
Einsteigöffnung f manhole, culvert; (e. Kupolofens:) cleaning door
EINSTELL ~, *(mach.)* (~bock, ~dorn, ~fehler, ~genauigkeit, ~griff, ~hebel, ~knopf, ~kasten, ~organ, ~ring, ~scheibe, ~schraube, ~skala, ~spindel, ~vorrichtung) setting ~
einstellbar a. adjustable
Einstellehre f setting gage
einstellen v.t. adjust; (Schneidwerkzeuge:) position, set; (Walzen:) line; (e. Zündung:) time; *(radio)* modulate
Einstellgewindelehre f thread setting gage
Einstellplan m (e. Werkzeugmaschine:) tooling diagram
Einstellung f adjustment, setting; positioning; timing; *s.a.* einstellen
Einstellupe f focusing glass

Einstellwinkel *m* (e. Meißels:) entering angle; *(UK)* plan angle

Einstellzeichnung *f (mach.)* tooling diagram

Einstellzeit *f* (e. Meßgerätes:) response time

einstemmen *v.t.* calk; *(UK)* caulk; (Holz:) mortise

Einstich *m* 1. plunge-cut; 2. infeed; 3. recess; *(rolling mill)* pass; *s.a.* Einstech ~

Einstichseite *f (rolling mill)* entering side

einstöpseln *v.t.* plug in

Einstößelräummaschine *f* single-ram broaching machine

Einstoßvorrichtung *f (rolling mill)* pusher

einstrahlen *v.t.* irradiate

Einstrahlung *f* irradiation

einströmen *v.i.* flow in, pass in

Einströmung *f* inflow, influx

Einströmventil *n* inlet valve

Einstufenzerkleinerer *m* single-passage crusher

einstufig *a.* single-stage

eintauchen *v.t.* immerse; *(grinding)* plunge

Eintauchschmierung *f* flood lubrication

einteilen *v.t.* divide; graduate, scale; classify

eintragen *v.t.* enter; record; plot

Eintritt *m* inlet; ingress; entry

Eintrittsschlitz *m* inlet port

Eintrittsseite *f* entry side

Eintrittsstutzen *m* inlet pipe connection

Eintrittswinkel *m* entering angle; *(opt.)* angle of incidence

Einwaage *f* amount weighed in

EINWEG(E) ~, *(mach.)* (~ausführung, ~bohrmaschine, ~maschine, ~verstärker) one-way ~

Einweggleichrichter *m* half-wave rectifier

Einwegschalter *m* single-way switch, one-way switch

Einwegwähler *m (tel.)* uniselector

einwirken *v.t.* act; affect; effect

Einwurftrichter *m* feed hopper, track hopper

EINZAHN ~, (~bewegung, ~kupplung, ~strehlen, ~teilen, ~zustellung) single-tooth ~

Einzahnmeißel *m* single-point cutting tool

einzäunen *v.t.* fence

einzeichnen *v.t.* plot; mark

Einzelantrieb *m* separate motor drive

Einzelaufhängung *f (auto.)* independent suspension

Einzelerdschluß *m (electr.)* single earth

Einzelfertigung *f* single-piece work

Einzelleiter *m (electr.)* single conductor

Einzelpunktschweißung *f* individual spot welding

Einzelradaufhängung *f (auto.)* independent wheel-suspension

Einzelschalter *m* separate switch

Einzelteil *n* single part, component part

Einzelteilungsfehler *m* single indexing error

Einzelteilzeichnung *f* component drawing

Einzelverzahnung *f* single-cycle method

Einzelzeichnung *f* detail drawing

einziehen *v.t.* draw-in; *(cold work)* close-in; (e. Schraube:) tighten

EINZWECK ~, (~drehmaschine, ~fräsmaschine, ~maschine, ~werkzeugmaschine) single-purpose ~

Einzylindermotor *m* single-cylinder engine

Eisblumenbildung *f (painting)* reticulation

Eisen *n* iron; cast iron; *s.a.* Stahl

EISEN ~, (~erz, ~erzlager, ~gießerei,

~gießereiwesen, ~industrie, ~kern, ~kernspule, ~kerntransformator, ~kitt, ~oxid, ~Portlandzement) iron ~

Eisenbahn f (US) railroad; (UK) railway

EISENBAHN ~, (~bau, ~brücke, ~damm, ~fahrzeug, ~kesselwagen, ~knotenpunkt, ~kreuzung, ~netz, ~schiene, ~schranke, ~schwelle, ~signalwesen, ~strecke, ~technik, ~überführung, ~übergang, ~unfall, ~unterbau, ~verbindung, ~verkehr, ~verwaltung, ~waggon) railroad ~; railway ~

Eisenbahnoberbau m permanent way, railway superstructure

Eisenbahnoberbaumaterial n permanent way material

Eisenbauindustrie f structural engineering industry

Eisenbegleiter m (met.) any element other than iron

Eisenbeton m cf. Stahlbeton

eisenerzeugende Industrie, iron producing industry

eisengekapselt a. iron-clad

Eisenguß m cast iron

Eisenhüttenbetrieb m ironworks

Eisenhütteningenieur m metallurgical engineer

Eisenhüttenkunde f ferrous metallurgy

Eisenhüttenmann m metallurgist

eisenhüttenmännisch a. metallurgical

Eisenhüttenwerk n ironworks; metallurgical plant

Eisenkohlenstoffdiagramm n iron-carbon diagram

Eisenkohlenstofflegierung f iron-carbon alloy

Eisenmeßgerät n permeameter

Eisenmischer m (met.) hot metal mixer

Eisennadelinstrument n permanent-magnet moving-iron instrument

eisenverarbeitende Industrie, metalworking industry, iron and steel working industry

Eisenwaren fpl. iron-ware, hardware

Elastikbereifung f (auto.) solid rubber tire (or tyre)

elastisch a. elastic, flexible

Elastizitätsgrenze f (mat.test.) elastic limit

Elastizitätsmodul m modulus of elasticity; Young's modulus

Elastizitätsprüfer m elastometer

elektrifizieren v.t. electrify

Elektrifizierung f electrification

Elektriker m electrician

elektrisch a. electric, electrical

elektrisieren v.t. & v.r. electrise, electrify

Elektrizität f electricity

Elektrizitätswerk n generating station, power station

Elektrizitätszähler m electric [supply] meter

Elektroantrieb m electric drive

Elektroausrüstung f electrical equipment

elektrobearbeiten v.t. spark-machine

Elektrobohrmaschine f electric drill

Elektrode f electrode

elektrodynamischer Lautsprecher, electrodynamic loudspeaker

Elektroenergie f electrical energy

Elektroerosion f spark erosion

elektroerosive Metallbearbeitung, electro-erosion machining; spark machining

elektroerosives Verfahren, electrospark process, electro-erosion process

Elektrofahrzeug n electric vehicle

Elektroflaschenzug m electric pulley block; electric hoist

Elektrogabelstapler *m* electric fork lift truck
Elektrogerät *n* electrical appliance
Elektrogewindebohrmaschine *f* electric tapper
Elektrohandbohrmaschine *f* electric hand drilling machine
Elektrohandwerkzeug *n* portable electric tool
Elektroherdofen *m* electric-hearth furnace
Elektrohochhubkarren *m* industrial truck with tiering attachment
Elektroindustrie *f* electrical industry
Elektroingenieur *m* electrical engineer
Elektroinstallateur *m* electrician
Elektrokarren *m* electric freight truck, motor driven truck
Elektrokarren mit Kippmulde, dump truck
Elektrokontaktanlage *f* *(electroerosion)* electro-contact unit
Elektrokontaktfühler *m* *(copying)* electro-contact tracer
Elektrokrankarren *m* electric crane truck
Elektrolamellenkupplung *f* electromagnetic clutch
Elektrolyse *f* electrolysis
Elektrolyt *m* electrolyte
Elektrolyteisen *n* electrolytic iron
Elektrolytgleichrichter *m* electrolytic rectifier
elektrolytisch *a.* electrolytic
elektrolytische Metallisierung, electroplating
Elektrolytkondensator *m* electrolytic capacitor
Elektrolytkupfer *n* electrolytic copper
Elektromagnet *m* electromagnet
elektromagnetische Kupplung, electromagnetic coupling
Elektromagnetventil *n* solenoid-controlled valve
Elektrometallurgie *f* electrometallurgy
Elektromotor *m* electric motor
elektromotorische Kraft, electromotive force
Elektron *n* electron
ELEKTRONEN ~, (~auffang, ~austritt, ~bahn, ~beugung, ~beschleuniger, ~entladung, ~fluß, ~hülle, ~kamera, ~konzentration, ~ladung, ~mikroskop, ~röhre, ~rückstoß, ~schale, ~vervielfacher, ~überschuß, ~zerfall) electron ~
Elektronenausstrahlung *f* emission of electrons
Elektronenaustritt *m* emission of electrons
Elektronenlehre *f* electronics
Elektronenröhrensteuerung *f* electronic control
Elektronenröhrentechnik *f* electronics
Elektronenschalter *m* electronic switch
Elektronenstrahlschweißen *n* welding by electron beam
Elektronik *f* electronics, electronic equipment
elektronische Datenverarbeitung, electronic data processing
elektronisches Meßgerät, electronic instrument
elektronische Steuerung, electronic control
elektronisch gesteuerte Maschine, computer-controlled machine
Elektroofen *m* electric furnace
Elektroplattierung *f* electro-deposition, electroplating
Elektrophysik *f* electro-physics

Elektropreßspan *m* electrical laminated fibre sheet
Elektro-Roheisen *n* electric pig iron
Elektrorüttelstampfer *m* electric vibrating tamper
Elektrosäge *f* electric saw
Elektroschachtofen *m* electric shaft furnace
Elektroschaltschrank *m* electrical switchgear cabinet
Elektroschleifmaschine *f* electric bench grinder
Elektroschlepper *m* electric tractor
Elektroschraubenzieher *m* electric screwdriver
Elektroschweißung *f* arc welding, electric welding
Elektrospannfutter *n* electrically operated chuck
Elektrostahl *m* electric steel
Elektrostahlerzeugung *f* electric steel production
Elektrostahlwerk *n* electric steel plant
elektrostatisch *a.* electrostatic
Elektrotechnik *f* electrotechnics; electrical engineering
elektrotechnische Industrie, electrical industry
Elektrotischbohrmaschine *f* electric bench drill
Elektrovulkanfiber *n* electrical vulcanized fiber
Elektrowerkzeugkasten *m* electric tool chest
Elektrozugkarren *m* tructractor
Element *n* (chem., phys.) element; (electr.) cell, battery; (mach.) component
Elementarteilchen *n* fundamental particle, elementary particle

Elevator *m* elevator
ELEVATOR ~, (~becher, ~behälter, ~gurt, ~kette, ~gehäuse) elevator ~
Elfenbeinschwarz *n* bone black
Ellira-Verfahren *n* (welding) Unionmelt process, submerged arc welding
Eloxalüberzug *m* anodic coating
Eloxalverfahren *n* anodic treatment (in oxalic acid), electrolytic oxidation process
eloxieren *v.t.* anodize
elysieren *v.t.* (electroerosion) machine electrolytically
Emaille *f* enamel
Emaillelack *m* enamel varnish
Emaillelackdraht *m* enamel-insulated wire
Emailleüberzug *m* vitreous enamel coating
emaillieren *v.t.* enamel
Emaillierofen *m* enamelling oven
Emaillierüberzug *m* enamel coat
Emission *f* (electron.) emission
EMISSIONS ~, (~konstante, ~spektrum, ~strom, ~temperatur) emission ~
Emissionsfähigkeit *f* emissivity
emittieren *v.t. & v.i.* emit
Empfang *m* (radio.) reception
empfangen *v.t.* receive
Empfänger *m* (radio) radioreceiver; (tel.) telegraph receiver
EMPFÄNGER ~, (~abgleich, ~abstimmung, ~antenne, ~ausgangsleistung, ~ausgangsspannung, ~rauschen, ~röhre, ~verstärker) receiver ~
EMPFANGS ~, (~anlage, ~antenne, ~apparat, ~gerät, ~gleichrichter, ~instrument, ~kreis, ~lautstärke, ~ort, ~röhre, ~station, ~störung, ~verstärker, ~welle) receiving ~

Empfangsleistung f (radio) received power
Empfangsspannung f (radio) signal voltage
empfindlich a. (von Instrumenten:) sensitive
Empfindlichkeit f (e. Meßgerätes:) sensitiveness
Empfindlichkeitsmessung f sensitometric measurement
Emulgator m emulsifier
Emulsion f emulsion
Emulsionsbildung f emulsification
emulsionsfähig a. emulsifiable
Emulsionsfett n emulsifiable grease
Emulsionsschmiermittel n emulsion lubricant
Endauflager n end support
Endausschalter m limit switch
Endausschlag m (mach.) end stop, limit stop
endbearbeiten v.t. (spanlos:) finish work; (zerspanend:) finish machine
Enddrehzahl f top speed
Enddruck m end thrust
Ende n (e. Spindel:) nose
Endfläche f end surface
Endglied n (math.) final term
Endkontakt m terminal contact
Endlage f end position, final position
Endlager n end bearing; end journal
endliche Zahl, finite number
endlos a. (Kette, Riemen, Seil:) endless
Endmaß n gage block, end measure
Endmast m (telegr., tel.) terminal pole
Endpentode f (radio) output pentode
Endröhre f power output tube, high-power tube

Endschalter m limit switch
Endspannung f terminal voltage
Endspiel n end play
Endstellung f final position, extreme position
Endstufe f (radio) power output stage
Endtemperatur f final temperature
Endtriode f (radio) output triode
Endverstärker m (tel.) terminal repeater; (radio) output amplifier
Endverteiler m (electr.) block terminal, terminal box
Energie f energy, power
ENERGIE ~, (~abfall, ~aufwand, ~ausstrahlung, ~austausch, ~dichte, ~erhaltung, ~gewinn, ~gleichung, ~umformer, ~verlust, ~verschwendung) energy ~
Energieniveau n quantum state
Energiequant n quantum of energy
Energiequelle f source of power
Energieverbrauch m power comsumption
Energieversorgung f (electr.) power supply
Energiewirtschaft f power economy
eng a. narrow; tight; close
enge Passung, snug fit, close fit
Engpaß m bottleneck
engtoleriert p.a. close-tolceranced
entbrummen v.t. dehum, buck the hum
Entbrummkondensator m antihum condenser
entfärben v.t. discolor
entfernen v.t. remove; eliminate
Entfernung f removal; elimination
Entfernungsmesser m distance meter
entfetten v.t. degrease
Entfettungsmittel n degreasing agent
entflammbar a. inflammable, flammable

Entfroster *m (auto.)* windscreen defroster; *(UK)* windshield defroster
entgasen *v.t.* degasify: extract gas (by distillation)
entgegengesetzt *a.* opposed, opposite
Entgrateinrichtung *f* deburring device (*or* attachment)
entgraten *v.t.* deburr; (Schmiedeteile:) trim
enthärten *v.t.* soften
entionisieren *v.t.* deionise
entkohlen *v.t. (met.)* decarburize
Entkohlung *f* decarburization
entkupfern *v.t.* (Bleiraffination:) extract copper
Entkupferung *f* copper extraction
entkuppeln *f.t.* uncouple, unclutch
ENTLADE ~, (~kreis, ~spannung, ~strom, ~widerstand) discharge ~
Entladebrücke *f* material handling bridge
Entladegreifer *m (automatic lathe)* unloading gripper
entladen *v.t.* unload; – *v.i. (electr.)* discharge
Entladevorrichtung *f (mach.)* unloading device
Entladung *f (electr.)* discharge
ENTLADUNGS ~, (~bahn, ~funke, ~kanal, ~kreis, ~lampe, ~spannung, ~stoß, ~verzug, ~widerstand) discharge ~
entlasten *v.t.* relieve; unload; (e. Ventil:) balance
Entlastung *f* release, relief; unloading
entlüften *v.t.* ventilate, deaerate; *(auto.)* (Kurbelwanne:) breath; *(hydraulics)* bleed
Entlüfter *m (auto.)* ventilator; (Kurbelwanne:) breather; bleeder

Entlüfterschlauch *m (auto.)* bleeding tube
Entlüfterstutzen *m* air vent
Entlüftung *f* ventilation; deaeration; *(hydraulics)* bleeding
Entlüftungshahn *m* air relief cock
Entlüftungsrohr *n (auto.)* breather tube
Entlüftungsschacht *m* ventilation shaft
Entlüftungsscheibe *f (auto.)* ventilation window
Entlüftungsschraube *f (auto.)* air vent screw; bleeder screw
Entlüftungsstopfen *m* ventilation plug
Entlüftungsventil *n* air-relief valve
entmagnetisieren *v.t.* demagnetise, de-energise
Entmagnetisierung *f* demagnetisation
entmodulieren *v.t.* demodulate
Entnahme *f* (Dampf:) extraction, bleeding
entnehmen *v.t.* (Dampf:) extract, bleed
Entphosphorung *f* dephosphorization, removal of phosphorus
entriegeln *v.t.* unlock
entrosten *v.t.* de-rust
Entsalzungsanlage *f* desalination plant, demineralization plant
Entsalzungsverfahren *n* desalination process, demineralization process
entschlacken *v.t.* remove slag, draw-off slag
Entschlackung *f* removal of slag, removal of cinder
entschwefeln *v.t.* desulphurize
entspannen *v.t.* unstress, relieve stresses; *(mech.)* release; (e. Feder:) slacken; (Werkzeuge:) season
Entspannungsglühen *n* stress-relieving anneal
Entstaubung *f* dust-exhaust, dust collection, dust removal

Entstaubungsanlage f dust arrester, dust collecting equipment
entstören v.t. (radio) suppress interferences
Entstörer m (radio) interference suppressor
Entstörgerät n anti-interference device
Entstörung f (electr., radio) interference suppression, noise suppression
Entstörungskondensator m suppression capacitor, anti-interference capacitor
entwachsen v.t. (painting) dewax
entwässern v.t. drain
Entwässerung f drainage
entwerfen v.t. design
entwickeln v.t. develop; (Gase:) generate; – v.r. liberate, evolve
Entwickler m (photo.) developer
Entwicklung f development; (von Gasen:) generation; (Dampf:) formation
Entwurf m draft; layout; project, plan
entzerren v.t. (radio) neutralize disturbing effects
Entzerrer m (radio) [attenuation] compensator
Entzerrung f attenuation compensation
entzündbar a. inflammable, flammable
entzünden v.t. ignite, kindle
entzundern v.t. descale
Entzunderungsofen m wash heating furnace
Entzunderungswärme f (met.) wash heat, cinder heat, sweating heat
Entzündung f ignition
Entzündungspunkt m ignition point; burning point
ERD ~, (~anschluß, ~antenne, ~arbeit, ~bauwerk, ~klemme, ~peilgerät, ~rückleitung, ~verbindung) earth ~

Erdalkali n alkaline earth, alkali-earth
Erdanziehung f earth's attraction
Erdarbeiter m navvy
Erdbeschleunigung f acceleration due to gravity
Erdbewegung f (civ.eng.) earthmoving
Erdboden m ground, soil
Erdbohrmaschine f soil boring machine
Erde f (civ.eng.) earth, soil; (electr.) earth; (US) ground
erden v.t. earth; (US) ground
Erder m earth connection; (US) ground connection
Erdfarbe f mineral pigment
Erdfehler m (electr.) earth fault
Erdgas n natural gas
Erdkabel n underground cable
Erdleiter m (electr.) earthing conductor
erdmagnetisch a. geomagnetic
Erdöl n crude oil, crude naphtha, petroleum
Erdölbenzin n naphtha
Erdölbitumen n petroleum asphalt
Erdpech n asphalt, asphaltum, mineral pitch
Erdplanumfertiger m (civ.eng.) subgrade planer
Erdprüfer m (electr.) ground detector
Erdschluß m dead earth; earth connection
Erdschlußprüfer m earth-leakage indicator
Erdschlußstrom m earth current
Erdung f (electr.) earthing, grounding, earth connection
ERDUNGS ~, (~anschluß, ~draht, ~klemme, ~punkt, ~schalter, ~stöpsel, ~trennschalter, ~widerstand) earthing ~

Erdungsmesser m earth resistance meter
Erdverlegung f underground laying

Erdwachs *n* mineral wax
Ergänzungsfarbe *f* complementary color
erhaben *a.* convex
Erhaltung *f* conservation
erhärten *v.t. & v.i.* harden
Erhebung *f (surface finish)* buckle, projection
erhitzen *v.t.* heat up
erhöhen *v.t.* raise, increase; elevate
erkalten *v.i.* cool down; (Leim:) chill
erliegen *v.i.* (Schneidwerkzeug:) fail
erlöschen *v.i.* (Lichtbogen:) extinguish
Ermittlung *f* (Kosten:) finding
ermüden *v.i.* (Stahl:) fatigue
Ermüdung *f (techn.)* fatigue
Ermüdungsbruch *m* fatigue failure
Ermüdungsgrenze *f* fatigue limit
Ermüdungsriß *m* fatigue crack, endurance crack
Ermüdungsversuch *m* endurance test, fatigue test
erneuern *v.t.* renew; replace
Erneuerung *f* renewal; (von Gleisanlagen:) relaying
erodieren *v.t.* sparkmachine
Erosion *f (metal cutting)* erosion, sparkmachining
EROSIONS ~, (~elektrode, ~gerät, ~krater, ~rückstände) erosion ~
erproben *v.t.* try [out]
erregen *v.t.* excite; *(US)* energize
Erreger *m (electr.)* exciter
ERREGER ~, (~feld, ~kreis, ~spannung, ~spule, ~strom, ~wicklung) exciting ~
Erregung *f (electr.)* excitation
errichten *v.t.* erect; install
Ersatz *m* replacement; (≈ Ersatzteil:) spare part

ERSATZ ~, (~leitung, ~reifen, ~sicherung, ~teil) spare ~
Ersatzschaltung *f* equivalent circuit
Ersatzstoff *m* substitute material
Ersatzteildienst *m* spare parts service
erschmelzen *v.t.* (Rohstoffe:) smelt; (Metall:) melt
erschütterungsfrei *a.* vibrationless
ersetzen *v.t.* replace; substitute
erstarren *v.i. (met.)* solidify; (Fett:) congeal; (Flüssigkeiten:) freeze; (Zement:) set
Erstarrungspunkt *m* freezing point, solidification point; congealing point; setting point; *s.a.* erstarren
erwärmen *v.t.* heat, warm up
erweichen *v.t. & v.i.* soften, fuse
Erweichungspunkt *m* fusion point, softening point
erweitern *v.t.* expand; (Löcher:) enlarge, widen
Erweiterung *f* enlargement; expansion
Erz *n* ore
ERZ ~, (~ader, ~aufbereitung, ~bergbau, ~bergwerk, ~brecher, ~bunker, ~frischreaktion, ~gicht, ~greifer, ~grube, ~lager, ~möller, ~röstofen, ~rutsche, ~schmelzofen, ~trübe, ~verladeanlage, ~zerkleinerungsanlage) ore ~
erzeugen *v.t.* produce; *(electr.)* generate
Erzeuger *m* (Dampf:) generator; (Gas:) producer
Erzeugung *f* production, manufacture; (Gas, Zahnräder:) generation
Erzfrischverfahren *n* direct process, ore process
erzwungene Schwingung, *(metal cutting)* forced vibration
Estrich *m* flooring

Evolvente *f* involute
EVOLVENTEN ~, (~fläche, ~meßgerät, ~prüfer, ~prüfung, ~rad, ~schnecke, ~wälzfräser, ~zahn) involute ~
explodieren *v.i.* explode, burst
Explosion *f* explosion, burst
Explosionsenergie *f* explosive energy
Explosionsgefahr *f* explosion hazard
explosionsgeschützt *a.* explosion-proof
Explosionsklappe *f* (e. Gichtgasleitung:) bleeder valve
Explosionsumformung *f* metal-forming with explosives, metalworking with explosives
Explosivstoff *m* explosive
Explosivumformung *f* explosive forming
extrahieren *v.t.* extract
extrudieren *v.t.* extrude
Exzenter *m* eccentric
EXZENTER ~, (~bewegung, ~bolzen, ~getriebe, ~hebel, ~kurve, ~presse, ~verstellung, ~welle, ~zapfen, ~ziehpresse) eccentric ~
Exzentrizität *f* eccentricity

F

Fabrik f factory, works, plant
Fabrikat n make; product
Fabrikationsbetrieb m production shop
Fabrikationskosten pl. cost of manufacture
Fabrikationslänge f manufacturing length
Fabrikationslehre f production gage
Fabrikationsnummer f serial number
Fabrikationsteil n production part
fabrikfertig a. ready-made
fabrikmäßig hergestellt, factory-produced
Fabrikzeichen n trade mark
Fach n branch; profession, line
Fächerscheibe f serrated lockwasher
fachgerecht a. workmanlike
FachIngenieur m specialist engineer
fachkundig a. competent
Fachmann m expert
Fachmesse f trade fair
Fachwerk n framework; lattice work; (e. Martinofens, Cowpers etc.:) checker work, chequer work
Fachwerkhängebrücke f lattice suspension bridge
Fachwerkkonstruktion f truss construction
Fachwerksmauerung f checkering, chequering
Faden m thread; (electr.) filament
Fadenkreuz n hairline cross, line cross, cross spiderline
Fadenlampe f filament lamp
Fadenstrich m spider line
Fading n (radio) fading

Fadingregelung f (radio) automatic gain control
Fähigkeit f capability, ability, quality
Fahne f (Schablonenformerei:) strickle; (e. Akkumulators:) lug
Fahrbahn f (auto.) track; roadway, driveway; (e. Krans:) runway
fahrbar a. portable; (auto.) manoeuvrable
fahrbare Reparaturwerkstatt, mobile repair shop
fahrbare Schmieranlage, mobile lubricating equipment
Fahrdraht m trolley wire
Fahrdrahtaufhängung f catenary suspension
Fahrdrahtweiche f trolley frog
Fahreigenschaften fpl. (auto.) roadability
fahren v.i. (auto.) drive; (motorcycle) ride; (e. Anlage:) run; ($\approx$ verfahren:) traverse
Fahrer m (auto.) driver; (motorcycle) rider
Fahrerhaus n (auto.) driver's cab
Fahrersitz m (auto.) driver's seat
Fahrgast m passenger
Fahrgestell n (auto.) chassis
Fahrgestellrahmen m (auto.) chassis frame
Fahrkran m travelling crane
Fahrleitung f bus-bar line; trolley line, overhead conductor
Fahrmotor m traction motor
Fahrpraxis f (auto.) driving experience
Fahrpreisanzeiger m (auto.) taximeter
Fahrrad n bicycle
FAHRRAD ~, (~beleuchtung, ~dynamo, ~kette, ~motor) bicycle ~

Fahrradschlüssel *m* cycle spanner
Fahrschalter *m* traction switch
Fahrstuhl *m* *(US)* elevator; *(UK)* lift
Fahrstuhlschacht *m* lift-shaft
Fahrstuhlsteuerung *f* lift control
Fahrtrichtungsanzeiger *m* *(auto.)* direction indicator
Fahrtschreiber *m* *(auto.)* time recorder
Fahrzeug *n* vehicle
Fahrzeugbatterie *f* traction-battery
Fall *m* fall, drop
Fallbügelinstrument *n* instrument with locking device
Fallbügelregler *m* hoop drop relay
fällen *v.t.* precipitate
fallender Guß, direct casting, top casting, down-hill casting, top pouring
fallend gießen *v.t.* top-pour, top-cast
Fallgeschwindigkeit *f* falling speed, velocity of fall
Fallgewicht *n* drop weight
Fallhammer *m* drop hammer
Fallhammerprüfung *f* *(mat.test.)* falling weight test
Fallhammerschmieden *n* drop forging
Fallhärteprüfung *f* drop hardness test
Fallhöhe *f* (e. Gesenkes:) height of drop
Fallkoks *m* *(Kupolofenbetrieb:)* coke recovered
Fallprobe *f (US)* drop test; *(K)* falling weight test
Fallschmierung *f* gravity lubrication
Fallschnecke *f* *(mach.)* drop worm
FALLSCHNECKEN ~, (~auslösung, ~gehäuse, ~hebel, ~lager, ~welle) drop worm ~
Fallstrom *m* downdraft
Fallstromvergaser *m* *(auto.) (US)* downdraft carburetor; *(UK)* downdraught carburetter
Fallwähler *m* *(tel.)* drop selector
Fallzerreißversuch *m* tensile impact test
Falschluft *f* infiltrated air, secondary air
Falte *f* fold, wrinkle
Fältelung *f* (als Gießfehler:) fold; (als Walzfehler:) lap
Fältelungsriß *m* *(surface finish)* lamination
Faltversuch *m* folding test, doubling test, bend-over test
Falz *m* *(mach.)* bead; seam; (Holz:) rebate; (shop term:) rabbet
falzen *v.t.* *(mach.)* fold, bead; seam; (Holz:) rebate; (shop term:) rabbet
Falzhobel *m* rebate plane
Falzung *f* *(mat.test.)* folding
Falzziegel *m* interlocking roofing tile
Fangdamm *m* cofferdam
Fanggitter *n* *(electron.)* suppressor grid
Fangleiste *f* (e. Gesenkes:) die lock
Fangschale *f* (für Späne:) pan, tray
Fangschaltung *f* *(electr.)* interception circuit
Farbanstrich *m* paint coat
färbbar *a.* colorable
Farbbottich *m* dye vat
Farbe *f* (streichfertig:) paint; color; (≈ Farbkörper:) pigment; (helle:) tint; (dunkle:) shade
farbecht *a.* fast-dyed
färben *v.t.* paint, color; (Holz:) stain; (Textilien:) dye
FARBEN ~, (~abstufung, ~chemie, ~libelle, ~mischung, ~skala, ~tafel, ~umschlag, ~zerlegung) color-~
farbenempfindlich *a.* color-sensitive
Farbenlehre *f* science of color, chromatics

Farbenmesser *m* chromatometer; colorimeter

Farbenmessung *f* colorimetry

Färber *m* dyer

Färbevermögen *n* coloring power

Farbfernsehen *n* color television

Farbglas *n* colored glass

farbig *a.* colored

farbiger Lack, enamel, lacquer

Farbkörper *m* pigment

Färbkraft *f* coloring power

Farblack *m* lake

Farbmesser *m* chromoscope

Farbreibemaschine *f* color grinding mill, paint roller mill

Farbschreiber *m* color trace recorder

Farbspritzanlage *f* paint-spraying equipment

Farbspritzpistole *f* paint-spraying gun

Farbstift *m* colored pencil

Farbstoff *m* (löslicher:) dye, dyestuff; (unlöslicher:) pigment, coloring matter, coloring substance, colorant

Farbton *m* color; shade; tint; hue

Farbtönung *f* (Arbeitsvorgang:) coloring; (Arbeitsergebnis:) shade

Farbumschlag *m* color change

Färbung *f* coloration; coloring; (Holz:) staining; *(plastics:)* dyeing

Färbungsmittel *n* coloring agent

Fase *f* chamfer; bevel

Faser *f* fiber; *(UK)* fibre; (Schlacke:) wool

Faserasbest *m* fibrous asbestos

faserig *a.* fibrous, fibriform

Faserkunstleder *n* fibrous artificial leather

Faserstoff *m* fibrous material

Faserstreifen *m* (metallo.) ghost line

Faserstruktur *f* (metallo.) fibering, fibrous structure

Faserung *f* texture; *(metallo.)* fibering; (Holz:) grain

Faserverlauf *m* grain flow, run of the grain

Faßabschmiergerät *n* *(auto.)* drum-type lubricating pump

Fassade *f* *(building)* façade, front

Fassadenbeleuchtung *f* front lighting

fassen *v.t.* hold; grip

Fassonarbeit *f* forming work, profiling work

Fassonautomat *m* automatic forming machine

Fassonblech *n* re-sheared sheet

fassondrehen *v.t.* form turn, profile

fassonpressen *v.t.* shape, form

Faßpumpe *f* *(auto.)* drum pump

Fassung *f* lamp holder; mounting; – pl. fittings; (Diamant:) mount

Fassungsschalter *m* socket switch

Fassungssteckdose *f* plug adapter

Fassungsstecker *m* lamp-holder plug

faulbrüchig *a.* short-brittle

Faulbrüchigkeit *f* rottenness, shortness, brittleness

Faulschlamm *m* digested sludge

Feder *f* (≈ Tragfeder:) spring; (≈ Paßfeder:) feather key; (e. Keilwelle:) spline; *(molding)* rib; *(woodworking)* tongue

FEDER ~, (~blatt, ~gehäuse, ~härte, ~kontakt, ~kraft, ~nute, ~stahl, ~waage, ~zirkel) spring ~

Federbolzen *m* spring-loaded bolt

Federgehänge *n* spring suspension

Federhammer *m* *(forge)* spring power hammer

Federhebel *m* spring-loaded lever

Federkeil *m* feather

Federlochtaster *m* inside spring caliper

federn *v.t.* spring-load; (Holz:) tongue; – *v.i.* spring, be resilient

federnd *p.a.* spring-loaded, spring-cushioned; resilient
federnde Zahnscheibe, lock washer
Federprüfgerät *n* spring tester
Federring *m* spring washer
Federrollenlager *n* flexible roller bearing
Federscheibe *f* washer
Federung *f* spring; elasticity; resilience; flexibility
Federungsvermögen *n* flexibility; resilience
Feder und Nut, *(mach.)* slot and key; *(woodworking)* tongue and groove
Fehlboden *m (building)* deadened floor
Fehler *m* defect; *(electr.)* fault
fehlerfrei *a.* free from defects; correct
fehlerhaft *a.* defective; wrong, faulty
fehlerhaft entgraten, (e. Schmiedestück:) mistrim
Fehlerstrom *m* fault current, leakage current
Fehlersuchgerät *n (electr.)* fault detector
Fehlgriff *m* faulty manipulation
fehlgriffsicher *a.* foolproof
Fehlgußstück *n* waste casting; misrun casting
fehlkantig *a.* (Holz:) waney-edged
Fehlschaltung *f (mech.)* faulty operation; (*or* engagement); *(electr.)* faulty switching
Fehlschmelze *f (steelmaking)* off-heat
Fehlstelle *f* (im Guß:) defect, fault
Fehlzündung *f (auto.)* misfiring
Feile *f* file
FEILEN ~, (~angel, ~bürste, ~härte, ~hauer, ~heft, ~stahl, ~zahn) file ~
Feilkloben *m* hand vise
Feilmaschine *f* filing machine
FEIN ~, (~bohrmaschine, ~bohrwerk, ~flächenschleifmaschine, ~gewinde, ~mechanik, ~mechanische Industrie, ~schleifen) fine ~
Feinabstimmkondensator *m* vernier capacitor
Feinabstimmung *f (radio)* sharp tuning
feinbearbeiten *v.t.* finish
Feinbearbeitung *f* finishing; fine machining
Feinblech *n* light sheet, thin gage plate, sheet
Feinblechgerüst *n* sheet mill stand
Feinblechstraße *f* sheet rolling train
Feinblechwalze *f* sheet roll
Feinblechwalzwerk *n* sheet rolling mill
feinbohren *v.t.* fine-bore, finish-bore, precision-bore
feindrehen *v.t.* precision-turn, fine-turn
Feindrehmaschine *f* precision lathe
Feindruckmesser *m* micro-pressure gage
Feineinstellschraube *f* micrometer screw
Feineinstellung *f* fine adjustment, micrometer adjustment
Feinflächenschleifmaschine *f* precision surface grinder
feinfühlig *a.* sensitive
Feinfühligkeit *f* sensitivity
feingängige Schraube, fine-pitch screw
Feingefühl *n* sensitive feel
Feingestalt *f* (e. Oberfläche:) microgeometric surface pattern
feingestuft *p.a.* (Drehzahlbereich:) sensitively adjustable
Feingewebe *n* fine fabric
feinhonen *v.t.* micro-hone, fine-hone
Feinkies *m* fine gravel
feinkopieren *v.t. (metal cutting)* finish-copy
feinkörnig *a.* fine-grained, close-grained

feinmahlen v.t. pulverize, powder
Feinmaßstab m precision scale
Feinmeßeinrichtung f precision measuring equipment
Feinmeßgerät n micrometer measuring instrument
Feinmeßokular n micrometer eyepiece
Feinmeßschieblehre f vernier caliper gage
Feinmeßschraublehre f micrometer caliper gage
Feinmeßtiefenlehre f micrometer depth gage
Feinmessung f precision measurement
Feinmeßwaage f precision spirit level
Feinsand m fine sand
feinschleifen v.t. precision-grind, fine-grind
Feinschliff m (Holz:) fine finish sanding
Feinsicherung f (electr.) fine-wire fuse
Feinstahl-Walzwerk n light-section rolling mill, small-section rolling mill
feinstbearbeiten v.t. precision-machine, superfinish
Feinstbearbeitungsmaschine f superfinishing machine
feinstbohren v.t. precision-bore, superfinebore; (UK) fine-bore
Feinstbohrwerk n superfine boring machine
Feinstdrehmaschine f high-precision lathe, super-finishing lathe
Feinstellscheibe f micrometer dial
Feinstellschraube f micrometer adjusting screw
Feinstgewebe n extra fine fabric
Feinstraße f sheet rolling train
Feinstruktur f microstructure, fine structure

Feinstschleifen n superfine grinding
Feinstschliff m superfine surface finish
feinstufig a. sensitive
feinstufiges Getriebe, selective speed gear mechanism
Feinstziehschleifen n superfinishing, microfinishing
Feinwaage f micrometric balance
Feinwerktechnik f precision engineering
Feinzerkleinerung f fine crushing
Feinziehschleifen n finish honing
Feinzink n high-grade zinc
Feinzug m (e. Ziehbank:) finishing block; (als Arbeitsvorgang:) finishing pass
Feinzugmatrize f sizing die
Feinzustellung f micrometer adjustment
Feld n (electr., magn.) field; (Stecker:) board, panel
FELD ~, (~änderung, ~erregung, ~magnet, ~regelung, ~regler, ~spule, ~stärke, ~telefon, ~verteilung, ~verzerrung, ~wicklung, ~wirkung, ~zerfall) field ~
Feldbahn f narrow-gage railway
Feldbahnlokomotive f field locomotive
Feldstärkemesser m field intensity meter
Felge f (auto.) rim
Felgenband n (auto.) rim tape
Felgenkurbel f brace rim wrench
Fensteraussteller m (auto.) window regulator
Fensterbank f window sill
Fensterbeschlag m window fitting
Fenstersteller m (auto.) window fastener
Fensterglas n window glass
Fensterkitt m glazier's putty
Fensterkurbel f (auto.) window crank handle

Fensterkurbelapparat *m (auto.)* window winding gear

Fernamt *n (US)* toll exchange; *(UK)* trunk exchange

Fernanruf *m* trunk call; long-distance call

Fernbedienung *f* remote control

Fernbus *m* motor-coach

Ferndrucker *m* teleprinter

Fernempfang *m* long-distance reception

Fernfahrer *m* truck driver

Ferngas *n* long-distance gas, grid gas

Ferngasnetz *n* gas grid

Ferngasversorgung *f* grid gas supply

Ferngespräch *n* trunk call; telephone call

ferngesteuert *p.a.* (Raketen:) remote-guided

Fernhörer *m* telephone receiver

Fernkabel *n* long-distance cable, trunk cable

Fernlastverkehr *m* long-distance road haulage

Fernleitung *f (tel.) (US)* long-distance line; *(UK)* trunk line; *(electr.)* transmission line

Fernleitungskabel *n* long-distance cable, trunk(-line) cable

Fernlicht *n* full headlight beam

Fernlichtkontrolleuchte *f (auto.)* main beam warning lamp

FERNMELDE ~, (~kabel, ~leitung, ~techniker) communication ~

Fernmeldewesen *n* telecommunication

Fernmeßanlage *f* telemetering system

Fernmeßeinrichtung *f* telemetering device

Fernmeßinstrument *n* telemeter

fernmündlich *a.* telephonic

Fernschaltung *f* remote control

Fernschreibamt *n* teletype exchange

fernschreiben *v.t.* teleprint

Fernschreiber *m* telewriter

FERNSEH ~, (~bild, ~empfang, ~empfänger, ~kanal, ~sender, ~technik) television ~

fernsehen *v.t.* televise

Fernsehen *n* television

Fernsehfunk *m* television

Fernsehsendung *f* video transmission

FERNSPRECH ~, (~amt, ~anlage, ~kabel, ~leitung, ~technik, ~verkehr, ~vermittlung) telephone ~

Fernsprechanschluß *m* subscriber's line

Fernsprechautomat *m* coin-box phone, pay station

Fernsprechbetrieb *m* telephony

Fernsprecher *m* telephone

Fernsprechzelle *f* telephone box, telephone booth

Fernsteuerung *f* remote control

Fernübertragung *f* teletransmission

Fernverbindung *f (tel.)* trunk call; trunk connection

Fernverkehr *m (traffic)* trunk traffic; long-distance communication

Fernverkehrsstraße *f* trunk road

Fernvermittlung *f (tel.)* trunk exchange

Fernvermittlungsamt *n* tandem office

Fernversorgung *f* long-distance supply

Fernzähler *m* telecounter

ferritisch *a.* ferritic

Ferritstreifen *mpl.* lines of ferrite, bands of ferrite

FERRO ~, (~chrom, ~legierung, ~mangan, ~nickel, ~silizium) ferro ~

Fertigbehandlung *f* finishing operation

Fertigbeton *m* ready-mixed concrete, transit-mixed concrete

fertigblasen *v.t. (met.)* blow full

fertigdrehen *v.t.* finish-turn

Fertigdrehmaschine f finishing lathe
fertigen v.t. manufacture, make, fabricate
Fertigerzeugnis n finished product
fertigfräsen v.t. finish-mill
Fertiggerüst n (rolling mill) finishing stand of rolls
Fertiggesenk n finisher
Fertigkaliber n (rolling) finishing groove
Fertigkeit f skill
fertigmachen v.t. (s. Schmelze:) finish; (Kupfer:) pole tough pitch
Fertigmaß n finished size
Fertigreibahle f finishing reamer
Fertigschlacke f refining slag, finishing slag
fertigschleifen v.t. finish-grind
fertigschlichten v.t. (mach.) fine-finish; (rolling mill) planish, polish
Fertigschlichtkaliber n last planishing pass, finishing pass
Fertigschlichtstich m last finishing pass
Fertigschliff m finish grinding
Fertigschmieden n finish-forging
Fertigschnitt m final cut
Fertigstauchstempel m (Schraubenfabrikation) header
Fertigstich m (rolling) finishing pass, shaping pass
Fertigstraße f (rolling) finishing mill train, finishing train
Fertigteil n finished part
Fertigung f production, manufacture
FERTIGUNGS ~, (~aufgabe, ~betrieb, ~ingenieur, ~kosten, ~kostenstelle, ~teil, ~toleranz, ~vorgang, ~zeit) production ~
Fertigungslohn m direct labor cost
Fertigungsplan m (work study) process chart

Fertigungsstandard m (cost accounting) production standard
Fertigungsstraße f transfer line
Fertigwalzen n finish rolling
Fertigzug m (Ziehtechnik) finishing pass
fest a. (stabil:) strong; (starr:) rigid; (massiv:) solid; (feststehend:) stationary; (festsitzend:) dead; (sicher:) secure, tight
Festbeton m hardened concrete
festbrennen v.i. (Sand am Guß:) sinter
festfressen v.r. seize, jam
festgelagert a. fixed; stationary
festhaftend a. firmly adhering
Festigkeit f strength; rigidity; sturdiness; stability
Festigkeitseigenschaften fpl. mechanical properties, strength properties
Festigkeitsguß m high-strength cast iron
Festigkeitsschweißung f strength weld
festkleben v.i. stick
Festkörper m solid
festlaufen v.i. jam
Festlehre f fixed gage
Festlünette f steadyrest
Festnaht f (welding) strength weld
Festscheibe f (belt drive) tight pulley
Festsitz m force fit; (≈ Preßsitz:) interference fit
festspannen v.t. fasten, clamp, lock; tie
feststehend a. stationary; fixed
feststehender Setzstock, plain steadyrest
Feststellbremse f parking brake
feststellen v.t. (mech.) secure, lock; clamp
Feststellschraube f locking screw
Feststellvorrichtung f locking device
Festzeitgespräch n fixed time call
festziehen v.t. (e. Schraube:) tighten

Fett *n* fat; *(techn.)* grease; (Riemen:) dressing

FETT ~, (~büchse, ~kanal, ~nippel, ~presse, ~spritze) grease ~

fetten *v.t.* grease

fettig *a.* greasy

Fettsäure *f* fatty acid

feucht *a.* moist, humid

feuchtempfindlich *n.* sensitive to humidity

Feuchtigkeit *f* moisture; (Luftfeuchtigkeit:) humidity

feuchtigkeitsdurchlässig *a.* permeable to moisture

Feuchtigkeitskorrosion *f* aqueous corrosion

Feuchtigkeitsmesser *m* hygrometer; psychrometer

Feuchtraum *m* damp room

Feuer *n* fire; (e. Kohlebürste:) sparking

FEUER ~, (~löscher, ~pumpe, ~melder, ~meldung, ~rost, ~tür) fire ~

feuerbeständig *a.* fire-resistant

Feuerbuchskupfer *n* firebox-copper

feuerfest *a.* fire-proof, fire-resistant, heat-proof; (Steine:) refractory

feuerfester Baustoff, refractory

feuerfester Mörtel, refractory mortar

feuerfester Ton, fireclay

feuerfester Ziegel, firebrick

Feuerfestigkeit *f* (Steine:) refractoriness

feuergefährlich *a.* inflammable, flammable

Feuerlöschkraftfahrzeug *n* fire engine

feuerlöten *v.t.* (hart:) muffle-braze; (weich:) sweat

feuerschweißen *v.t.* forge-weld

Feuerung *f* heating, firing; (als Anlage:) fireplace, firebox, fire; furnace; hearth

Feuerungsniederschläge *mpl.* furnace deposits

feuerverbleien *v.t.* lead-coat

feuerverkupfern *v.t.* copper-coat

feuerverlöten *v.t.* sweat together

feuerverzinken *v.t.* hot galvanize, pot galvanize

feuerverzinnen *v.t.* tin coat

Feuerwehrleiter *f* fire escape

Feuerwehrleiter-Anhänger *m* fire escape trailer

Feuerwehrschlauch *m* fire hose

Feuerwehrwagen *m* fire-fighting vehicle

Filter *m* filter

FILTER ~, (~drossel, ~einsatz, ~kondensator, ~packung, ~spule, ~tuch) filter ~

Filz *m* felt

FILZ ~, (~dichtung, ~einlage, ~nagel, ~unterlage, ~unterlegscheibe) felt ~

Finanzbuchhaltung *f* financial accounts department

Finanzkonto *n* finance account

Findling *m* erratic block

Finger *m* *(lathe),* dog

Firnis *m* linseed oil

First *m* *(building)* ridge

Fischleim *m* fish glue

fixieren *v.t.* fix, locate, secure

Fixierstift *m* locating pin

flach *a.* flat; (≈ eben:) plane; (Gewinde:) square; (Kegel:) long; (Mulde:) shallow; (Winkel:) slight

Flachbagger *m* surface digging machine

Flachbaggerung *f* surface digging, shallow cut digging

Flachbahnanlasser *m* faceplate starter

Flachbahnführung *f* *(lathe)* square guide

Flachbettfelge *f* *(auto.)* flat base ring

Flachdach *n* flat roof

Flachdichtung f gasket
Fläche f surface; area; plane
Flächendruck m surface pressure
Flächengewichtswaage f area weight balance
Flächenmaß n measure of area
Flächenpressung f surface pressure
Flächenschleifmaschine f surface grinder
Flachgewinde n square thread
Flachkaliber n *(rolling mill)* box pass
Flachkeil m parallel key
Flachknüppel m *(met.)* slab billet
Flachlehre f flat gage
Flachmeißel m square-nose tool
Flachnaht f *(welding)* flush weld
Flachrevolverdrehmaschine f flat turret lathe
Flachrevolverkopf m flat turret head
Flachrundkopf m mushroom head
Flachschleifmaschine f face grinder
Flachschweißung f flush welding
Flachstahl m flat steel; – *pl.* flats
Flachstich m *(rolling mill)* flat pass
Flachstromvergaser m *(auto.)* crossdraught carburetter, cross-draft carburetor
Flachwinkel m *(tool)* millwrights' steel square
Flachwulststahl m flat bulb steel; – *pl.* beaded flats
Flachzange f flat-nose plier
Flammenführung f flame baffling
Flammenrückschlag m *(welding)* flashback, back firing
Flammgrundieren n flame priming
flammhärten v.t. (obs.) cf. Randschichthärten
flammlöten v.t. flame-braze, flame solder and braze

Flammofen m reverberatory furnace, air furnace, gas furnace
Flammofenschlacke f air-furnace slag
Flammofenschmelzen n reverberatory smelting
flammpolieren v.t. flame-polish
Flammpunkt m flash point
Flammpunktprüfer m flash-point tester
flämmputzen v.t. deseam
Flammrohrkessel m flue boiler
Flammschockspritzen n flame plating
flammspritzen v.t. flame-spray
flammstrahlen v.t. flame-clean
Flanke f *(gearing, threading)* flank
FLANKEN ~, (~anlage, ~berührung, ~fläche, ~form, ~krümmung) flank ~
Flankendurchmesser m pitch diameter; *(UK)* effective diameter
Flankeneinbrand m *(welding)* fusion penetration
Flankenkehlnaht f flank fillet weld
Flankenlinie f *(gears)* tooth trace
Flankenspiel n (notwendiges:) flank clearance; (unerwünschtes:) backlash
Flankenweite f (e. Fugennaht:) half angle of throat
Flankenwinkel m (e. Doppelkehlnaht:) angle of bevel; *(gears)* flank angle; *(threading)* thread angle
Flansch m flange
Flanschbefestigung f flange-mounting
Flanschendichtung f flanged joint
Flanschlager n flange-mounted bearing
Flanschmotor m flange-mounted motor
Flanschträger m flanged beam
Flanschverbindung f flange joint
Flanschwelle f flanged shaft
Flasche f *(gas)*, cylinder

Flaschengas *n* cylinder gas
Flaschenhalskokille *f* bottle top mold
Flaschenzug *m* [chain] block, tackle block
flattern *v.i.* (Ventil:) knock, flatter
Fleck *m* stain; *(op., nucleon.)* spot
Fleckspachtel *m* spot primer
Fleischseite *f* (e. Riemens:) flesh-side
flicken *v.t.* repair; (Ofenfutter:) patch; (Autoreifen:) mend
fliegend *a.* overhung
fliegende Schere, flying shears
Fliehkraft *f* centrifugal force
FLIEHKRAFT ~, (~anlasser, ~antrieb, ~kupplung, ~moment, ~regler, ~schalter, ~wirkung) centrifugal ~
Fliese *f* tile
Fließband *n* assembly line
Fließbandmontage *f* progressive assembly
Fließdrehmaschine *f* metal flow-turning lathe; (≈ Drückbank:) spinning lathe
fließen *v.i.* flow; (Werkstoffe:) yield
Fließfiguren *fpl.* flow lines, stretcher strains
Fließgrenze *f (material testing)* yield point
Fließpresse *f* extrusion press
fließpressen *v.t. (cold work)* (Buchsen, Stirnräder, Kegelräder:) cold-extrude
Fließpressen *n* [cold] extrusion
Fließpreßverfahren *n* cold extrusion process
Fließpreßwerkzeug *n* cold extrusion die
Fließpunkt *m* (Öl:) pour point
Fließreihe *f* transfer line
Fließsand *m* quick sand
Fließscheide *f* (e. Gesenkes:) neutral flow plane
Fließspan *m* flow chip, continuous chip

Fließspannung *f (material testing)* yield stress
Flocken *fpl.* (Fehlstelle in legiertem Stahl:) flakes
Flockenempfindlichkeit *f* (e. Stahls:) susceptibility to flakes
flockenhaltig *a. (met.)* flaky
Flockenriß *m* flake
flockenrissig *a.* flaky
fluchteben *a.* flush
fluchten *v.t. & v.i.* align
fluchtgerecht *a.* flush
flüchtig *a.* volatile
Fluchtlinie *f* alignment, straight line
Flüchtung *f* alignment
Flugasche *f* flue dust
Flugbahn *f* trajectory
Flügel *m* wing; blade; vane; (e. Schraube:) thumb drive
Flügelmutter *f* wing nut, thumb nut
Flügelschraube *f* wing, screw, thumb screw
Flügelventil *n* fly-valve
Flugkreisdurchmesser *m (woodw.)* cutting track circle
Flugmotor *m* aircraft engine
Flugzeugbenzin *n* aviation gasoline
Flugzeugmotor *m* aircraft engine
Fluoreszenz *f* fluorescence
FLUORESZENZ ~, (~lampe, ~licht, ~schirm, ~strahlung) fluorescent ~
Fluoreszenzmeßgerät *n* fluorometer
Flurebene *f* floor level
Fluß *m (electr., magn.)* flux
flüssig *a.* fluid; liquid; (schmelzflüssig:) molten
flüssiger Einsatz, *(met.)* liquid metal charge
flüssiges Eisen, molten metal

Flüssigkeit f liquid, fluid
FLÜSSIGKEITS ~, (~antrieb, ~bremse, ~druck, ~getriebe, ~motor) hydraulic ~
Flüssigkeitskupplung f fluid clutch
Flüssigkeitsmechanik f hydromechanics
Flüssigkeitsreibung f fluid friction
Flußmittel n flux, fluxing agent
Flußspat m fluor spar, fluorite
Flußstahl m ingot steel, soft steel, mild steel
Flußstahlblech n mild sheet steel
Flußstahlblock m mild-steel ingot block
fokussieren v.t. focus
Fokussierung f focusing
Folge f effect; consequence; (≈ Reihenfolge:) sequence
Folgeschnitt m successive operation
Folgesteuerung f sequence control
Folgewerkzeug n follow-on tool
Folie f foil; thin sheet, film
Folienklebung f foil bonding, bonding of foils
Folienkunstleder n artificial leather plastic foil, leatherette
Förderanlage f conveying machinery
Förderband n belt conveyor, conveyor belt
Förderbremse f hoist brake
Förderer m (techn.) conveyor
Fördergefäß n conveyor bucket
Fördergerüst n hoist structure
Fördergurt m conveyor belt
Förderkohle f run-of-mine coal
Förderkübel m conveyor bucket, elevator bucket
Förderleistung f conveying capacity
Fördermittel n conveying appliance
fördern v.t. convey; deliver; transport; (Flüssigkeiten:) pump

Förderpumpe f feed pump; (auto.) fuel supply pump
Förderrinne f conveyor trough
Förderseil n winding rope, hauling rope; hoisting rope
Förderung f conveyance; haulage; transfer; output; delivery
Förderwagen m transport car, delivery car
Form f form, shape; contour; profile; (≈ Matrize:) die; (Gieß-, Preßform:) mold; (Windform:) tuyere
Formänderung f deformation
Formänderungsfestigkeit f resistance to [plastic] deformation
Formänderungsgeschwindigkeit f (forging) rate of deformation
Formänderungskraft f (forging) forming force; deformation force
Formarbeit f (founding) molding operation, molding; (mach.) contouring operation
Formatschnitt m (woodw.) dimension sawing
Formätzen n chemical milling
Formbeständigkeit f dimensional stability
Formbildung f shaping
Formblech n shaped plate
Formdraht m section wire
Formdreheinrichtung f profile-turning attachment
Formdrehmaschine f form-turning lathe
Formdrücken n stamping, forming; (specif.) shallow forming
Formeinrichtung f (founding) molding equipment
Formel f formula
formen v.t. form, shape, profile, contour; (≈ gestalten:) construct; (founding) mold
Formenbau m die- and mold-making
Former m (founding) molder

Formerei | 100 | **fortschrittlich**

Formerei f *(founding)* molding shop
Formereieinrichtung f *(foundry)* mold-making equipment
Formerwerkzeug n molder's tool
formfräsen v.t. profile-mill, form-mill, contour-mill
Formfräser m formed milling cutter
Formfräsmaschine f profile miller, contour miller
formgeben v.t. profile, shape
Formgebung f forming, shaping; (spanlos:) non-chip forming; (zerspanend:) shape cutting
formgepreßtes Rohr, *(plastics)* compression molded tube
Formgraviermaschine f form engraving machine
Formgußteil n die-casting
Formhälfte f *(founding)* mold part; (untere:) drag; (obere:) cope; *(shell molding)* shell
Formhohlraum m *(founding)* mold cavity
Formkaliber n *(rolling mill)* shaping pass, section groove
Formkasten m *(founding)* flask
formkopieren v.t. copy
Formlineal n *(metal cutting)* template, former
Formmaschine f molding machine
Formmaske f *(founding)* shell mold
Formmaskenguß m shell-molded casting
Formmaskenverfahren n *(founding)* shell molding process, Croning molding process
Formmasse f molding compound, molding composition
Formmeißel m form cutting tool
Formpressen n *(plastics)* compression-mold

Formrändel n shaped knurl
formsägen v.t. contour-saw
Formsand m molding sand
Formsandaufbereitungsanlage f molding sand preparation plant
Formschalenverfahren n *(founding)* cf. Formmaskenverfahren
formschlichten v.t. finish-form
formschlüssig a. form-fitting
formschlüssige Passung, form fit
formschlüssig verbunden, form fitting
formschmieden v.t. precision-forge
formschruppen v.t. rough-form
Formstahl m structural steel, sectional steel; sections
Formstahlwalzwerk n section rolling mill
formstanzen v.t. (obs.) cf. gesenkziehen
Formstempel m forming die
Formstich m *(rolling mill)* shaping pass
Formstoff m *(founding)* mold material; *(plastics)* molded material
Formstück n fitting
Formtechnik f *(plastics)* molding technique
Formteil n *(plastics)* molded part, molded article
Formtrockenofen m *(foundry)* mold drying oven
Form- und Schraubenautomat m automatic screw machine
Formung f shaping, forming
Formwalze f *(rolling mill)* grooved roll
Formwalzen n section rolling
Formziegel m purpose-made tile
Formzyklus m cycle of deformations
fortpflanzen v.r. propagate
Fortpflanzung f (z. B. von Schallwellen:) propagation
fortschrittlich a. progressive

fortspülen *v.t.* wash away
Fotokopie *f* photostat
Fotomontage *f* photomontage
Fotozelle *f* photocell; photoelectric cell
FRÄS ~, (~antrieb, ~arbeit, ~bereich, ~dorn, ~einrichtung, ~gang, ~leistung, ~maschine, ~messer, ~motor, ~schnitt, spindel, ~support, ~technik, ~vorrichtung, ~vorschub, ~werkzeug) milling ~
Fräsanschlag *m (woodw.)*. molding fence
Fräsautomat *m* automatic milling machine
Fräsdornring *m* spacing collar
fräsen *v.t.* mill; (Gesenke:) die-sink; (Keilwellen:) spline; (Naben:) spotface; (nach dem Wälzverfahren:) hob; *(woodworking)* shape
Fräser *m* milling cutter; (in Wortverbindungen kurz:) cutter; *(woodw.)* molding cutter
FRÄSER ~, (~abhebung, ~aufnahme, ~dorn, ~einstellehre, ~futter, ~prüflehre, ~schärfmaschine, ~schleifmaschine, ~zahn) cutter ~
Fräserei *f* milling shop
Fräsfutter *n* cutter chuck
Fräskopf *m* (e. Fräsmaschine:) cutter head; (≈ Messerkopf:) cone-type face milling cutter
Fräslager *n* cutter spindle bearing
Frässchlitten *m* cutter slide, spindle slide, milling head; (e. Wälzfräsmaschine:) hob saddle
Frässtichel *m (engraving)* single-lip cutter
Fräs- und Bohrmaschine *f* milling, drilling and boring machine
Fräswerk *n* horizontal drilling, boring, and milling machine; jigmil

freiarbeiten *v.t. (metal cutting)* relieve, undercut
Freifallmischer *m (concrete)* rotary drum mixer
Freifläche *f* (e. Drehmeißels:) flank; (e. Fräsers:) back
freiformschmieden *v.t.* hand-forge, hammer forge
Freiformschmieden *n* open die forging, hand forging, smith forging, hammer forging
Freiformschmiedestück *n* smith forging
Freigabe *f* release
freigeben *v.t.* release
freigeformtes Schmiedestück, smith forging, hand forging
freihandschleifen *v.t.* grind offhand
Freilauf *m (auto.)* free-wheeling
Freilaufbremse *f (bicycle)* back-pedalling brake
Freilauffelge *f (auto.)* free-wheel rim
Freilaufkupplung *f* free-wheel clutch, overrunning clutch
Freilaufnabe *f (auto.)* free-wheel hub
Freileitung *f (electr.)* overhead [transmission] line
FREILUFT ~, (~anlage, ~schalter, ~station, ~umspanner) outdoor ~
Freiluftschaltanlage *f (electr.)* switchyard
freischleifen *v.t.* relief-grind
freischneiden *v.t.* relieve, back off, cut free
freischwebend *p.a.* freely supported
Freistich *m* undercut
freistrahlen *v.t.* shot-blast
Freistrahlgebläse *n* hose sandblast gun
freitragen *v.i.* overhang; self-support
freitragend *p.a.* self-supporting, overhung
Freiwinkel *m* (e. Meißels:) clearance angle

Freizeichen n (tel.) free line signal, dial tone
fremdbelüften v.t. ventilate separately
Fremdbelüftung f (Motor:) separate ventilation, fan cooling
fremderregte Schwingungen, (metal cutting) forced vibrations
Fremderregung f (electr.) separate excitation
Fremdkörper m foreign substance, foreign particle
Fremdlüftung f separate ventilation
Fremdstrom m parasitic current; (Strom aus Fremdbezug:) outside power
Frequenz f frequency
FREQUENZ ~, (~abweichung, ~anzeiger, ~band, ~gang, ~messer, ~modulation, ~regler, ~steuerung, ~teiler, ~umformer, ~umsetzer, ~verdopplung, ~vervielfacher, ~verschiebung, ~wandler, ~wiedergabe) frequency ~
fressen v.t. & v.i. (Ofenfutter:) wear, erode; (chem.) corrode; (Schlacke:) cut; (Lager:) seize; (metalcutting) score
friemeln v.t. (cold work) (Wellen, Rohre:) cross-roll; (US) reel
Friktionsschmiedepresse f friction forging press
frisch a. (Holz:) green
Frischdampf m live steam
Frischeisen n refined iron
frischen v.t. refine; oxidize; (Windfrischen:) convert, blow; (Blei:) reduce
Frischerz n oxidizing ore
Frischfeuereisen n charcoal hearth iron
Frischherd m refining hearth
Frischherdverfahren n refinery process
Frischluftheizung f (auto.) interior heater
Frischluftzufuhr f admission of fresh air

Frischöl n fresh oil
Frischperiode f (met.) oxidizing period
Frischreaktion f (met.) boil
Frischschlacke f oxidizing slag
Frischverfahren n refining process; oxidizing process
Frontantrieb m (auto.) front wheel drive
Frontlader m front-end loader
Fronträumer m bulldozer
Frosch m (mech.) clip; (railw.) frog
Froschklemme f (tel.) wire grip
Froschplatte f (power press) bolster plate, bolster
Frostriß m frost shake
Frostschutzmittel n anti-freeze mixture, anti-freezer
Frostschutzpumpe f anti-freeze pump
Frostschutzscheibe f (auto.) demister screen
Frühzündung f pre-ignition, premature ignition
Fuchsschwanz m (tool) hand saw
Fuge f (≈ Stoßfuge:) joint; (≈ Naht:) seam; (≈ Spalt:) interstice; (≈ Schlitz:) slot; (welding) gap, open joint, edge; (Nut:) groove
Fügemaschine f jointer
fügen v.t. join, joint; mate; assemble
Fugenflanke f groove flank
Fugenhartlöten n open joint brazing
Fugenhobeln n joint gouging
Fugenlänge f joint length
Fugenlöten n open joint brazing
Fugenverguß m (concrete) joint sealing
Fugenvergußarbeit f (building) joint grouting work
Fugenvorbereitung f (welding) edge preparation
Fühler m (copying) feeler, tracer

fühlergesteuert *a.* tracer-controlled
Fühlerlehre *f* thickness gage
Fühlersteuerung *f* tracer control [mechanism]
Fühlfinger *m (copying)* tracer
Fühlhebelschraublehre *f* indicating micrometer
führen *v.t.* guide; (Strom:) carry
Führer *m* (Kran:) operator, driver
Führersitz *m (auto.)* driver's seat
Fuhrpark *m* motor-vehicle fleet
Führung *f* guide; *(mach.)* guideway; *(met.)* (e. Ofens:) operation; (e. Schmelze:) working; (e. Kabels:) run
FÜHRUNGS ~, (~kasten, ~rohr, ~schiene, ~säule, ~stange, ~stift, ~streifen, ~stück, ~trommel, ~zylinder) guide ~
Führungsbahn *f* (e. Werkzeugmaschine:) guide-way
Führungseigenschaft *f* (e. Lagers:) locating characteristic
Führungsfase *f* (an e. Schneide:) heel, margin
Führungslager *n* locating bearing; (e. Bohrspindel:) pilot bearing; (e. Fräsdornes:) arbor support
Führungsleiste *f (lathe)* taper gib
Führungsrolle *f (rolling mill)* feed roller
Führungsschlitten *m (boring mill)* pilot carriage
Führungszapfen *m* pilot [guide]
Füllkitt *m* filling cement
Füllkoks *m (founding)* bed coke
Füllluftmesser *m (auto.)* air inflation indicator
Füllsand *m (Formerei:)* body sand, backing sand
Füllschraube *f* plug screw, filler plug
Füllstoff *m* filler

Fülltrichter *m* hopper
Fundament *n* foundation, base
FUNDAMENT ~, (~aushub, ~beton, ~bolzen, ~platte, ~sockel, ~sohle, ~schraube) foundation ~
Fundamentierung *f* foundation
Fünferalphabet *n* five unit code
Fünfkantrevolverkopf *m* five-face turret head
Fünfpolröhre *f* pentode, five-electrode valve
Fünfspindelautomat *m* automatic five-spindle screw machine
Fünfspindelfutterautomat *m* five-spindle chucking automatic
FUNK ~, (~bake, ~empfänger, ~feuer, ~gerät, ~meßtechnik, ~nachrichtenkanal, ~navigation, ~ortung, ~peiler, ~peilung, ~sender, ~sendung, ~sprechgerät, ~sprechverkehr, ~streifenwagen, ~technik, ~telefonie, ~telegrafie, ~trägerfrequenz, ~turm, ~verbindung, ~wellen) radio ~
Funkbild *n* photoradiogram
Funkbildübertragung *f* radiophotography
Funke *m* spark
Funkenbildung *f* sparking
Funkenerosion *f* spark machining
funkenerosiv *a.* electroerosive
Funkenfolgefrequenz *f (electroerosion)* group frequency
Funkenfrequenz *f* spark frequency
funkenhärten *v.t.* toughen by sparks
Funkenhärtung *f* spark discharge toughening
Funkenlöschkondensator *m* spark quenching condenser
Funkensender *m* spark transmitter
funkensicher *a.* non-sparking

Funkentstörung f radio interference suppression
Funkenstrecke f spark gap
Funkenüberschlag m *(electroerosion)* sparkover
Funker m wireless operator
Funkspruch m radiogram
Funktelegramm n radiogram
Funkübertragung f wireless transmission
Furnier n veneer
furnieren v.t. veneer
Furnierplatte f veneer panel
Fuß m (e. Maschine:) leg; *(gearing, threading)* root; (e. Schiene:) base, flange
Fußabblendschalter m *(auto.)* foot dip switch
Fußanlasser m *(motorcycle)* kickstarter; *(auto.)* foot starter
Fußanlasserschalter m *(auto.)* foot-operated starting switch
Fußausrundung f *(gears)* fillet
Fußbodenbelag m floor covering
Fußbodenbrett n floor board
Fußbodenlack m floor varnish
Fußbremshebel m *(auto.)* foot brake pedal
Fußflanke f *(gears)* flank
Fußgashebel m *(auto.)* gas pedal
Fußhebel m foot-pedal
Fußkreis m *(gearing)* root circle
Fußlinie f *(gearing)* root line
Fußmatte f *(auto.)* matting
Fußmotor m foot-mounted motor
Fußraste f *(auto.)* foot rest
Fußschalter m foot-pedal switch
Fußspindelpresse f foot press
Fußtiefe f *(gears)* dedendum
Fußwinkel m *(gears)* dedendum angle
Futter n (e. Ofens, Lagers:) lining; (≈ Spannfutter:) chuck
Futteraufspannung f chucking
Futterautomat m chucking automatic
füttern v.t. *(techn.)* line
Futterrevolverautomat m turret-type chucking automatic
Futterschlüssel m chuck wrench

G

Gabel f fork; *(für Getrieberäder:)* shifter fork; *(tel.)* hand-set
Gabelhebel m fork lever
Gabelkopf m *(motorcycle)* fork crown
gabeln v.r. bifurcate
Gabelpfanne f *(founding)* bull ladle, shank ladle
Gabelpleuelstange f *(auto.)* forked connecting rod
Gabelrohr n *(auto.)* bifurcated tube
Gabelschlüssel m engineers' wrench
Gabelstapler m fork truck
Gabelung f bifurcation
galvanisch a. galvanic
galvanische Metallisierung, electrodeposition, metal-plating, electro-plating, electrolytic deposition
galvanisches Element, galvanic cell, primary cell
galvanische Verchromung, chrome-plating
galvanische Vernickelung, nickel-plating
galvanische Verzinkung, electro-galvanizing, cold galvanizing
galvanisch verkupfern, copper-plate
galvanisch verzinken, cold-galvanize, electro-galvanize
galvanisch verzinnen, tin-plate
galvanisieren v.t. electroplate
Galvanisierung f electroplating, electrodeposition
Galvanisierwerkstatt f electroplating shop
Galvanometer m galvanometer
Galvanotechnik f electro-deposition, electro-plating practice
Gang m *(auto.)* speed, gear; (e. Gewindes:) turn, thread; (e. Schmelzofens:) run; operation; (e. Schnecke:) start; (e. Schmelze:) heat; (e. Motors:) speed; (e. Maschinentisches:) traverse; *(rolling)* pass; (≈ Weg:) path, passage; (≈ Bewegung:) movement, motion; (Frequenzen:) response; (e. Spindel:) rotation
Ganganzeiger m *(motorcycle)* gear position indicator
Gangart f *(met.)* gangue
Gängigkeit f (e. Gewindes:) number of threads; (e. Schnecke:) number of starts
Ganghöhe f (e. Gewindes:) lead
Gangrichtung f (e. Gewindes:) direction of thread; (e. Schneidwerkzeuges:) direction of hand
Gangschalter m *(auto.)* gear shift lever
Gangschaltung f *(auto.)* gear change, gear shifting
Gangwechsel m *(auto.)* gear shift
Ganzautomat m full-automatic lathe
gar a. *(met.)* refined; (Koks:) carbonized
Garage f garage
Garagengeräte npl. garage equipment
garen v.t. (Stahl:) refine; (Kohlen:) coke
Garkupfer n refined copper
Garn n yarn
Garungszeit f *(Kokerei:)* coking period, coking time
GAS ~, (~abzugsrohr, ~anzünder, ~aufkohlung, ~batterie, ~dichte, ~dichtemesser, ~druck, ~einschluß, ~entladung, ~entladungsröhre, ~erzeuger, ~feuerung, ~flasche, ~hahn, ~herd,

~kohle, ~koks, ~leitung, ~motor, ~pedal, ~preßschweißen, ~reinigungsanlage, ~rohr, ~schweißbrenner, ~schweißen, ~schweißung, ~turbine, ~werk) gas ~

Gasabzug *m* gas issue; (Rohrleitung:) gas offtake; (e. Hochofens:) downcomer

Gasblase *f (founding)* blowhole; *(welding)* gas pocket

Gasbrennschweißen *n* constant-temperature pressure welding

Gasdrehgriff *m (motorcycle)* throttle twistgrip

Gaserzeugergas *n* producer gas

Gasfang *m* gas collector; (Gichtverschluß:) gas seal bell

Gasfeuerung *f* gas firing

gasförmig *a.* gaseous

Gasfußhebel *m* accelerator pedal, gas pedal

Gasgewindebohrer *m* pipe tap

Gashebel *m (auto.)* throttle hand lever, gas pedal

Gasmaske *f* respirator

Gasnitrieren *n (heat treatment)* dry nitriding

Gasrohrschlüssel *m* 'bull dog' wrench

Gasverschluß *m* gas seal; (e. Hochofengicht:) bell-and-hopper arrangement

Gaswäscher *m* gas washer, scrubber

Gas-Wulstschweißen *n* pressure gas welding

gattieren *v.t. (met.)* calculate a charge

Gattieren *n* mixture-making, mixing

Gattierungswaage *f* charging scales; *(blast furnace)* burden balance

Gattierungswagen *m* charging car

Gaze *f* gauze

Geber *m (telegr.)* telegraph transmitter

Gebinde *n* container

Gebläse *n (auto.)* fan, blower; *(welding)* blow torch

Gebläsekies *m* blast grit

Gebläselampe *f* blow torch

Gebläsemaschine *f* blowing engine, blower engine

Gebläsemotor *m* blast-injection engine

Gebläsesand *m* sandblasting sand

Gebläseschachtofen *m* blast furnace

Gebläsevorverdichter *m (auto.)* blower-type supercharger

Gebläsewind *m* air blast, blast

gebrannter Kalk, calcined lime, quick lime

Gebrauchsanweisung *f* direction for use

gebrauchsfertig *a.* I ready for use

gebrauchsfertiges Gemisch, ready-for-use mixture

Gebrauchsmusterschutz *m* legal protection of registered designs

Gebrauchsspannung *f* service voltage

gebrochenes Härten, interrupted quenching

Gebührenanzeiger *m (tel.)* call-fee indicator

gebührenpflichtiger Anruf, chargeable call

gedrungen *a.* compact, sturdy

Geigerzähler *m* Geiger-Müller counter

Gefahrenschalter *m* emergency switch

Gefäß *n* vessel; container; receptacle

Geflecht *n (Draht:)* netting, cloth

Gefrierschutzmittel *n* anti-freezing agent, anti-freeze

Gefüge *n (met.)* structure

Gefügeaufbau *m (cryst.)* structural constitution

Gefügeausbildung f crystalline structure
Gefügebeständigkeit f structural stability
Gefügebestandteil m structural constituent
Gefügeumwandlung f structural transformation
Gegendruckturbine f back-pressure turbine
Gegendruckventil n back pressure valve
gegenelektromotorische Kraft, counter-electromotive force
Gegenflansch m counter-flange
Gegenführung f (e. Werkzeugmaschine:) steady
Gegengewicht n counterweight
Gegenhalter m (miller) overarm; (hobber) counterstay; (turret lathe) tail center; (e. Fräsdornes:) outer bearing; (riveting) dolly bar
Gegenhalterstütze f (miller) outer brace
Gegeninduktivität f mutual inductance
Gegenkolbenmotor m (auto.) opposed-piston engine
Gegenkopplung f negative feedback
Gegenlager n (miller) outer bearing
Gegenlagerständer m end support
Gegenlauffräsen n up-cut milling
Gegenlauffräsmaschine f conventional milling machine
gegenläufige Kolben, opposite-stroke pistons
Gegenlehre f mating gage
Gegenlichtblende f anti-dazzle screen
Gegenmutter f lock nut, check nut
Gegenphase f reverse, phase, phase opposition
Gegenrad n (gears) mating gear
Gegenschlaghammer m (forge) counter-blow hammer

Gegensprechanlage f two-way telephone equipment
Gegensprechbetrieb m duplex operation
Gegensprechtelefonie f duplex telephony
Gegenständer m (boring mill) boring stay
Gegenstrom m countercurrent; reverse current
Gegenstromprinzip n countercurrent principle
Gegenstromvergaser m cross-draft carburetor; (UK) cross-draught carburetter
Gegenstück n counterpart, companion part
GEGENTAKT ~, (~ablenkung, ~gleichrichter, ~mikrophon, ~schaltung, ~stufe, ~transformator, ~verstärker) push-pull ~
Gegenverbundmotor m differential compound wound motor
Gegenverkehr m (tel.) duplex operation
gegenwirken v.t. counteract
Gehalt m content
Gehänge n suspension attachment; (e. Gießpfanne:) bail
Gehäuse n casing, housing; compartment; chamber; (e. Batterie, e. Ventils:) box; (gears:) case, box, housing
Geheimschalter m secret-position switch
gehren v.t. miter; (UK) mitre
Gehrmaß n (Meßzeug:) T-bevel
Gehrung f miter
Gehrungskreissäge f miter and bevel saw
Gehrungssäge f miter box saw
Gehrungsschnitt m mitering cut
Gehrungsschweißung f miter weld
Gehrungsstoß m miter joint
Gehrungswinkel m miter angle; (Meßzeug:) miter square
gekapselt a. enclosed, encased

Gekrätz n waste, refuse, discard; dross; slag
gekreuzter Riemen, crossed belt
gekröpft p.a. offset; cranked; (Meißel:) goose-necked
gekröpfte Achse, crank axle
gekröpfter Meißel, goose-neck tool
gekröpfter Meißelhalter, knee turning toolholder
Gelände n territory
Geländefahrzeug n cross-country vehicle
geländegängiger Wagen, cross-country car
Geländegängigkeit f (auto.) cross-country mobility
Geländelastwagen m cross-country truck
Geländer n railing, banister
Gelbgießerei f brass and bronze foundry
Gelbguß m yellow brass
Gelbkiefer f yellow pine, pitchpine
Gelbkupfer n yellow brass, brass
Gelbmetall n brass and bronze
Geleise cf. Gleis
Geleitfahrzeug n escort vehicle
Geleitzug m convoy (of trucks)
Gelenk n joint
gelenkartig a. articulated
Gelenkfahrzeug n articulated vehicle
gelenkig aufhängen, suspend pivotally
Gelenkkette f sprocket chain
Gelenkkopf m (auto.) swivel head
Gelenkleuchte f flexible arm lamp
Gelenkmutter f swivel nut
Gelenkrohrleitung f (auto.) articulated pipe
Gelenkschraube f swing bolt
Gelenkspindelbohrmaschine f multiple-spindle drilling machine with universally adjustable spindle

Gelenkstück n universal joint
Gelenkstulpen m (auto.) universal joint sleeve
Gelenkwelle f universal joint shaft; (auto.) propeller shaft
gelöschter Kalk, slaked lime, slack lime
gelten v.i. apply (to)
Gemeinkosten pl. overhead cost, general cost
Gemeinschaftsanschluß m (tel.) party-line
Gemeinschaftsantenne f communal antenna (or aerial)
Gemeinschaftswelle f (radio, telev.) shared channel
Gemisch n (auto.) gasoline-benzol mixture
gemischtes Hüttenwerk, iron and steel works, integrated steel plant
genau a. accurate, exact, true
Genauguß m precision casting
Genauigkeit f accuracy; precision
GENAUIGKEITS ~, (~gewinde, ~lager, ~prüfung, ~schalter, ~teilgerät, ~wasserwaage) precision ~
genaulaufend p.a. true-running
Generalüberholung f general overhaul
Generator m generator
Generatorgas n producer gas
Gepäck n luggage, baggage
Gepäckbrücke f (auto.) trunk rack
Gepäckgallerie f luggage rack
Gepäcknetz n (auto.) luggage net
Gepäckraum m (auto.) trunk room, luggage compartment
Gepäckträger m (motorcycle) luggage carrier
gerade a. straight; (Zahl:) even
Gerade f straight line
geraderichten v.t. straighten

Geradfräsen n straight-line milling
Geradheit f straightness
geradlinig a. straight-lined
geradverzahnt a. straight-toothed
Geradverzahnung f straight-tooth system
Geradzahn m straight tooth
Geradzahnkegelrad n straight-tooth bevel gear
Geradzahnrad n straight-toothed gear
Gerät n appliance, device; instrument; utensil; *(radio)* set; – *pl. (building)* plant; (Batterie:) set
Geräteanschlußschnur f flexible lead
Gerätedose f plug box
Geräteschalter m plug switch
Gerätesteckdose f utensil socket
Gerätestecker m coupler plug, utensil plug
Gerätetafel f (e. Werkzeugmaschine:) control panel
Gerätewagen m pick-up truck
Geräusch n noise
geräuscharm a. silent, quiet
Geräuschdämpfer m *(auto.)* exhaust silencer, exhaust muffler
Geräuschpegel m noise level
gerben v.t. (Leder:) tan
geriffelt a. grooved, serrated
Geröll n rubble
Gerüst n *(building)* scaffold; (rolling mill) stand
Gerüstbalken m scaffold beam
Gerüstbau m scaffolding
Gerüstklammer f scaffolding cramp
Gerüststange f scaffold pole
GESAMT ~, (~belastung, ~gewicht, ~leistung, ~spannung, ~spin, ~strahlung, ~strahlungspyrometer, ~verbrauch, ~verlust, ~widerstand, ~wirkung) total ~

Gesamtansicht f general view
Gesamtfehler m *(gears, metrol.)* cumulative error
Gesamthöhe f height overall
Gesamtlänge f length overall
Gesamtsumme f total amount, total
Gesamtwirkungsgrad m overall efficiency
geschichteter Preßstoff, laminated molded plastics
Geschirrspüler m dish-washer
Geschirrtrockner m dish-dryer
Geschirrziehpresse f reducing press
geschmeidig a. ductile
Geschoß n *(building)* floor
geschränkter Riemen, *(belt drive)* crossed belt
Geschützbronze f gunmetal, gun bronze
Geschwindigkeit f speed; velocity; rate
Geschwindigkeitsmesser m speedometer
Geschwindigkeitsregler m speed governor
Geschwindigkeitsschreiber m tachograph
Geschwindigkeitsstufung f (von Drehzahlen:) speed rate increment
Geschwindigkeitszähler m tachometer
Gesenk n die; (≈ Gesenkblock:) die block; (als Schmiedegerät:) swage
GESENK ~, (~amboß, ~arbeit, ~bau, ~fräser, ~gravur, ~herstellung, ~stahl) die ~
Gesenkbearbeitung f *(milling)* die-sinking
Gesenkdrücken n swaging
Gesenkdrückmaschine f swaging machine
Gesenkformfräsmaschine f die-sinker
Gesenkfräsen n die-sinking
Gesenkfräser m die-sinking cutter

Gesenkfräsmaschine f die-sinking machine, duplicator
Gesenkfräs- und Graviermaschine f die-sinking and engraving machine
Gesenkführung f side match line
Gesenkmacherei f die shop
Gesenkplatte f (Lochplatte:) swage block
Gesenkpresse f die-stamping press
Gesenkpreßteil n die-pressed part
Gesenkschlosser m die-maker
Gesenkschmied m drop forger
Gesenkschmiede f drop forge
gesenkschmieden v.t. drop-forge
Gesenkschmiedepresse f drop forging press
Gesenkschmiedestück n drop forging
Gesenkschräge f draft
Gesenkstahlblock m die Block
Gesenkziehen v.t. die-drawing
Gesetzmäßigkeit f regularity; (jur.) legality
Gesichtsmaske f (welding) face shield
Gesims n cornice
Gespann n (e. Kokille:) bottom plate
Gespannfahrzeug n pneumatic-tired car
Gespannguß m group teeming (of steel ingots)
Gespannplatte f group teeming plate, bottom pouring plate
Gespräch n (tel.) call
Gesprächszähler m (tel.) call-meter
gestaffelte Walzstraße, staggered rolling train
Gestalt f shape; profile; contour; (e. Oberfläche:) configuration, structure
gestalten v.t. design; construct
Gestaltfestigkeit f strength depending on shape (or design), rated fatigue limit
Gestaltung f design
Gestänge n (auto.) linkage

Gestehungskosten pl. first cost
Gesteinsmehl n mineral powder
Gesteinsschotter m broken rock
Gesteinstrümmer pl. rock debris
Gestell n frame, stand; (e. Batterie:) rack; (Hochofen:) hearth; (tel.) bay; (telegr.) frame
Gestellpanzer m (Hochofen:) hearth jacket, hearth casing
gesteppte Matte, quilted mat
gesteuert p.a. controlled
gesundheitsschädlich a. health-impairing
Getriebe n (Zahntrieb:) gear transmission, gear unit, gearing; (als Triebwerk:) drive mechanism; (hydraulics) transmission
GETRIEBE ~, ~aggregat, ~block, ~bremse, ~fertigung, ~kupplung, ~raum, ~schmierung, ~schrank, ~schutz, ~untersetzung, ~welle, ~zug) gear ~
Getriebeabdeckung f (auto.) gearbox cover
Getriebebild n gearing diagram
Getriebeblockzahnrad n cluster gear
Getriebebolzen m gearbox stud
Getriebefett n transmission grease
Getriebefüllapparat m (auto.) gearbox lubrication equipment
Getriebegehäuse n gear housing, gear case; gearbox
Getriebeglied n gearing component
Getriebekasten m gearbox, transmission case
Getriebeluftpumpe f (auto.) gear-driven air pump
Getriebemotor m geared motor
Getriebeöl n transmission oil

Getriebeölpumpe f (auto.) transmission oil pump
Getriebeplan m gearing layout, gearing diagram
Getrieberad n transmission gear; gear wheel
Getriebeschema n gearing layout
Getriebeübersetzung f transmission ratio
Getriebezahnrad n transmission gear
Gewähr f guarantee
gewährleisten v.t. guarantee
Gewebe n (Draht:) netting, cloth, gauze
Geweberiemen m fabric belt
Gewerbe n craftsmen trade, handicraft
Gewerbesteuer f trade tax
Gewicht n weight
Gewichtsvorschub m weight-operated feed
Gewinde n thread, screw thread
GEWINDE ~, (~achse, ~auslauf, ~drücken, ~entgrateinrichtung, ~flanke, ~form, ~fräsen, ~fräser, ~fräsmaschine, ~gang, ~kopieren, ~läppen, ~läppmaschine, ~lücke, ~profil, ~prüfung, ~rachenlehre, ~reihe, ~rille, ~rollenrichtung, ~rollen, ~schälmaschine, ~schleifeinrichtung, ~schleifen, ~schleifscheibe, ~schleifmaschine, ~schneideinrichtung, ~schneidmaschine, ~schneidwerkzeug, ~steigung, ~strehlen, ~strehler, ~teilung, ~walzbacke, ~walzen, ~wälzen, ~wälzfräser, ~wirbelmaschine, ~wirbeln) thread ~
Gewindebohreinrichtung f tapping attachment
gewindebohren v.t. tap
Gewindebohrer m tap
Gewindebohrmaschine f tapper; (UK) tapping machine
Gewindedrehmaschine f screw-cutting lathe
Gewindeeinstechschleifen n plunge-cut thread grinding
Gewindeflansch m threaded flange
Gewindegrund m root (or bottom) of a thread
Gewindekappe f screwed cap
Gewindelehrdorn m screw plug gage
Gewindelehre f screw thread gage
Gewindeleitbacke f follower
Gewindeleitpatrone f leader
Gewindeloch n tapped hole
Gewindemeißel m threading tool
Gewindemeßkomparator m optical screw thread measuring machine
Gewindemuffe f screwed socket
Gewindenippel m screwed nipple
Gewindepatrone f leader
Gewinderäderkasten m screwcutting gearbox
Gewinderille f thread groove
Gewinderillenfräser m multiple thread milling cutter
Gewindeschaft m threaded shank
Gewindeschneidanzeiger m thread dial indicator
Gewindeschneideisen n threading die
gewindeschneiden v.t. cut a thread, screw
Gewindeschneiden n thread-cutting, screw-cutting; (Gewindebohren:) tapping
Gewindeschraublehre f screw thread micrometer, caliper
Gewindespindel f screw spindle
Gewindestift m grub screw; (mit Innensechskant:) set screw
Gewindestopfen m screw plug
Gewindestrehlen n thread chasing

Gewindeteilung f pitch
Gewindeuhr f thread indicator, dial indicator
Gewindewalzbacke f thread rolling die
Gewindewälzfräser m thread milling hob
Gewindewalzmaschine f thread rolling machine
Gewinn m (radio) gain
Gewinnung f (chem.) extraction; (Benzol:) recovery
Gewölbe n vault, arch
Gewölbeteil n (e. Flammofens:) bung
Gicht f (Hochofen:) furnace throat; (≈ Begichtung:) charge
Gichtaufzug m furnace hoist
Gichtbühne f charging platform
Gichtgas n blast furnace gas
Gichtgasabzugsrohr n downcomer
Gichtgasleitung f blast furnace gas main
Gichtgasstaub m flue dust
Gichtglocke f furnace-top bell
Gichtkübel m charging bucket
Gichtöffnung f furnace throat
Gichtstaub m flue dust
Giebel m (building) gable
Giebelsims m gable molding
Gießbarkeit f pourability
Gießbühne f pouring, platform, teeming platform
gießen v.t. cast; (Blöcke:) teem; (allgemein:) pour
Gießen im Gespann, group casting, bottom casting
Gießen mit Aufsatz, hot topping
Gießen mit verlorener Gießform, investment casting
Gießer m founder; (specif.) caster
Gießerei f foundry
GIESSEREI ~, (~betrieb, ~kupolofen, ~mann, ~ofen, ~roheisen, ~sand, ~technik) foundry ~
Gießereiwesen n founding, foundry practice
Gießfähigkeit f pourability
Gießform f mold; (Druckgußtechnik) die
Gießgeschwindigkeit f rate of pouring
Gießgrube f casting pit, foundry pit; (für Gußblöcke:) teeming box
Gießhalle f casting bay, pouring bay
Gießhaut f skin
Gießkopf m feeder head, sinkhead
Gießlöffel m pouring cup
Gießpfanne f casting ladle, foundry ladle
Gießpfanne mit Schnauzenausguß, top-pour ladle
Gießpfanne mit Stopfenausguß, bottom-pour ladle, teeming ladle
Gießpfannengabel f ladle shank
Gießpfannenkran m ladle crane
Gießpfannenwagen m ladle truck
Gieß-Preßschweißverfahren n molten metal pressure welding
Gießrinne f pouring spout
Gießschnauze f (e. Pfanne:) pouring lip, pouring nozzle
Gießschweißen n molten metal welding
Gießtrichter m pouring gate, gate; downgate, sprue; runner
Gießwagen m casting bogie
Gießwanne f tundish
Giftstoff m toxic substance
Gipskalk m plaster lime
Gipsmörtel m gypsum mortar
Gipsspat m sparry gypsum
Gitter n (constr., math., cryst.) lattice; (e. Regenerativkammer:) checker; (electr., electron.) grid
GITTER ~, (electr.) (~draht, ~drossel,

~gleichrichter, ~kondensator, ~kreis, ~spannung, ~strom, ~vorspannung, ~widerstand) grid ~

Gitterbatterie f C-battery, grid bias battery
Gitterkammer f (e. Ofens:) checker chamber
Gittermast m lattice tower; lattice mast
Gitterstein m checker brick
Gitterträger m lattice girder
Glanz m gloss, brilliancy, lustre
Glas n glass
Glaser m glazier
Glaserkitt m glazier's putty
Glasgewebe n woven glass
glasig a. vitreous
glasklar a. transparent
Glaskolben m glass bulb
Glasmaßstab m glass rule
Glaspapier n glass paper
Glasprüfmaß n optical flat gage
Glasschneider m glass cutter
Glasspannungsprüfer m glass strain tester
Glasur f glaze, gloss
Glaswolle f glass wool
glatt a. smooth; (≈ eben:) plane; (≈ bündig:) flush
glattdrücken v.t. planish
glätten v.t. smooth, planish, polish; burnish; (Rohre:) reel
glatthobeln v.t. (woodworking) finish
Glättkelle f (building) plastering trowel
Glattputz m (building) fair faced plaster
Glättung f (electr.) smoothing
Glättungskondensator m smoothing capacitor
Glättwalzwerk n smoothing rolls; (Rohre:) reeling machine
gleichachsig a. coaxial; concentric

Gleichdick n spherical triangle
Gleichdruck-Axialturbine f axial flow impulse turbine
Gleichdruckbrenner m balanced pressure torch
Gleichdruck-Kondensationsturbine f impulse condensing turbine
Gleichdruckverbrennung f (auto.) constant-pressure combustion
gleichgerichtet a. parallel; (electr.) rectified
Gleichgewicht n equilibrium; balance
gleichlauffräsen v.t. down-cut mill, climb-cut mill
Gleichlauffräsmaschine f down-cut milling machine, downfeed milling machine
Gleichlaufmotor m synchronous motor
Gleichlaufvorrichtung f synchronizing mechanism
gleichpolig a. homopolar
gleichrichten v.t. (electr.) rectify
Gleichrichter m rectifier
GLEICHRICHTER ~, (~anode, ~elektronenröhre, ~element, ~instrument, ~kathode, ~kreis, ~meßgerät, ~röhre, ~schaltung) rectifier ~
Gleichrichtung f (electr.) rectification
gleichschalten v.t. coordinate; synchronize
gleichschenkliger Winkelstahl, equal-sided angle steel; – pl. equal angles
gleichseitig a. (geom.) equilateral
Gleichspannung f direct-current voltage, d.c. voltage
Gleichstrom m direct current; D. C., d-c
GLEICHSTROM ~, (~betrieb, ~empfänger, ~generator, ~kabel, ~motor, ~netz, ~schweißstromgenerator, ~schweißumformer, ~transformator,

Gleichstromerzeuger — 114 — **Glockenschneidrad**

~übertragung, ~umformer, ~widerstand) direct-current ~
Gleichstromerzeuger m dynamo, direct-current generator
Gleichstromregelmotor m variable-speed D. C. motor
Gleichung f equation
gleichwinklig a. equiangular
Gleis n [railway] track, rail track, railroad track
GLEIS ~, (~erneuerung, ~hebewinde, ~joch, ~meßwagen, ~relais, ~schotter, ~stopfer, ~stopfmaschine, ~stromrelais, ~tafel, ~verbindung, ~verlegung, ~wasserwaage) track ~
Gleisanlage f railway trackage, permanent way and fixed installations
Gleisanschluß m siding
Gleisarbeiten fpl. railway trackwork
Gleisbagger m rail-mounted excavator
Gleisbau m track construction; track laying
Gleisbaumaschinen fpl. track laying machinery
Gleisbettung f sleeper bedding
Gleisbettungswalze f permanent way roller
Gleisbremse f rail-brake car retarder
Gleisdreieck n Y-track
Gleiskette f crawler, creeper track, chain track
Gleiskettenantrieb m crawler drive
Gleiskettenfahrzeug n crawler-type vehicle
Gleiskettenschlepper m crawler-tractor, caterpillar tractor
Gleiskettenschlepperkran m crawler tractor crane
Gleiskettenzugmaschine f crawler-tractor
Gleiskreuzung f railroad crossing

Gleissperre f scotch block
Gleisüberführung f fly-over
Gleiswaage f wagon weighbridge
GLEIT ~, (~backe, ~keil, ~kontakt, ~kupplung, ~lager, ~sitz, ~widerstand) sliding ~
Gleitbahn f slideways, sliding ways; guideways
Gleitebene f (cryst.) slip plane, gliding plane
gleiten v.i. slide, slip, glide
Gleitfläche f (cryst.) slip plane; (e. Bremse:) friction surface
Gleitlinie f (forging) slip line
Gleitschiene f (e. Stoßofens:) skid
Gleitschuh m sliding shoe; (e. Zylinders:) crosshead slipper
Gleitschutz m anti-skid protection
Gleitschutzblech n anti-slip plate
Gleitschutzkette f (auto.) anti-skid chain
gleitsicher a. non-slip, skid-proof
Glied n (als Bauelement:) element, member; (≈ Teilstück:) component; (e. Kette:) link
Gliederkessel m sectional boiler
Gliederkette f link chain
Gliedermaßstab m folding rule, zigzag rule
Gliederriemen m link belt
Gliederung f classification, (Kosten:) subdivision
Gliederwelle f articulated shaft
GLIMM ~, (~entladung, ~entladungslampe, ~röhre) glow ~
glimmen v.i. glow
Glimmer m mica
Glocke f (blast furnace) bell-type distributing gear
Glockenschneidrad n hub-type shaper cutter

Glockenwinde f *(blast furnace)* bell hoist, bell operating gear
Glühbehandlung f annealing (operation)
Glühbirne f bulb
Glühelektrode f hot-cathode
glühen v.t. *(heat treatment)* anneal
Glühfaden m filament
Glühfadenlampe f incandescent filament lamp
Glühfadenpyrometer m disappearing filament pyrometer
glühfrischen v.t. malleablize
Glühkasten m *(founding)* annealing box
Glühkathode f hot cathode
Glühkathodengleichrichter m hot cathode rectifier
Glühkerze f *(auto.)* heater plug
Glühkopf m hot bulb
Glühkopfmotor m mixed cycle engine, semi-diesel, hot-bulb engine
Glühkopfzündung f hot-bulb ignition
Glühkörper m incandescent body
Glühlampe f filament lamp; incandescent lamp
Glühlampenfassung f incandescent lamp socket
Glühlicht n incandescent light
Glühofen m annealing furnace; (Stahlwerk:) reheating furnace
Glühspuk m flash figures
Glühtopf m (für Temperzwecke:) annealing pot
Glühung f annealing, anneal
Glühverlust m ignition loss
Glühzunder m mill scale
Glutfestigkeit f resistance to incandescence
Goldschaum m *(Zinkentsilberungsverfahren)* gold crust

Goldscheideanstalt f gold refinery
Goldscheidung f gold refining, gold parting
Graben m trench, ditch
Grabenbagger m ditch digger, trench excavator
Grabenpflug m trenching plough
Grad m degree
Gradeinteilung f degree graduation
Granalien fpl. shot, granulated metal
Granaliennickel n nickel shot
Granit m granite
Granulationsanlage f granulating plant
granulieren v.t. granulate
Graphit m graphite; (mineralischer:) plumbago
Graphitausscheidung f *(founding)* separation of graphite
Graphitglühen n graphitizing
graphithaltig a. graphitiferous
graphitisieren v.t. & v.i. graphitize
Graphitschmiermittel n graphite lubricant
Graphittiegel m graphite crucible
Grat m *(building)* hip; *(forging)* flash
Gratbahn f (e. Gesenkes:) land
grätig a. (Walzgut:) overfilled
Gratmulde f (e. Gesenkes:) gutter, flush gutter
graues Roheisen, gray pig iron
Graugießerei f grey iron foundry
Grauguß m grey [cast] iron
Graukalk m magnesian limestone
GRAVIER ~, (~anstalt, ~arbeit, ~fräser, ~maschine, ~schablone, ~werkzeug) engraving ~
gravieren v.t. engrave
gravierfräsen v.t. engrave
Gravierspindel f cutter spindle

Gravur f (e. Gesenkes:) impression, cavity; (operation:) engraving
Greifbagger m grab buckel crane, grab excavator
Greifer m (Verschiebevorrichtung im Walzwerk:) tappet, finger; (e. Krans:) grab
Greiferkette f grab chain
Greiferkorb m grab bucket
Greiferkran m grab excavator
Greiffähigkeit f gripping power
Greifvorrichtung f gripping appliance
Greifwinkel m angle of contact
Greifzirkel m calipers pl.
Grenzdrehzahl f limiting speed
Grenze f limit; (metallo.) boundary
Grenzfläche f boundary surface; contact surface; (cryst.) cleavage plane
Grenzflächenspannung f interfacial tension
Grenzfrequenz f limiting frequency
Grenzlehrdorn m limit screw plug gage
Grenzlehre f limit gage
Grenzlinie f boundary line
Grenzmaß n limiting size
Grenzrachenlehre f limit snap gage
Grenzreibung f boundary friction
Grenzschicht f boundary layer
Grenzstein m border stone
Grenzwiderstand m critical resistance
Griff m (≈ Heft:) handle; knob; lever; (fig.) grip, touch
Griffkreuz n star handle
Griffmutter f knurled nut
griffnahe a. within easy reach
Griffschale f (auto.) handle escutcheon
Griffzeit f (time study) handling time
Grobblech n heavy plate, plate
Grobblechlehre f plate gage
Grobblechschere f plate shears

Grobblechwalze f plate roll
Grobblechwalzwerk n plate rolling mill
Grobbrecher m coarse crusher
Grobeinstellung f coarse adjustment
Grobgefüge n macrostructure
Grobgestalt f macro-geometric contour
Grobgewinde n coarse thread
Grobkies m coarse gravel
Grobkorn n (metallo.) coarse grain
Grobkornglühen n full annealing, coarse grain annealing
grobkörnig a. coarse-grained
grobmaschig a. wide-meshed
Grobpassung f loose fit; coarse fit
Grobsand m coarse sand
grobschleifen v.t. rough-grind, snag
grobschlichten v.t. rough-finish
grobschmieden v.t. rough-forge
Grobsitz m loose fit; coarse fit
Grobsplitt m coarse stone chipping
Grobstraße f (rolling mill) breaking down mill train
Grobwalzwerk n blooming mill
Grobzerkleinerungsmühle f crushing mill
Grobzug m (Drahtfabrikation:) roughing block
Grobzustellung f coarse adjustment
Großbaustelle f large-scale project site
Größe f size; quantity; magnitude; degree; factor; rate
Großgerät n (building) heavy-duty construction equipment
Großserienfertigung f long-run production
großstückig a. lumpy, large-sized
Großtankstelle f service station
Größtmaß n maximum dimension (or size); (Passungen:) upper limit
Größtspiel n maximum clearance

Großzahlforschung f frequency statistics
Grübchen n (Oberflächenfehler:) pit, stain
Grübchenbildung f pitting
Grube f pit
Grubenhobelmaschine f pit planer
Grubenholz n mine timber
Grubenlampe f miner's lamp
Grubenlokomotive f mine locomotive
Grubenstempel m pit prop
Grund m ground; foundation; base; bottom; cause, reason; *(threading, welding)* root
Grundanstrich m prime coat, priming coat, primer, flat coat
Grunddrehzahl f basic speed
Grundfarbe f prime color, flat color
Grundfläche f floor space
Grundfrequenz f fundamental frequency
Grundgetriebe n basic gearing
grundieren v.t. prime
Grundierfarbe f flat paint, flat color, primer
Grundierung f priming coat
Grundkreis m *(gearing)* base circle
Grundloch n bottom hole, blind hole
Grundlohn m base rate
Grundlohnsatz m base rate earnings
Grundmaß n basic size
Grundmasse f *(metallo.)* matrix
Grundmetall n base metal, parent metal, backing metal
Grundplatte f baseplate
Grundraupe f *(welding)* bottom layer [of a weld]
Grundreibahle f rose chucking reamer
Grundriß m plan view
Grundschaltung f basic circuit
Grundstandard m *(cost accounting)* basic standard

Grundteilung f *(gears)* base pitch
Gründung f *(civ.eng.)* foundation
Gründungspfahl m foundation pile
Gründungsspiegel m ground water table
Grundwasser n ground water
Grundwelle f fundamental wave
Grundwerkstoff m *(welding)* base metal, parent metal; base material
Grundzeit f floor-to-floor time
grüner Sand, green sand
Grünsandform f *(founding)* green sand mold
Grünsandformerei f green-sand molding
Grünspan m copper rust
Gruppenantrieb m group drive
Gruppenbohrmaschine f gang driller
Gruppenschalter m gang switch
Gruppenschmierung f one-shot lubrication
Gruppenwähler m group selector
Grus m breeze, [coke] fines
Gummi m & n rubber
GUMMI ~, (~band, ~belag, ~dichtung, ~handschuh, ~isolierung, ~kabel, ~kleber, ~lösung, ~matte, ~reifen, ~riemen, ~schlauch, ~schnur, ~stopfen, ~tülle) rubber ~
Gummihammer m rubber mallet
Gurt m *(conveying)* belt
Gurtförderer m belt conveyor
Gurtung f *(building)* chord
Gurtungsversteifung f *(building)* chord bracing
Guß m cast iron; (≈ Gußstück:) casting; (≈ Gießen:) casting, pouring, pour
Gußasphalt m mastic asphalt
Gußasphaltestrich m mastic asphalt floor
Gußbeton m cast concrete
Gußblock m ingot

Gußbruch m grey iron scrap
Gußeisen n cast iron
Gußeisenschrott m grey iron scrap
gußeisern a. cast-iron
Gußglas n cast glass
Gußgrat m fin
Gußhaut f skin (of a casting)
Gußmetall n cast metal
Gußnaht f seam, feather, fin
Gußputzerei f foundry cleaning room
Gußrohr n cast iron pipe
Gußschrott m cast iron scrap
Gußstahl m cast steel
Gußstahlwerk n cast steel plant
Gußstück n casting
Gußwaren fpl. cast-iron ware

Gut n ($\approx$ Material, Werkstoff:) material
Güte f quality
Güteeigenschaft f quality characteristic
Gütefaktor m quality factor
Gütekontrolle f quality control
Güterfernverkehr m long-distance road haulage
Güterschuppen m freight house
Güterverkehr m goods traffic, freight traffic
Güterwagen m freight car; *(UK)* railway wagon
Gütevorschrift f quality specification
Gütezahl f quality index figure
Gutgewindelehrdorn m 'go' plug screw gage
Gutlehre f 'go' gage

H

Haarlineal *n* knife-edge straightedge
Haarnadelkurve *f (traffic)* hair-pin bend
Haarriß *m* micro-flaw, hairline crack, tiny crack
Haarseite *f* (e. Riemens:) grain side
Haarwinkel *m* bevelled edge square
Hacksäge *f* power hacksaw
Haftelektrode *f* adherent electrode
haften *v.i.* adhere, stick to
Haftfähigkeit *f* adhesive property
Haftgrundmittel *n (painting)* wash-primer
Haftpflicht *f (auto.)* third party liability
Haftpflichtversicherung *f* liability insurance
Haftreibung *f* static friction
Haftsitz *m* wringing fit
Haftspannung *f* adhesive stress
Haftung *f (phys.)* bond
Hahn *m* cock; tap
Hahnküken *n* plug of the cock
Hahnschlüssel *m* box wrench
Haken *m* hook
Hakengeschirr *n* hook fittings
Hakenmeißel *m* recessing tool
Hakenschlüssel *m* hook spanner
Hakenschraube *f* T-head bolt, Tee-bolt
Halbautomat *m* semi-automatic lathe
halbautomatisch *a.* semi-automatic
Halbfabrikat *n* semi-finished product
Halbholz *n* scantling
halbiert *a.* (Roheisen:) mottled
halbkontinuierliche Walzstraße, semi-continuous rolling train
Halbkreis *m* semicircle
Halbleiter *m* semi-conductor
Halbmesser *m* radius

Halbrundkopf *m* button head
Halbrundniet *m* spherical head rivet
Halbrundräumen *n* radius broaching
halbselbsttätige Drehmaschine, semi-automatic lathe
Halbwertzeit *f* half-life period
Halbzeug *n* semi-finished products, semi-finished steel
Halbzeugstraße *f* semi-finishing mill train
Halde *f* waste-dump, dump, yard
Haldenschlacke *f* dump slag
Haldenschüttung *f* stockpiling
Halslager *n* journal bearing
Halssenker *m* counterbore
Haltbarkeit *f* life, durability, stability
Haltepunkt *m (metallo.)* thermal critical point
Haltepunktsdauer *f* critical range, critical interval
Halter *m* holder; handle; clamp; *(scaffolding)* lashing
Halterung *f* mounting; clamping fixture
Haltesignal *n (auto.)* stop sign
Haltestein *m* (e. Gesenkes:) dowel
Haltestelle *f* stop
Halteverbot *n (auto.)* no waiting
Haltezeichen *n (auto.)* stop sign
Haltezeit *f* (e. Maschinentisches:) dwell
Hämatitroheisen *n* hematite pig iron
Hammerbahn *f* hammer face
hämmerbar *a.* (warm:) forgeable; (kalt:) malleable
Hämmerbarkeit *f* (warm:) forgeability; (kalt:) malleability
Hammerbrecher *m* hammer crusher

Hammerfeder f hammer strap
Hammergarmachen n (Kupfer:) poling tough pitch
Hammerlötkolben m hatchet soldering copper
hämmern v.i. hammer; (e. Schweißnaht:) peen
Hammerpinne f hammer pane
Hammerschlag m hammer blow; (Zunder:) hammer scale
Hammerschweißung f forge welding
Hammerwerk n hammer mill
Handamt n manual exchange
Handbeil n hatchet
Handblechschere f tinners' snip
Handbohrmaschine f hand drill; breast drill
Handbremse f hand brake
Handbremshebel m hand brake lever
Handbuch n manual
Handelsbaustahl m commercial structural steel
Handelsblech n steel sheet of commercial quality
Handelsfeinblech n commercial light-gage sheet
Handelsgüte f commercial grade (or quality)
Handelsstabstahl m merchant bar
Handfeuerlöscher m fire extinguisher
Handformerei f hand molding; hand molding shop
Handformmaschine f hand-operated molding machine
Handgashebel m (auto.) hand accelerator
Handgewindebohrer m hand tap
Handgriff m handle; manipulation
handhaben v.t. manipulate

Handhebelbohrmaschine f sensitive drill press
Handhebelfettpresse f hand lever-operated grease gun
Handhebelsteuerung f manual lever control
Handkabelwinde f hand cable winch
Handkreuz n capstan wheel, star wheel
Handkurbel f hand crank
Handlampe f portable lamp
Handlanger m helper
Handlauf m (building) handrail
Handleuchte f inspection lamp
Handrad n handwheel
Handreibahle f hand reamer
Handschalthebel m (auto.) change speed lever
Handschaltung f manual shifting
Handschleifarbeit f free-hand grinding
Handschuh m glove
Handspannfutter n hand-operated chuck
Handspindelpresse f hand screw press
Handstampfer m tamper
Handvermittlung f (tel.) manual exchange
Handverstellung f manual adjustment
Handwahlschalter m hand selector switch
Handwerker m craftsman
Handwerkzeug n small tool, hand tool
Handzeit f (work study) handling time
Hanf m hemp
Hanfpackung f hemp packing
Hanfriemen m hemp belt
Hanfseele f (e. Seils:) hemp core
Hanfseil n (US) manila rope; (UK) cotton rope
Hängebahn f aerial ropeway, overhead trolley; suspension railway
Hängebahngießpfanne f trolley ladle

Hängebahnkreuzung f trolley conveyor crossing
Hängebahnkübel m ropeway bucket, swing bucket
Hängebahnlaufkatze f overhead conveyor trolley
Hängebahnwaage f balance for overhead chain and trolley conveyor
Hängebahnweiche f trolley rail switch
Hängebrücke f suspension bridge
Hängedruckknopftafel f pendant push-button panel
Hängekübel m swing bucket, suspension bucket
Hängelager n drop hanger
Hängeleiter f suspension ladder
hängen v.t. & v.i. suspend, hang; (der Gicht:) scaffold
Hängeschiene f overhead rail
Hängeschloß n padlock
Hängetafel f pendant control panel
hantieren v.t. handle, manipulate
Hartauftragschweißung f hard surfacing
härtbar a. (heat treatment) hardenable; (plastics) thermosetting
härtbare Formmasse, thermosetting material
härtbarer Kunststoff, thermosetting plastics
Härtbarkeit f hardenability
Hartbeton m granulithic concrete
Härte f hardness; (Naturhärte e. Stahles:) temper; (e. Schleifscheibe:) grade
Härteanlage f heat treating equipment
Härtebeständigkeit f retentivity of hardness
Härtebildner m (metallo.) hardening constituent
Härtekessel m (concrete) curing vessel

härten v.t. harden
Härteofen m hardening furnace
Härteprüfer m hardness tester
Härteprüfmaschine f hardness testing machine
Härteprüfung f hardness test
Härter m (plastic) setting agent
Härterei f hardening room, hardening shop, heat treating department
Härteriß m heat treatment crack, hardening crack
Härteschicht f (Einsatzschicht:) hardened case
Härtespannung f hardening stress; hardening strain
Härtetiefe f (heat treatment) depth of hardening zone; (Einsatzhärtung:) depth of case
Härteverzug m hardening distortion
Härtezustand m temper
Hartfaserplatte f hard wood fiber
Hartgewebe n laminated fabric, fabric-base laminate
Hartgewebezahnrad n faboid gear
Hartgummi n hard rubber; vulcanite
Hartguß m chilled cast iron
Hartgußwalze f chilled roll
Hartholz n hardwood
Hartlot n brazing solder, hard solder, spelter
hartlöten v.t. braze
Hartmatte f glass-mat-base laminate
Hartmetall n sintered hard carbide, cemented metal carbide; cutting alloy
HARTMETALL ~, (~auflage, ~bestückt, ~bestückung, ~einlage, ~einsatz, ~legierung, ~meißel, ~plättchen, ~schneide, ~werkzeug) carbide ~

Hartpapier *n* kraft paper, paper-base laminate, laminated paper; Manila paper

Härtung *f* hardening

Hartverchromung *f* hard chrome plating

Hartverzinkung *f* spelterizing

Hartwalze *f* chilled roll

Hartzerkleinerungsmaschinen *fpl.* crushing and grinding equipment

Hartzink *n* spelter

Hartzinn *n* pewter

Harz *n* resin; *(UK)* rosin

Harzbitumen *n* resinous bitumen

Harzesterlack *m* gum lacquer

Harzöl *n* resin oil

Haspel *f* reel

Haube *f (auto.) (US)* hood; *(UK)* bonnet; *(met.)* (e. Blockform:) feeder head, dazzle

Haubenglühofen *m* bell type annealing furnace

Hauer *m* (Feilen:) cutter

HAUPT ~, (~antriebsmotor, ~arbeitsspindel, ~schalter, ~schalthebel, ~schalttafel, ~schnittdruck, ~sicherung, ~spindel, ~ständer, ~stromkreis, ~verteiler, ~vorschub, ~wellenleitung) main ~

Hauptabmessungen *fpl.* (e. Werkzeugmaschine:) specifications

Hauptanschluß *m (tel.)* main station, subscriber's station

Hauptleitung *f (electr.)* mains; *(tel.)* trunk line

Hauptschlitten *m (lathe)* saddle; *(US)* carriage

Hauptschlußmotor *m* series-wound motor

Hauptschneide *f* (Meißel:) primary cutting edge

Hauptschnittfläche *f (metal cutting)* work surface

Hauptstrommotor *m* series-wound motor

Hauptstrom-Ölfilter *m (auto.)* full-flow oil filter

Haupttelegrafenamt *n* central telegraph office

Hauptverkehrsstraße *f* trunk road, main road

Hauptverkehrszeit *f* peak-traffic hour

Hauptwelle *f (auto.)* mainshaft gear shift lever shaft

Hauptzeit *f (time study)* machining time, cutting time

Haus *n* building; (für Kesselanlage:) room

Hausanschluß *m (tel.)* residence telephone

Hausbrand *m* domestic fuel

Haushalttarif *m* domestic tariff

Haushaltszähler *m (electr.)* house service meter

Hausinstallation *f (electr.)* home wiring

Haustelefon *n* interphone

Hautleim *m* hide glue

Hebebock *m* lifting jack

Hebebühne *f* lifting platform

Hebekopf *m* (e. Pochstempels:) tappet

Hebel *m* lever

HEBEL ~, (~arm, ~griff, ~knarre, ~lochstanze, ~schere, ~stanze) lever ~

Hebelschalter *m* knife switch

Hebelübersetzung *f* leverage

Hebelumschalter *m* throw-over switch

Hebelwaage *f* beam scale

Hebelwirkung *f* leverage

heben *v.t.* raise, elevate; lift; (Lasten:) hoist

Hebetisch *m* (Walzblöcke:) tilter

Hebewinde *f* lifting jack

Hebezeug *n* hoist

Heck n (auto.) rear deck
Heckantrieb m (auto.) rear-engine drive
Heckklappenschloß n (auto.) boot lock
Heckmotor m rear-mounted engine
Heft n handle
heften v.t. (welding, riveting) tack
Heftnaht f (welding) tack weld
Heftniet m tacking rivet
Heftnietung f tack riveting
Heftschweiße f tack weld
Heftschweißen f tack welding, stitch welding
Heftzapfen m tang
Heftzwecke f thumb tack
Heißdampf m superheated steam
Heißeisensäge f hot saw
Heißgasschweißen n hot-gas welding
heißlaufen v.i. run hot
Heißlufterhitzer m hot-air-heater
Heißteer m hot tar
Heißtemperaturzone f hot temperature zone; (e. Temperofens:) soaking zone
Heißwasserspeicher m hot water or high-temperature storage tank
Heißwind m hot-air blast, hot blast
Heißwindkupolofen m hot blast cupola
Heißwindleitung f hot-blast main
HEIZ ~, (~fläche, ~gas, ~kessel, ~kraft, ~mantel, ~ofen, ~rohr, ~spirale, ~spule, ~widerstand) heating ~
Heizbatterie f heating battery; (US) A-battery
Heizdraht m resistance wire, heating wire
Heizer m stoker
Heizfaden m filament; heater
Heizkissen n electric pad
Heizkörper m radiator
Heizöl n fuel oil

Heizplatte f hot plate
Heiztransformator m filament transformer
Heizung f (≈ Erwärmung:) heating; (≈ Feuerung:) firing
Heizungsanlage f heating system, heating installation
Heizungsgebläse n (auto.) heating blower
Heizwert m calorific power, thermal power
Helligkeit f brightness, luminosity
hemmen v.t. & v.i. retard; obstruct; hinder
Hemmung f obstruction; retardation
herabsetzen v.t. decrease, reduce
herausnehmbar a. removable; detachable, withdrawable
herausnehmen v.t. remove; withdraw
herausragen v.i. protrude, overhang
herausschlagen v.t. knock out
herausschrauben v.t. unscrew
herausstoßen v.t. eject
herausziehen v.t. pull out, withdraw; (e. Stift:) extract
Herd m hearth; fireplace; furnace bottom; (≈ Ofen:) furnace; range
Herdfrischverfahren n (steelmaking) open-hearth process
Herdwagenofen m bogie hearth furnace
herrichten v.t. prepare
herstellen v.t. manufacture, make, produce; fabricate
Herstellkosten pl. manufacturing cost
Herstelltoleranz f manufacturing tolerance
Hertz n cycles per second
herunterdrücken v.t. depress
herunterschalten v.t. (electr.) step down
heruntertransformieren v.t. step down
herunterwalzen v.t. roll down, rough down; (Blöcke:) cog down
hervorragen v.i. protrude, project
Herzklaue f lathe dog

Herzwelle f stud shaft

HF ~, (~behandlung, ~drossel, ~kreis, ~messung, ~schirm, ~spannung, ~übertragung) radio-frequency ~

HF-Gleichrichter m first detector

HF-Störung f (radar) jamming

HILFS- ~, (~antenne, ~ausrüstung, ~fläche, ~frequenz, ~kontakt, ~leitung, ~pol, ~schalter, ~spannung, ~spule, ~transformator) auxiliary ~

Hilfsamt n sub-exchange

Hilfsbetrieb m service department

Hilfseinrichtung f labor-aiding equipment, ancillary equipment

Hilfskraftlenkung f (auto.) servo-assisted steering gear

Hilfslenkung f (auto.) servo steering

Hilfsmaßnahme f remedial measure

Hilfsmittel n aid

Hilfsmotor m auxiliary motor, pony motor

Hilfsphase f (Motor:) phase splitting device

Hilfspolmotor m split-pole motor

Hilfspumpe f booster pump

Hilfsstoffe mpl. (cost accounting) operating supplies

hinaufspannen v.t. (electr.) step up

hinauftransformieren v.t. step up

Hindernis n obstacle, obstruction

Hinterachsantrieb m (auto.) rear axle drive, final drive

Hinterachsbrücke f back driving axle

Hinterachse f (auto.) rear axle, back axle

Hinterachsgehäuse n (auto.) rear axle casing

Hinterachsstrebe f (auto.) rear axle torque bar

Hinterachswelle f (auto.) rear axle shaft

hinterarbeiten v.t. (metal cutting) relieve, back off

Hinterbohreinrichtung f back drilling attachment

hinterbohren v.t. back-drill

Hinterdreheinrichtung f relieving attachment

hinterdrehen v.t. relief-turn

Hinterdrehmeißel m relief-turning tool

Hintereinanderschaltung f series connection

hinterfräsen v.t. relief-mill

hinterfüllen v.t. (e. Gießform:) back up

Hinterfüllmasse f (shell molding) back-up material

Hinterfüllung f (e. Gießform:) backing

Hinterkipper m end tipper

hinterlegen v.t. (techn.) back [up]

Hinterrad n rear wheel

Hinterradachse f (auto.) rear axle

Hinterradantrieb m (auto.) rear-axle drive

Hinterradwelle f (auto.) rear axle shaft

Hinterschleifeinrichtung f relief-grinding attachment

hinterschleifen v.t. relief-grind

Hinterschliff m relieving, backing off

Hinterschliffwinkel m clearance angle, relief angle

hinterschütten v.t. (e. Modellplatte:) back up

hinterstechen v.t. (metal cutting) recess

Hinterstechwerkzeug n back facing tool

Hinterwalzer m (rolling) catcher

Hin- und Herbewegung f reciprocating movement

Hinweiszeichen n (traffic) indication sign

Hitzdraht m (electr.) hot wire

Hitzdrahtinstrument n hot wire instrument

hitzebeständig *a.* (Stahl:) resistant to heat; (ff. Steine:) refractory
Hitzebeständigkeit *f* refractoriness; heat resisting quality
HM-Platte *f* carbide tip
Hobel *m* plane
Hobelbank *f* carpenters' bench
Hobelbett *n* planer bed
Hobelbreite *f* planing width; shaping width; *s.a.* hobeln
Hobeleinrichtung *f* planing attachment
Hobeleisen *n* plane iron, plane knife
Hobeleisenschleifmaschine *f* planing irion grinder
Hobelfutter *n* planer chuck
Hobelkamm *m (gear cutting)* rack-type cutter
Hobellänge *f* planing length; shaping length; *s.a.* hobeln
Hobelmaschine *f* planing machine, *planer*; (≈ Waagerechtstoßmaschine:) shaper; *(Zahnradherstellung nach dem Wälzverfahren:)* generator
Hobelmeißel *m* planing tool; shaping tool; *s.a.* hobeln
hobeln *v.t.* 1. (auf der Langtischhobelmaschine:) plane; 2. (auf der Waagerechtstoßmaschine:) shape; 3. (von Zahnrädern nach dem Wälzverfahren:) generate
Hobelschlitten *m* 1. crossrail slide; 2. tool carrier slide; *s.a.* hobeln
Hobelstahl *m cf.* Hobelmeißel
Hobelstrich *m* 1. planing stroke; 2. shaping stroke; *s.a.* hobeln
Hobelsupport *m* 1. planing head, crossrail head; 2. toolhead; *s.a.* hobeln
Hobel- und Fräsmaschine *f* planer-miller
Hobler *m cf.* Waagerechtstoßmaschine

Hochantenne *f* elevated antenna; *(UK)* outdoor aerial
Hochbau *m* building above ground; building construction
hochbocken *v.t.* jack up
Hochdruck *m* high pressure
HOCHDRUCK ~, (~brenner, ~dampf, ~dampfmaschine, ~gasleitung, ~hydrierung, ~kessel, ~leitung, ~pumpe, ~rohr, ~stufe, ~ventil, ~verdichter) high-pressure ~
Hochdruckexplosivstoff *m* high explosive
hochfester Stahl, high-tensile steel, high-strength steel
hochfeuerfest *a.* highly refractory
Hochfrequenz *f* high frequency; *(radio)* radio frequency
HOCHFREQUENZ ~, (~ausgleich, ~beschleuniger, ~feld, ~generator, ~induktionsofen, ~kondensator, ~leistung, ~ofen, ~röhre, ~sender, ~strom, ~stufe, ~technik, ~transformator, ~übertragung, ~verstärker, ~widerstand) high-frequency ~
Hochfrequenzspule *f* radio coil
hochgekohlter Stahl, high carbon steel
Hochgenauigkeitslager *n* high-precision bearing
Hochglanz *m* mirror finish
hochglanzpolieren *v.t.* burnish, finish bright
Hochglanzpolitur *f* high-mirror finish, bright luster
Hochheber *m (auto.)* high level jack
hochkantbiegen *v.t.* bend on edge
hochkanten *v.t.* (Bleche:) edge-raise, raise edges

hochkantstellen v.t. upend, place on edge, turn upside down
Hochlauf m (Motor:) full speed, run up
hochlegierter Stahl, high-alloy steel
Hochleistung f high duty, heavy duty
HOCHLEISTUNGS ~, (~drehmaschine, ~fräsmaschine, ~niethammer, ~schleifmaschine, ~stoßmaschine, ~werkzeug) high-duty ~
Hochleistungsmotor m high-power motor; high-power engine
Hochleistungsröhre f (radio) high-power tube
Hochleistungssicherung f quick-break fuse
Hochleistungstransformator m high power transformer
Hochleistungsumformung f (explosive metal-forming) high-energy forming method
Hochofen m blast furnace
HOCHOFEN ~, (~anlage, ~aufzug, ~begichtung, ~betrieb, ~gas, ~gestell, ~gicht, ~koks, ~panzer, ~rast, ~reise, ~schacht, ~schaumschlacke, ~schlacke, ~schlackenzement, ~werk) blast-furnace ~
Hochofenzement m blast furnace slag cement
Hochöfner m blast furnaceman
Hochschulterkugellager n deep groove ball bearing
Hochspannung f high voltage, high tension
HOCHSPANNUNGS ~, (~anlage, ~batterie, ~freileitung, ~kabel, ~kondensator, ~leitung, ~netz, ~schalteinrichtung, ~schalter, ~sicherung, ~technik, ~transformator, ~wicklung) high-voltage ~
HÖCHST ~, (~drehzahl, ~gewicht, ~last, ~leistung, ~stand, ~strom, ~temperatur, ~wert) maximum ~
hochstellen v.t. elevate, raise
Hochstraße f overhead roadway; skyway
Hochtemperaturlegierung f high-temperature alloy
hochtouriger Motor, (auto.) high-speed engine
Hochvakuumröhre f high-vacuum tube
Hochvakuumtechnik f high-vacuum engineering
hochwertiger Grauguß, high-test grey iron, high-strength cast iron
hochwertiger Stahl, high-tensile steel
hochwertiges Gas, rich gas
Höhe f height, altitude
Höheneinstellung f vertical adjustment
Höhenmaßstab m depth gage
Höhenmesser m altimeter
Höhenschieblehre f vernier height gage
Höhensupport m (planer) rail-head, tool-head
höhenverstellbar a. vertically adjustable
Höhenverstellgetriebe n elevating gear
Höhenverstellung f vertical adjustment
Höhenvorschub m vertical feed
hohl a. hollow; concave
Hohlbeitel m gouge
hohlbohren v.t. hollow out; trepan
Hohlbohrer m trepanning tool
Hohlbolzenkette f hollow pin chain
Hohlkegel m taper hole
Hohlkehle f fillet
Hohlkehlhobel m fluting plane
Hohlkehlnaht f (welding) concave fillet weld

Hohlkehlschweißung f concave fillet weld
Hohlkeil m saddle key
Hohlkern m *(founding)* hollow core
Hohlmeißel m gouge
Hohlpfanne f *(building)* hollow tile
hohlprägen v.t. *(cold work)* (Löffel, Gabel:) emboss
Hohlraum m cavity, hollow-space, void
hohlschmieden v.t. hollow-forge
Hohlwalzwerk n rotary piercing mill
Hohlwelle f hollow shaft
Hohlziegelstein m hollow brick
Holzbohrer m wood auger
Holzfaserbruch m fibrous structure, woody structure
Holzfaserplatte f wood fibre board
Holzhammer m mallet
Holzimprägnierung f wood preservation
Holzkeil m wooden wedge
Holzkohle f charcoal
Holzkohlenroheisen n charcoal pig iron
Holzleim m adhesive for wood
Holzmast m wooden pole
Holzmehl n wood flour
Holzöl n tung oil
Holzpfahl m wooden pile
Holzplatz m lumber yard
Holzsäge f wood-cutting saw
Holzschraube f wood screw
Holzspannplatte f wood chip panel
Holzteer m wood tar
Honahle f honing tool, hone
honen v.t. hone
Honmaschine f honing machine
Honstein m honing stick
hörbar a. audible
Hörbarkeit f audibility
Hörbereich m *(radio)* audio range

Horchgerät n intercept receiver
Horchortung f sound ranging
Horde f hurdle
Hordenwascher m hurdle type scrubber
Hörempfang m audio reception
Hörer m *(tel.)* receiver
Hörfrequenz f audible frequency, audio frequency
horizontal a. horizontal; *s.a.* waagerecht
Horizontalbiegepresse f bulldozer
Horn n *(Amboß)* horn
hornartig a. horny
Hub m lift; (Exzenter, Kurbel:) throw; (Kolben:) stroke; (Pumpe:) lift
Hubanzeiger m stroke indicator
Hubbalkenofen m walking beam heating furnace
Hubbereich m range of stroke
Hubbrücke f vertical lift bridge
Hubgeschwindigkeit f stroke rate
Hubgetriebe n elevating gear
Hubkette f hoisting chain
Hubkraft f lifting power; hoisting power
Hublager n (e. Kurbelwelle:) crankpin
Hubmagnet m lifting magnet
Hubmotor m elevating motor
Hubnocken m eccentric cam
Hubraum m *(auto.)* swept volume, piston displacement, cylinder capacity
Hubscheibe f eccentric disc
Hubscheibenrad n *(shaper)* bull gear
Hubschlitten m *(saw)* stroke slide
Hubspindel f elevating screw
Hubtransporter m forklift truck
Hubventil n lift valve
Hubvolumen n *(auto.)* cubic capacity, swept volume, cylinder capacity
Hubwagen m lift truck, industrial truck, jacklift

Hubwelle f eccentric shaft
Hubwerk n hoisting gear; (radial) elevating gear
Hubwinde f lifting winch
Hubzähler m stroke counter
Hubzapfen m (e. Kurbelwelle:) crankpin
Hufstollenstahl m grooved flats
Hülle f (electron.) shell
Hülse f sleeve; bushing; ($\approx$ Pinole:) quill; (e. Kette:) barrel
Hülsenkupplung f ring compression coupling
Hund m (rolling mill) guard; (mining) minecar
Hupe f (auto.) horn
hupen v.t. (auto.) sound, hoot
Hupensignal n hooting signal
Hutmutter f cap nut
Hütte f metallurgical plant; (specif.) blast furnace plant; (Schmelzhütte:) smelting plant
Hüttenaluminium n primary aluminium pig

Hüttenbims m foamed slag: expanded slag
Hüttenkoks m blast furnace coke
Hüttenkunde f metallurgy
Hüttenmann m metallurgist, blast furnaceman
hüttenmännisch a. metallurgical
Hüttenwerk n metallurgical works, metallurgical plant, ironworks
Hüttenwesen n metallurgical engineering
Hüttenzement m slag cement
Hydraulik f hydraulics
Hydraulikgetriebe n hydraulic transmission
Hydrauliköl n hydraulic oil
Hydraulikpumpe f hydraulic pump
hydraulische Pumpe, hydraulic press
hydraulischer Antrieb, hydraulic drive
hydrieren v.t. hydrogenate, hydrogenize
Hydrierung f hydrogenation
Hydrodynamik f hydrodynamics
hydrodynamisch a. hydrodynamic
Hysteresisschleife f hysteresis loop

I

Illuminationslampe *f* decorative lamp
illuminieren *v.t.* illuminate
imprägnieren *v.t.* impregnate; (Holz:) creosote
Imprägnierung *f* impregnation
Imprägnierungsmittel *n* preservative
Impuls *m* impulse; *(phys.)* momentum; *(wave mech.)* pulse
IMPULS ~, (~abstand, ~breite, ~dauer, ~energie, ~folge, ~frequenz, ~geber, ~modulation, ~schalter, ~sender, ~strahlung, ~umformer, ~verstärker, ~zähler) pulse ~
Impulspeilung *f* radar
Impulsschweißbetrieb *m* pulsed resistance welding
Impulssperrung *f (telev.)* gating
Impulszahl *f (electron.)* count
I-Naht *f (welding)* square butt weld
Inbusschlüssel *m (tradename)* hexagon socket wrench
Inbusschraube *f (tradename)* socket head cap screw
Index *m* (= Indexbolzen, Indexstift) index pin; *(math.)* subscript
INDEX ~, (~bolzen, ~hebel, ~kurbel, ~loch, ~nocken, ~raste, ~schalthebel, ~stellung, ~teilung) index ~, indexing ~
indexieren *v.t.* index
Indirektteilverfahren *n* indirect method of indexing
Induktion *f (electr.)* induction
INDUKTIONS ~, (~fluß, ~frequenz, ~funke, ~härtung, ~kreis, ~meßgerät, ~messer, ~motor, ~schweißen, ~spule, ~strom) induction ~

induktionsfrei *a.* non-inductive
Induktionsmotor *m* ohne Hilfsphase für Handanwurf, crank start motor (without phase-splitting device)
induktiver Blindwiderstand, inductance
induktiver Strom, inductive current
induktiver Widerstand, inductive reactance
induktives Erwärmen, induction heating
Induktivität *f* inductance
Induktivitätsmesser *m* inductance meter
Industriefahrzeug *n* commercial vehicle
Industrieofen *m* industrial furnace
induzieren *v.t.* induce
ineinanderpassen *v.t.* fit together
Inertschweißen *n* inert-gas arc welding
Informationsspeicher *m* information store
Informationsverarbeitung *f* data processing
Infrarotstrahlung *mpl.* infra-red rays
Infrarotstrahlung *f* infra-red radiation
ingangsetzen *v.t.* start
Ingenieurbauten *mpl.* civil engineering works
Ingenieurbüro *n* engineering office
Ingenieurhammer *m* machinists' hammer
Injektorbrenner *m* low pressure torch
INNEN ~, (~beleuchtung, ~dreheinrichtung, ~getriebe, ~gewindefräsen, ~kegel, ~kopierapparat, ~läppmaschine, ~paßteil, ~raum, ~räumen, ~räumnadel, ~räumwerkzeug, ~riß, ~rundschleifmaschine, ~rüttler, ~schleifmaschine, ~spannung, ~verzahnung) internal ~

innenausdrehen v.t. turn inside diameters, bore
Innenausstattung f (auto.) interior furnishing
innendrehen v.t. bore, turn inside diameters
Innendrehmeißel m inside turning tool; (≈ Bohrmeißel:) boring tool
Innendrehzahn m boring bit
Innendrücker m (auto.) inside door handle
Innendurchmesser m inside diameter
inneneinstechen v.t. recess internally
Innengewinde n female thread
Innengrat m (forging) wad
Innenkegel m taper hole, taper bore
innenkopieren v.t. copy internally
Innenlack m interior varnish
innenläppen v.t. lap internally
Innenlenker m (auto.) saloon car, sedan limousine
Innenmaß n inside dimension
innenplandrehen v.t. face internally
innenräumen v.t. broach internally
innenrundläppen v.t. lap round holes internally
innenschleifen v.t. grind internally
Innenschlichtmeißel m finish boring tool
Innenschräge f (Gesenk:) inside draft
Innenschraublehre f inside micrometer caliper
innenschweißen v.t. weld internally
Innensechskant m (e. Schraube:) hexagon socket, hexagon hole
Innensechskantschlüssel m hexagon head socket wrench
Innensechskantschraube f socket head cap screw
Innentaster m inside caliper
Innenvierkant m square drive hole

inner a. interior, internal, inner
innenbetriebliche Dienstleistung, internal service
Installateur m plumber, fitter
Installation f installation; plumbing; (electr.) wiring
Installationsplan m (electr.) wiring diagram
Installationsrohr n (electr.) thin-gage conduit
Installationsschalter m house wiring switch
Installationswerkzeug n plumbers' tool
installieren v.t. install
instandhalten v.t. maintain
instandsetzen v.t. repair
Instandsetzung f repair
Instandsetzungsbetrieb m repair shop
Instrument n instrument, apparatus; tool; device; contrivance
Instrumentenbrett n (auto.) instrument panel
Instrumentenleuchte f (auto.) dashboard light
Interferenzkomparator m interferometer
Interferenzstreifen m interference band
interkristalline Brüchigkeit, cleavage brittleness
interkristalline Korrosion, intercrystalline corrosion, intergranular corrosion
Invalidenfahrzeug n (auto.) invalid carriage
Inventur f inventory
Investition f investment
Investitionskosten pl. capital outlay
Investitionsplanung f investment planning
Ion n ion
IONEN ~, (~austausch, ~bündelung,

~dichte, ~entladung, ~fleck, ~gitter, ~strahl, ~zähler) ion ~
Ionenlehre f ionics
Ionenröhre f ionic valve
Ionenschleuder f ionic centrifuge
Ionisation f ionization
IONISATIONS ~, (~energie, ~kammer, ~spannung) ionization ~
ionisieren v.t. ionize
Ionisierung f ionization
I-Profil n I-beam section
Irrstrom m stray current, parasitic current
Isolation f (electr.) insulation
ISOLATIONS ~, (~fehler, ~prüfer, ~prüfung, ~schaden, ~schutz, ~widerstand) insulation ~
Isolationsstrom m leakage current
Isolator m insulator
Isolierband n insulating tape
isolieren v.t. (electr.) insulate

Isolierlack m insulating varnish
Isoliermittel n insulating material, insulating compound
Isolierrohr n insulating tube
Isolierschicht f (building) insulating course
Isolierschlauch m (auto.) insulated sleeving
Isolierung f insulation; (chem.) isolation
Isotop n isotope
Isotopenanreicherung f isotope enrichment
Isotopentrennung f isotope separation
Ist-Kosten pl. actual cost
Istmaß n actual size
Istmenge f actual quantity
I-Stoß m (welding) square butt joint
I-Träger m I-beam, I-girder
Izod-Pendelhammer m Izod pendulum hammer

J

Japanlack m lac varnish, japan varnish, japan lacquer
Japanschwarz n (painting) black japan
J-Naht f (welding) single-J butt weld
Joch n (magn.) yoke

junger Beton, green concrete
justieren v.t. adjust; fit; level
Justierschraube f levelling screw
Justierzange f flat nose plier

K

Kabel *n* cable
KABEL ~, (~abzweiger, ~anschluß, ~bewehrung, ~einführung, ~endverteiler, ~fehler, ~führung, ~graben, ~kanal, ~klemme, ~kralle, ~kran, ~leiter, ~leitung, ~mantel, ~mast, ~messer, ~muffe, ~netz, ~schacht, ~schelle, ~schrapper, ~schuh, ~seele, ~seil, ~stecker, ~stollen, ~trommel, ~verbinder, ~verbindung, ~verlegung, ~verteilungskasten, ~winde) cable ~
Kabeltelegramm *n* cablegram
Kabelzange *f* lineman's plier
Kabinenroller *m* cabine scooter
Kachel *f* glazed tile
Käfig *m* (lapping) workholder; (Lager:) cage
Käfigläufermotor *m* squirrel-cage motor
Kalfaterhammer *m* calking mallet
Kaliber *n* diameter of bore; *(rolling)* groove, pass
Kaliberanzug *m* (rolling mill) taper of a groove
Kalibergwerk *n* potash-mine
Kaliberpresse *f* sizing press
Kaliberwalze *f* grooved roll
Kalibreur *m* designer
kalibrieren *v.t.* calibrate; size; (Walzen:) groove
Kalibrierpresse *f* sizing press
Kalibrierung *f* roll designing, groove designing
Kalidüngesalz *n* potash salt
Kalilauge *f* caustic potash solution
Kalisalpeter *m* potassium nitrate, saltpeter
Kaliwasserglas *n* potassium silicate

Kalk *m* line
KALK ~, (~brennen, ~brennofen, ~farbe, ~grube, ~kitt, ~mergel, ~mörtel, ~ofen, ~stein) lime ~
  gebrannter ~, quicklime
  gelöschter ~, slaked lime
Kalkbeton *m* lime concrete
Kalkelend *n (blast furnace)* lime set
kälken *v.t.* (Draht:) lime-coat
Kalkmilch *f* whitewash
Kalksteinteermakadam *m* limestone tarmacadam
Kalkzuschlag *m* addition of lime
kalorisieren *v.t.* calorize, alitize
Kalotte *f* cup; spherical indentation
kalt-abbinden *v.i.* (Klebstoff:) cold set
Kaltarbeitsstahl *m* cold work steel
Kaltaushärtung *f* strain hardening
Kaltbeanspruchung *f* cold straining, cold working
Kaltbearbeitbarkeit *f* cold-working property
kaltbearbeiten *v.t.* cold work
Kaltbearbeitung *f* cold work, coldworking
Kaltbildsamkeit *f* ductility
Kaltbruch *m* cold-shortness, cold brittleness
kaltbrüchig *a.* cold-short
Kaltbrüchigkeit *f* cold shortness
kaltdrücken *v.t.* (lathe) spin; ($\approx$ kaltfließpressen:) cold-extrude
Kältebeständigkeit *f* anti-freezing property
kalteinsenken *v.t.* hob
Kalteinsenkpresse *f* hobbing press
kalt erblasen *v.t.* blow cold

kalterblasenes Roheisen, cold-blast pig iron
Kältetechnik f refrigerating technique
Kaltfestigkeit f (founding) (e. Formmaske:) cold strength
kaltfließpressen v.t. cold-extrude
Kaltfließpressen n cold extrusion
kaltformen v.t. cold work
Kaltformung f cold work
Kaltgesenkdrückmaschine f cold swaging machine
kalthämmerbar a. malleable
Kalthämmerbarkeit f malleability
Kalthärtbarkeit f strain hardenability
Kalthärte f strain hardness, wear hardness
kalthärten v.t. strain harden
Kalthärtung f strain hardening, work hardening; cold straining; room temperature precipitation hardening
kaltkleben v.t. cold bond
Kaltkreissäge f cold circular saw
Kaltleim m cold-setting adhesive
kaltlochen v.t. perforate
Kaltmatrize f cold die
Kaltmeißel m cold chisel
kaltnachziehen v.t. redraw cold
Kaltnachzug m cold redrawing
kaltpressen v.t. (Rohre:) sink; (plastics) cold-mold
Kaltpreßschweißen n cold pressure welding
Kaltpreßstempel m hob
kaltrecken v.t. cold strain
Kaltreckung f cold straining
Kaltschlagmatrize f cold heading die
Kaltschlagwerkzeug n cold heading tool
Kaltschweiße f cold shut, spill
Kaltsprödigkeit f cold brittleness
Kaltstart m cold starting

Kaltstauchmatrize f cold upsetting die
Kaltstreckgrenze f yield point at normal temperature
Kaltverfestigung f strain hardening
Kaltverformbarkeit f cold forming property
kaltverformen v.t. cold-work, cold-form
Kaltverformung f cold working, cold forming
kaltwalzen v.t. cold roll
Kaltwalzwerk n cold rolling mill
Kaltwasserprüfdruck m cold water test pressure
kaltziehen v.t. cold-draw
Kaltziehmatrize f cold drawing die
Kaltzug m cold draw
Kalzinierofen m calcining kiln
Kamera f camera
Kamin m chimney, stack
Kammbrenner m comb-type burner
Kammeißel m rack-type cutter
kämmen v.t. (gearing) mesh; – v.i. mate
Kammer f (techn.) chamber
Kammerofen m chamber furnace; batch-type furnace; (Kokerei:) by-product oven
Kammlager n collar thrust bearing
Kammrad n cog wheel
Kammwalze f pinion
Kammwalzgerüst n pinion housing
Kampheröl n camphor oil
Kanal m channel; conduit; duct; (e. Blockform:) runner; (Kabel:) tunnel; troughing, duct; (Kathode:) canal; (telev.) channel
Kanalbau m (civ.eng.) sewer construction
Kanalisation f canalization, sewerage
Kanalisationsrohr n sewer pipe
Kanalisationstunnel m sewer tunnel
Kanalofen m tunnel kiln

Kanalstein *m* (Kokillenguß:) runner brick
Kanalwähler *m (telev.)* channel selector
Kanister *m* fuel can
kannellieren *v.t.* (Blech:) bead; (Holz:) flute
Kannellierhobel *m* fluting plane
Kantbrett *n (scaff.)* guard plank
kanten *v.t.* (Walzgut:) edge, turn over; (Holz:) square
kantengerade *a.* straight-edged
Kantenkehlmaschine *f* edge molder
Kantenlänge *f*(e. Würfels:) length of side
Kantenpressung *f* edge pressure
kantenschleifen *v.t. (woodworking)* square
Kanter *m (rolling mill)* edger; manipulator
Kantholz *n* squared timber
kantigdrehen *v.t.* square
Kapazität *f (electr.)* capacity; capacitance
Kapazitätsmesser *m* capacitance meter
kapazitiver Widerstand, capacitive reactance
Kappe *f* cap; hood; dome; (e. Walzenständers:) top; (e. Zinkdestillierofens:) arch
Kappnaht *f (welding)* sealing run, backing run
Karabinerhaken *m* snap hook
Karbid *n* carbide
karbonitrieren *v.t.* carbonitride
Karborund *n* carborundum
karburieren *v.t. (met.)* carburize; *(chem.)* carburet
Kardanantrieb *m (auto.)* universal-shaft drive
Kardanaufhängung *f* cardanic suspension
Kardangehäuse *n* cardan casing
Kardangelenk *n* universal joint, flexible joint
Kardankranz *m (auto.) universal joint ring*
Kardanrohr *n (auto.)* tubular transmission shaft
Kardanwelle *f* propeller shaft, cardan shaft, universal shaft
karmesin *a. (painting)* crimson
Karosserie *f* auto body
Karosserieblech *n* auto body sheet, automobile body sheet
Karosserielack *m* coachwork lacquer
Karusselldrehmaschine *f* vertical turning and boring mill; (≈ Drehwerk:) vertical turret lathe
Karussellfräsmaschine *f* rotary milling machine
Kaskade *f* cascade
KASKADEN ~, (~anlasser, ~motor, ~schaltung, ~umformer, ~verstärker, ~wandler) cascade ~
Kasten *m* box, case; (e. Batterie:) container; *(rolling mill)* cage; (für Werkzeuge:) kit
Kastenbett *n* (e. Werkzeugmaschine:) box-section bed
kastenförmig *a.* box-shaped, of box design
Kastenfuß *m (lathe)* box-section leg
Kastengesims *n* boxed cornice
kastenglühen *v.t.* box anneal, pot anneal, close anneal
Kastenglühofen *m* box annealing furnace, pot annealing furnace
Kastenglühung *f* pot annealing, box annealing, close annealing
Kastenkaliber *n (rolling mill)* box pass, box groove

kastenlose Formtechnik, *(founding)* boxless molding
Kastenständer *m* (e. Werkzeugmaschine:) box-type column
Kastentisch *m* box-type table
Kastenträger *m (building)* box girder
Kastenwagenaufbau *m (auto.)* box body
Kathode *f* cathode
KATHODEN ~, (~fall, ~klemme, ~kreis, ~leitwert, ~modulation, ~niederschlag, ~oszillograph, ~potential, ~sprung, ~strahl, ~strom, ~vorspannung, ~widerstand) cathode ~
Kathodenröhre *f* cathode-ray tube
Kathodenstrahloszillograph *m* cathode-ray oscillograph
Katze *f* (= Laufkatze) trolley
Katzenauge *n (auto.)* cat's eye
Katzenbahn *f* trolley track
Katzenkopf *m (lathe)* cathead
katzfahren *v.i.* trolley
Kausche *f* (e. Seils:) dead-eye
Kautschuk *m* caoutchuc, India rubber
Kegel *m (mach.)* taper; (≈ Vollkegel:) taper shank; (≈ Verjüngung:) conicity; *(geom.)* cone
KEGEL ~, (~bohrung, ~flansch, ~gewinde, ~lehrdorn, ~lehre, ~leiste, ~lineal, ~passung, ~schleifen, ~sitz, ~winkel) taper ~
Kegelantenne *f* cone antenna; *(UK)* cone aerial
Kegeldrehautomat *m* automatic lathe for taper turning
kegeldrehen *v.t.* taper turn
Kegeldrehvorrichtung *f* taper turning attachment
Kegeldruckversuch *m* cone-thrust test, cone indentation test

Kegelfeder *f* conical spring
kegelförmig *a.* tapering, tapered, conical
Kegelfräseinrichtung *f* taper milling attachment
Kegelgetriebe *n* bevel gearing; bevel gear drive
Kegelhülse *f* taper bushing; taper sleeve
kegelig *a.* taper, tapering, conical
kegeligsenken *v.t.* countersink
Kegelkuppe *f* chamfered end
Kegelrad *n* bevel gear
Kegelradgetriebe *n* bevel gear drive
Kegelradhobelmaschine *f* bevel gear planer; (Wälzverfahren:) bevel gear generator
Kegelradschleifmaschine *f* bevel gear grinder
Kegelradumlaufgetriebe *n* planetary-type bevel gearing
Kegelradwälzfräser *m* bevel gear hob
Kegelradwälzhobelmaschine *f* bevel gear generator
Kegelreibahle *f* taper shank reamer
Kegelrollenlager *n* taper roller bearing; *(US)* Timken bearing
Kegel-Schmiernippel *m* conical head lubricating nipple
Kegelsenker *m* rose-bit
Kegelsenkschraube *f* tire bolt
Kegelstumpf *m* frustum of a cone
Kehle *f* (Holz:) groove, flute; (Hohlkehle:) fillet; (am Schweißstoß:) throat
kehlen *v.t. (woodworking)* mold; channel
Kehlhobelmaschine *f* molding machine
Kehlkopfmikrofon *n* throat microphone, laryngophone
Kehlnaht *f (welding)* fillet weld
Kehlnahtschweißung *f* fillet welding
Kehlschweißung *f* fillet welding

Kehlwinkel *m* (e. Kehlnaht:) included angle

Kehrmaschine *f* road-sweeping machine

Kehrwert *m* reciprocal

Keil *m* (als Maschinenelement:) key; (als Spalt- oder Stellkeil:) wedge

Keilbahn *f* keyway, keyseat; (e. Keilwelle:) splineway

keilförmig *a.* tapering, conical, wedge-shaped

Keilleiste *f* taper gib

Keilnabe *f* splineway, female spline

Keilnut *f* keyway, keyseat; (e. Bandbremse:) V-groove

keilnuten *v.t.* keyway

Keilnutenlehre *f* keyway gage

Keilnutenmeißel *m* keyway tool

Keilnutenräummaschine *f* keyway broaching machine

Keilnutenstoßen *n* keyseating

Keilnutenstoßmaschine *f* keyseater

Keilriemen *m* Vee-belt, V-belt

Keilriemenscheibe *f* V-belt pulley

Keilriemenstufenscheibe *f* Vee-belt cone pulley

Keilspannfutter *n* collet bar chuck

Keilverzahnung *f* splining

Keilverzahnungsfräsmaschine *f* spline milling machine

Keilwelle *f* splineshaft

Keilwellenfräsmaschine *f* splineshaft milling machine

Keilwellenschleifmaschine *f* splineshaft grinder

Keilwinkel *m* (e. Meißels:) *(US)* lip angle; *(UK)* wedge angle

Keilzahn *m* (e. Keilwelle:) spline-tooth

Kennbuchstabe *m* identification letter, index letter

Kennlinie *f* characteristic

Kennzeichen *n* feature, characteristic

Kennzeichenleuchte *f* *(auto.)* number plate lamp

Kennzeichenschild *n* number plate

Kennziffer *f* index figure

Keramikschneidwerkzeug *n* ceramic-cutting tool

Kerbbiegeprobe *f* notched bar bend test

Kerbe *f* notch; (≈ Aussparung:) recess; (≈ Schlitz:) slot; (≈ Nut:) groove

Kerbempfindlichkeit *f* notch sensitivity

Kerbnut *f* V-notch

Kerbprobe *f* notched specimen

Kerbschlagbiegeprüfung *f* notched bar impact bending test

Kerbschlagprobestab *m* notched test bar

Kerbschlagprüfung *f* notched bar impact test, Izod test

Kerbschlagzähigkeit *f* Izod impact strength, notch toughness, notched bar impact strength

Kerbstift *m* notched taper pin

kerbverzahnen *v.t.* serrate

Kerbverzahnung *f* serration

Kerbwirkung *f* notch offect

Kerbwirkungszahl *f* notch factor

Kerbzähigkeit *f* *(incorr.)* cf. Kerbschlagzähigkeit

Kerbzahnwelle *f* serrated shaft

Kerbzugfestigkeit *f* notched tensile property

Kern *m* *(founding, magn., heat treatment)* core; *(cryst.)* nucleus; (Holz:) central pith; (e. Schraube:) body

KERN ~, (~anregung, ~chemie, ~dichte, ~drehung, ~energie, ~ladung, ~physik, ~reaktion, ~reaktor, ~spaltung,

~teilchen, ~umwandlung, ~zahl, ~zertrümmerung) nuclear ~

Kernansatz m (e. Schraube:) half dog point

Kernbindemittel n (founding) core binder

Kernblaseinrichtung f (founding) core-blowing equipment

Kernblasmaschine f core-blowing machine

kernbohren v.t. trepan

Kernbohrer m core drill

Kernbohrkopf m trepanning head

Kerndurchmesser m (e. Gewindes:) minor diameter; inside diameter

Kernfestigkeit f core strength

Kernformmaschine f core molding machine

Kernhaltestift m (founding) dabber

Kernholz n heartwood

Kernkasten m core box

Kernloch n core hole, cored hole

kernloser Induktionsofen, coreless induction furnace

Kernlunker m pipe

Kernmasse f (nucl.) nuclear mass

Kernnagel m (founding) core nail, chaplet

Kernofen m (founding) core oven

Kernrachenlehre f core snap gage

kernrissig a. (Holz:) shaky

Kernsand m core sand

Kernschrott m high-grade melting scrap

Kernstütze f (founding) chaplet

Kerntrockenofen m core-baking oven

Kerntrockenschrank m (founding) core baking oven

Kernzerschmiedung f forging burst

Kerze f (auto.) spark plug

Kerzenschlüssel m (auto.) plug spanner

Kessel m boiler

KESSEL ~, (~anlage, ~armaturen, ~bekohlungsanlage, ~blech, ~boden, ~druck, ~feuerung, ~haus, ~leistung, ~mantel, ~naht, ~reinigung, ~rohr, ~rost, ~speisepumpe, ~stein, ~wasser) boiler ~

Kesselschmied m boiler maker

Kesselwagen m tank car

Kette f (mech.) chain; (electr.) cell; (Gewebe:) warp

Ketten ~, (~antrieb, ~becherwerk, ~bolzen, ~brücke, ~flaschenzug, ~förderer, ~führungsrolle, ~glied, ~haken, ~hülse, ~lasche, ~last, ~leitrolle, ~molekül, ~reaktion, ~säge, ~spanner, ~strang, ~teilung, ~trieb, ~trommel, ~winde, ~wirbel) chain ~

Kettenfahrzeug n crawler-type vehicle, tracked vehicle

Kettenkranz m (bicycle) sprocket

Kettennaht f (welding) intermittent weld

Kettennuß f sprocket wheel

Kettenpunktschweißung f double-row spot welding

Kettenrad n sprocket [wheel], chain wheel

Kettenradfräser m sprocket cutter

Kettenschlepper m crawler-tractor, caterpillar tractor

Kettenschutzkasten m (auto.) chain-case

Kies m gravel, grit; (Kupfer:) pyrites; (blasting) shot

Kiesabbrände mpl. iron pyrites

Kiesasphalt m gravel asphalt

Kiesbeton m gravel concrete

Kieshinterfüllung f (shell molding) gravel-backup

Kilohertz n kilocycle per second

Kilometerzähler m (auto.) mileage recorder

Kilowattstundenzähler m kilowatthour meter
Kimme f rear sight
Kinematik f kinematics
Kinetik f kinetics
kinetische Energie, kinetic energy
KIPP ~, (~anhänger, ~bühne, ~ofen, ~pfanne, ~tisch, ~vorrichtung, ~werk) tilting ~, tipping ~
Kippackerwagen m agricultural tipper
Kippaufzug m skip hoist
kippen v.t. & v.i. tilt, tip, dump
Kippen m dumper, tipper, dump truck
Kippfrequenz f sweep frequency
Kippgefäß n (e. Maskenformmaschine:) dump box
Kippgefäßaufzug m skip hoist
Kippgefäßbegichtung f skip-charging, skip-filling
Kippgenerator m relaxation oscillator
Kippgetriebe n tipper gear
Kipphebel m tumbler lever; (auto.) rocker arm
Kipphebelschalter m tumbler switch
Kippkarren m dump-barrow, tipping cart
Kippkreis m (electr.) sweep circuit, relaxation circuit
Kippkübel m dump bucket, skip car
Kippkübelaufzug m bucket hoist
Kippmoment n (e. Motors:) pull-out torque
Kippmuldenaufzug m (building) [concrete] skip hoist
Kippschalter m toggle switch, tumbler switch
Kippständer m (motorcycle) prop stand
Kipptransformator m sweep transformer
Kippvorrichtung f tipping gear
Kippwagen m dump car, dump truck
Kippwerk n (e. Ofens:) tilting machinery
Kistenöffner m nail puller
Kitt m putty; cement
kitten v.t. cement; putty; lute; glue
Kittflüssigkeit f cement liquid
Kittpulver n cement powder
Klappbrücke f bascule bridge
Klappe f (Ventil:) flap, (tel.) calling drop
Klappenscharnier n flap hinge
Klappenschloß n flap lock
Klappenschrank m (telegr.) drop-type switchboard
Klappenventil n flap valve
Klappfenster n pivoted sash
Klappsitz m (auto.) folding seat
Klappverdeck n (auto.) folding top
Klärgrube f (building) cesspool
Klarlack m clear varnish, pale varnish
Klarlackanstrich m clear varnish coat
Klärschlamm m sludge
Klarsichtscheibe f (auto.) windscreen demister; (UK) windshield demister
Klartext m running text; plain language
Klärung f (Abwasser:) clarification
klassieren v.t. classify, sort, grade; (nach Korngröße:) screen, size
Klassiersieb n classifying screen, grizzly
Klassierung f (Aufbereitung) classification, sorting, grading; (nach Korngröße:) screening, sizing
klauben v.t. (≈ auslesen, aussortieren) pick
Klaubetisch m picking table, sorting table
Klaue f (e. Hammers:) claw; (≈ Kloben:) jaw; (zum Spannen:) dog
Klauenbeil n claw hatchet
Klauenkupplung f jaw clutch coupling
Klauenöl n neatsfoot oil
Klebband n adhesive tape

kleben *v.t.* bond; (≈ leimen:) glue; (≈ kitten:) cement; – *v.i.* stick
Kleber *m* adhesive
Klebestreifen *m* adhesive tape
Klebezettel *m* adhesive label
Klebfläche *f* adhesion surface area
Klebfolie *f* adhesive foil
Klebkitt *m* adhesive cement
Kleblack *m* adhesive varnish
klebrig *a.* sticky
Klebrigkeit *f (plastics)* tackiness
Klebsand *m* plastic refractory clay
Klebschraubstock *m* bench vise
Klebstoff *m* adhesive
Kleeblatt *n (rolling mill)* wobbler
Kleinautomat *m (electr.)* automatic cut-out
Kleinbetrieb *m* small-sized factory
Kleinbildkamera *f* miniature camera
Kleinbus *m* small omnibus
Kleineisenindustrie *f* small hardware industry
Kleineisenwaren *fpl.* small ironware, small hardware
Kleinkonverter *m* small converter, baby converter
Kleinlastkraftwagen *m* small truck
Kleinmotor *m* fractional H. P. motor
Kleinserienfertigung *f* small batch production
Kleinspannung *f* low voltage
Kleinstmaß *n* minimum size
Kleinstmotor *m* pilot motor
Kleinwagen *m* light car
Kleister *m* paste
Klemmbacke *f* clamping jaw, gripping jaw
Klemmbrett *n (electr.)* terminal board
Klemme *f (mach.)* clamp; grip; *(electr.)* terminal; (für Kabel:) binding screw

klemmen *v.t.* clamp; grip; – *v.r.* jam, bind, seize
Klemmenbrett *n* terminal board
Klemmenkasten *m* terminal box
Klemmenspannung *f* terminal voltage
Klemmfutter *n* clamping chuck
Klemmlager *n* adapter bearing
Klemmschraube *f* clamp bolt; binding screw
Klemmung *f* locking; (als Bauteil:) locking mechanism
Klemmutter *f* lock nut
Klemmvorrichtung *f* clamping device
Klempner *m* tinsmith, plumber; fitter
Klempnerarbeit *f* plumbing work
Klempnerschere *f* tinners' snips
Klempnerwerkzeug *n* plumbers' tool
Klimaanlage *f* air conditioning plant
Klinge *f* blade
Klingel *f (electr.)* bell, ringer
Klingeldraht *m* ringing wire
klingeln *v.i.* (Motor:) pink
Klingelstrom *m* ringing current
Klingeltransformator *m* bell transformer
Klinke *f* (e. Gesperres:) pawl; (e. Tür:) handle; *(tel.)* jack; (≈ Riegel:) latch
Klinkenrad *n* ratched wheel
Klinkenvorschub *m (planer)* ratchet and pawl feed
Klinker *m* (≈ Klinkerstein) clinker [brick]
Klirrfaktor *m (radio)* distortion factor
Kloben *m* (Planscheibe:) jaw; (Flaschenzug:) block
Klopffestigkeit *f* (Benzin:) anti-knock properties, octane rating
Klotz *m* block
Kluppe *f* screw plate stock, stock and die
knacken *v.i. (tel.)* crackle, click
Knagge *f* cam; dog

K-Naht f (welding) double-bevel butt weld
Knallgasflamme f (welding) oxyhydrogen flame
Knarre f ratchet stock
Knarrenhebel m ratchet lever
Knarrenkluppe f ratchet pipe stock
Knarrenschlüssel m ratchet wrench
knattern v.i. (auto.) backfire
Knebel m locking handle
Knebelgriff m T-handle, locking handle, tommy bar
Knebelschalter m jack switch
Knebelschraube f tommy screw
Kneifzange f carpenters' pincers
kneten v.t. knead; squeeze
Knetlegierung f wrought alloy
Knetmischer m pug-mill
Knick m kink; crease; sharp bend; break
knicken v.t. & v.i. buckle; break; kink; nick
Knickfestigkeit f buckling strength
Knickspannung f buckling stress
Knickversuch m buckling test, crippling test
Knie n (Rohr:) elbow, bend
Kniegelenk n toggle joint
Kniehebel m toggle
Kniehebelbreitziehpresse f toggle drawing press
Kniehebelnietmaschine f toggle-joint riveter
Kniehebelprägepresse f knuckle-joint embossing press
Kniehebelpresse f toggle-joint press; knuckle joint press
Kniehebelziehpresse f toggle drawing press
Knierohr n elbow, bend
Knochenleim m bone glue
Knochenöl n bone oil

Knopf m button; knob
Knoten m (electr., wave mech.) node
Knotenamt n (tel.) central exchange
Knotenblech n gusset-plate
Knotenpunkt m (railw.) center
Knotenstahl m (concrete) knobbled steel
Knüppel m (met.) billet
KNÜPPEL ~, (~gerüst, ~schere, ~straße, ~walze, ~walzwerk, ~wärmofen) billet ~
Knüppelabschnitt m (forging) slug
Kochherd m cooking range
Kochplatte f table range
Koerzitivkraft f coercive force
Kofferaufbau m (auto.) box-type van body
Kofferempfänger m portable receiver
Kofferraum m (US) luggage compartment, trunk room; (UK) boot
Kofferraumschloß n luggage boot lock
Kohle f coal; (chem.) carbon
KOHLE ~, (~bürste, ~elektrode, ~faden, ~lichtbogen, ~mikrofon) carbon ~; s. a. Kohlen ~
KOHLEN ~, (~aufbereitung, ~aufbereitungsanlage, ~bergbau, ~bergwerk, ~bunker, ~feuer, ~schaufel, ~staub, ~wäsche) coal ~
Kohlenoxid n carbon monoxide
Kohlensack m (blast furnace) bosh
Kohlenstaubbrenner m pulverized coal burner
Kohlenstaubfeuerung f powdered (or pulverized) coal firing
Kohlenstoff m carbon
KOHLENSTOFF ~, (~aufnahme, ~gehalt, ~stahl, ~werkzeugstahl) carbon ~
kohlenstoffreich a. rich in carbon, high-carbon
Kohlenzeche f colliery

Kohlungsmittel n *(met.)* carburizing agent, carburizer
Kokerei f coking plant, coke-oven plant
Kokereigas n coke-oven gas
Kokereiofen m coke-oven
Kokille f *(met.)* ingot mold
Kokillenguß m gravity die-casting
Kokillenschleuderguß m hot-mold centrifugal casting process
Kokosmatte f coconut matting
Koks m coke
KOKS ~, (~ausdrückmaschine, ~begichtungswagen, ~gabel, ~grus, ~kammer, ~löschwagen, ~ofen, ~ofenbatterie, ~ofengas) coke ~
Kokszeche f coke-oven plant
Kolben m piston; (≈ Tauchkolben:) plunger; (e. Glühlampe:) bulb; (Labor:) flask
KOLBEN ~, (~antrieb, ~dichtung, ~druck, ~fläche, ~gebläse, ~geschwindigkeit, ~hub, ~kopf, ~liderung, ~manschette, ~ring, ~ringzange, ~schieber, ~schleifmaschine, ~ventil, ~verdrängung) piston ~
Kolbenbolzen m *(auto.) (UK)* gudgeon pin; *(US)* wrist pin
Kolbenbolzenlager n *(auto.)* small-end bearing, wrist-pin bearing
Kolbenkompressor m reciprocating compressor
Kolbenmotor m reciprocating engine
Kolbenpumpe f plunger pump
Kolbenstange f piston rod; (Ventil:) piston stem
Kolbensteuerung f *(copying)* hydraulic tracer control
Kolbenverdichter m reciprocating compressor

Kolk m *(metal cutting)* crater, cup
Kolkverschleiß m wear by cratering
Kollektor m collector, commutator
Kollektorbürste f commutator brush
Kollektormotor m commutator motor
Kollergang m pan grinder, pug mill, edge mill
Kolonne f (Arbeiter:) crew; (Destillation:) column
Kolophonium n colophony
Kombinationskraftwagen m multipurpose vehicle
Kombinationszange f combination plier
Kombiwagen m *(US)* utility car; *(UK)* station car
Kombiwagenaufbau m multi-purpose van body
Kommando n command, signal, order
KOMMANDO ~, (~knopf, ~organ, ~stand, ~tafel) control ~
Kommandogabe f commanding, signalling
Kommandosignal n command signal
Kommutator m commutator
KOMMUTATOR ~, (~anker, ~bürste, ~gleichrichter, ~motor) commutator ~
Kommutierung f commutation
Kompaß m compass
Kompensator m *(electr.)* potentiometer
Kompoundmotor m compound wound motor
Kompression f compression
KOMPRESSIONS ~, (~druck, ~hub, ~pumpe, ~verhältnis, ~zündung) compression ~
kompressorlose Einspritzung, airless solid injection
Kompressormotor m supercharged engine, forced induction engine

Kondensat n condensate
Kondensator m capacitor; condenser
Kondensatorlautsprecher m capacitor loud-speaker
Kondensatormotor m capacitor motor
Kondensatorschweißung f stored energy welding
Kondenstopf m steam trap
Kondenswasser n water of condensation
Konferenzschaltung f conference circuit
Konservendosenlack m preserve can lacquer
Konsistenz f consistency
Konsolbauart f (miller) knee-and-column design
Konsole f bracket; (Winkeltisch:) knee-table
Konsolfräsmaschine f knee-and-column milling machine
Konsolkran m wall crane
Konsolager n bracket bearing
Konsolschlitten m (miller) saddle
Konsolspindel f (miller) elevating screw
konstruieren v.t. design
Konstrukteur m designer
Konstruktion f construction; design
Konstruktionsbüro n drawing office; engineering office
Konstruktionsfehler m faulty design
Konstruktionsmerkmal n constructional feature
konstruktionstechnisch a constructional
Konstruktionszeichnung f workshop drawing, engineering drawing
Kontakt m (electr.) contact
KONTAKT ~, (~bürste, ~druck, ~feder, ~fläche, ~hebel, ~potential, ~ring, ~rolle, ~schiene, ~schleifmaschine, ~schraube, ~spannung, ~stück, ~widerstand, ~zone) contact ~
Kontaktgeber m contact maker
Kontaktkohle f (auto.) carbon brush
Kontaktstoff m catalyzer, catalyst
Kontenrahmen m model chart of accounts
Kontermutter f check nut, lock nut
kontern v.t. lock
kontinuierliches Drahtwalzwerk continuous rod mill
Konto n account
Kontrollampe f pilot lamp
Kontrolle f control, inspection
Kontrollehre f master gage
Kontrolleuchte f (auto.) indicator lamp
kontrollieren v.t. control; inspect; check
Kontrollprobe f (chem.) check
Kontrollstab m (Prüfwesen) standard test bar
Kontrolluhr f check clock
Konus m (gearing) cone; s. a. Kegel
Konverter m (met.) converter
KONVERTER ~, (~anlage, ~auskleidung, ~bauch, ~betrieb, ~bühne, ~futter, ~mündung, ~zapfen) converter ~
Koordinate f coordinate
KOORDINATEN ~, (~achse, ~bewegung, ~bohrmaschine, ~ebene, ~einstellung, ~messung, ~netz, ~system) coordinate ~
Koordinaten-Bohr- und Fräswerk n jig boring mill, jigmill
koordinieren v.t. coordinate
Kopallack m copal varnish
Kopf m (techn.) head; (e. S. M.-Ofens:) port; (e. Brenners:) tip; (e. Gußblockes:) top end, head; (e. Bohrmaschine:) drilling head; boring head
Kopfanziehschraube f screw

Kopfdrehmaschine *f* facing lathe, chuck lathe
Kopfhöhe *f (gearing)* addendum
Kopfhörer *m* head receiver; headphone
Kopfkreis *m (gearing)* tip circle, addendum circle
Kopfkreisdurchmesser *m* tip circle diameter
Kopfmaske *f (welding)* face shield
Kopfschleifscheibe *f* cup wheel
Kopfschutzhaube *f (welding)* helmet
Kopfsenker *m* counterbore
Kopfspiel *n (gearing)* crest clearance
Kopfstaucher *m (tool)* header
Kopfsteinpflaster *n* cobble stone paving
Kopfstempel *m* heading die
Kopfverband *m* (Mauerwerk) head bond
Kopfzahnflanke *f* tooth face
kopieren *v.t. (metal cutting)* copy; duplicate
KOPIER ~, (~arbeit, ~bereich, ~drehmeißel, ~einrichtung, ~fühler, ~schlitten, ~steuerung, ~support, ~taster, ~vorschub) copying ~
Kopierbock *m (mach.)* guide bracket
Kopierdrehautomat *m* automatic copying lathe
Kopierdrehmaschine *f* copy-turning lathe, duplicating lathe
Kopierfräseinrichtung *f* copy-milling attachment; (Gesenke:) die-sinking attachment
kopierfräsen *v.t.* copy-mill, contour mill; (Gesenke:) die-sink
Kopierfräsmaschine *f* copy-milling machine, contour milling machine, profiler
Kopierhobelmaschine *f* contouring shaper

Kopierlineal *n* form plate, master plate, [copying] template
Kopierstift *m* tracer [pin]
koppeln *v.t.* couple
Kopplung *f (radio)* coupling
KOPPLUNGS ~, (~kondensator, ~spule, ~transformator, ~widerstand) coupling ~

Korb *m* basket
Kordelgriff *m* knurled handle
Kordelmutter *f* diamond-knurled nut
kordeln *v.t.* diamond-knurl
Kordelung *f* spiral knurling
Kork *m* cork
Korkeiche *f* cork oak, cork tree
Kornbildung *f* crystallization
Körner *m (tool)* center punch
Körnernarbe *f* center mark
Körnerspitze *f* lathe center
Korngefüge *g* grain structure
Korngrenze *f (cryst.)* grain boundary
körnig *a.* granular, grained; gritty; globular
Körper *m* body; substance; particle; *(mach.)* element; (fester:) solid
korrodieren *v.i.* corrode
korrosionsbeständig *a.* corrosion-resistant, rust-resisting
korrosionsbeständiger Stahl, stainless steel
Korrosionsbeständigkeit *f* corrosion resistance
Korrosionsdauerfestigkeit *f* corrosion fatigue
korrosionsempfindlich *a.* corrodible, sensitive to corrosion
Korrosionsermüdung *f* corrosion fatigue
korrosionsfest *a.* corrosion resisting, rustproof

Korrosionsfestigkeit f corrosion resistance
Korrosionsnarbe f corrosion pit
Korrosionsschutz m rust protection
kosmische Strahlen, cosmic rays
Kosten pl. cost; (≈ Unkosten:) expenses
KOSTEN ~, (~aufwand, ~ermittlung, ~gliederung, ~konto, ~rechnung, ~schlüssel, ~stelle, ~träger, ~überschlag, ~umlage, ~verlauf, ~voranschlag) cost ~
Kostenanschlag m estimate
Kostenrechner m calculator
Kostensatz m unit cost
Kostenstelle f cost center
Kostenstelleneinsatz m cost center input
Kostenstellenschlüssel m *(accounting)* works cost and expense code
Kotflügel m *(US)* fender; *(UK)* mudguard
Kraft f force; energy; *(electr.)* power
KRAFT ~, (~anlage, ~anschluß, ~antrieb, ~bedarf, ~hammer, ~leistung, ~leitung, ~netz, ~quelle, ~spannung, ~station, ~steckdose, ~strom, ~übertragung, ~verbrauch, ~verstärker, ~werk) power ~
Kraftfahrer m motorist, driver
Kraftfahrtechnik f automotive engineering
kraftfahrtechnisch a. automotive
Kraftfahrwesen n motoring, motorcar industry, automotive engineering
Kraftfahrzeug n automobile, motorcar, motor vehicle
Kraftfahrzeugbau m automotive engineering
Kraftfahrzeughalter m car owner
Kraftfahrzeughandel m automobile trade
Kraftfahrzeughändler m automobile dealer
Kraftfahrzeugindustrie f motor-vehicle industry
Kraftfahrzeugingenieur m automotive engineer
Kraftfahrzeugkette f *(auto.)* vehicle chain
Kraftfahrzeugmotor m automobile engine, motorcar engine
Kraftfahrzeugsteuer f motor [vehicle] tax
Kraftfahrzeugtechnik f automotive engineering
kraftfahrzeugtechnisch a. automotive
Kraftfahrzeugverkehr m motor traffic
Kraftfahrzeugwesen n motoring, automotive engineering
Kraftfahrzeugzubehör n motor-car accessories
Kraftfluß m magnetic flux
Kraftheber m *(auto.)* power jack
kräftig a. sturdy, rugged, strong
Kraftlinie f *(electr., magn.)* line of force
Kraftmaschine f engine
Kraftmeßdose f pressure gage, standardizing box
Kraftmesser m dynamometer
Kraftrad n motorcycle
kraftschlüssige Verbindung, non-positive connection
Kraftspannfutter n power-operated chuck
Kraftstoff m fuel; (Diesel:) fuel oil
KRAFTSTOFF ~, (~behälter, ~einspritzung, ~hahn, ~kanister, ~messer, ~pumpe, ~reservetank, ~uhr, ~verbrauch) fuel ~
Kraftstoffschlauch m *(auto.)* flexible fuel tubing
Kraftverbrennungsmaschine f internal combustion engine
Kraftwagen m motorcar, motor vehicle
Kraftzange f pincers

Kraftzentrale f power station
Kragbalken m cantilever
Kragensteckdose f socket with shrouded contacts
Kralle f (Kabel:) grip
Krampe f staple
Kran m crane
KRAN ~, (~anlage, ~ausleger, ~bau, ~brücke, ~gerüst, ~gießpfanne, ~gleis, ~greifer, ~haken, ~katze, ~kette, ~last, ~magnet, ~pfanne, ~schiene, ~seil, ~waage) crane ~
Kranbagger m shovel crane, craneexcavator
Kranführer m crane operator, crane driver
Krankenkraftwagen m motor ambulance
Kranwagen m (auto.) crane truck, recovery vehicle, wrecking car
Kranz m (e. Rades:) rim
Kranzbrenner m ring burner
Kranzspannfutter n geared scroll chuck
Krappłack m madder lake
Krapprot n alizarine
Krater m (metal cutting) crater, pit
Kratzer m scratch
Kratzerförderer m drag-chain conveyor
Kratzfestigkeit f resistance to scratching
Kreide f chalk; (gemahlene:) whiting
Kreis m circle; (electr.) circuit
Kreisbewegung f rotary motion
Kreisbogen m circular arc
Kreisdurchmesser m diameter
Kreiselbrecher m gyratory crusher
Kreiselkompaß m gyro-compass
kreiseln v.i. gyrate
Kreiselpumpe f centrifugal pump, rotary pump
kreisen v.i. rotate, revolve; circulate

kreisförmig a. circular
Kreisfrequenz f (phys.) angular frequency; (US) radian frequency
Kreisgenauigkeit f cylindricity
Kreislauf m circulation; cycling; (hydr.) circuit
Kreismesser n circular slitting saw
kreisrund a. circular; (Bohrung:) cylindrical
Kreissäge f circular saw; (Maschine:) circular sawing machine
Kreissägeblatt n circular saw blade
Kreisschere f circular shear
Kreisumfang m circumference
Kreiswiderstand m (electr.) circuit resistance
Kresol n cresole
Kreuzbewegung f (e. Tisches:) compound motion
kreuzen v.r. cross; (≈ sich schneiden:) intersect
Kreuzgelenk n universal joint
Kreuzgelenkkupplung f universal joint coupling
Kreuzkopf m crosshead
Kreuzlibelle f cross-bubble
Kreuzloch n cross hole
Kreuzlochlehre f star gage
Kreuzlochmutter f cross hole nut, Phillips nut
Kreuzlochschraube f recessed head machine screw, Phillips screw
Kreuzmeißel m cape chisel
Kreuzschalter m joystick switch
Kreuzschieber m compound rest slide; (miller) saddle
Kreuzschlag m (e. Seils:) cross lay
Kreuzschlaghammer m straight pane hammer

Kreuzschlitten m *(lathe)* compound slide rest
Kreuzschlüssel m four-way rim wrench
Kreuzstrebe f diagonal strut
Kreuzstück n cross [piece]
Kreuzsupport m compound rest
Kreuzung f *(auto.)* crossroads, street crossing
KREUZUNGS ~, (~ebene, ~linie, ~weiche, ~winkel) crossing ~
Kreuzungsmast m *(tel.)* transposition pole
Kreuzungspunkt m junction
Kreuzverband m diagonal brace
kreuzweise a. crosswise
Kreuzwinkel m *(tool)* T-square
Kreuzzapfen m universal trunnion
Kriechdehnung f creep
kriechen v.i. creep
kriechfest a. creep-resistant
Kriechfestigkeit f creep strength
Kriechgang m *(mach.)* inching
Kriechgrenze f creep limit, limiting creep stress
Kriechspur f *(traffic)* creep lane
Kriechstrom m leakage current
Kriechstromfestigkeit f tracking resistance
Kriechvorschub m inching feed
Kriechweg m *(electr.)* leakage path
Kristall m crystal
Kristallachse f crystal axis
kristallartig a. crystalline
Kristallausbildung f crystalline structure
Kristallautsprecher m piezo-electric loudspeaker
Kristallehre f crystallography
Kristallfläche f crystal plane
Kristallisation f crystallization

Kristallisationsgeschwindigkeit f rate of crystalline growth
Kristallkorn n crystalline grain, crystal grain
Kristall-Lack m crystallizing lacquer
Kristallplastizität f *(explosive metalforming)* plastic flow in crystals
Kristallspiegelglas n polished plate glass
Kristallstruktur f crystalline structure
Kristalltonabnehmer m crystal pickup
Kristallverstärker m transistor
Kristallwachstum n crystal growth
kritisch a. critical
Krone f *(hydr. eng.)* crown; crest
Kronenmutter f castellated nut
kröpfen v.t. offset; crank
Kröpfmaschine f joggling machine
Kröpfung f offset; crank; (e. Maschinenbettes:) gap
krumm a. crooked, out-of-straight; bent; curved
Krümmer m (Rohr:) elbow, bend; *(auto.)* manifold
Krümmung f curvature; radius, bend; curve; bow; (seitliche:) camber; (senkrechte:) sweep
K-Stegnaht f *(welding)* double-bevel butt weld with root face
Kübel m bucket
Kübelaufzug m bucket elevator
Kübelbegichtung f bucket-charging
Kugel f *(mach.)* ball; *(geom.)* sphere; (e. Lampe:) globe
Kugeldruckhärte f ball thrust hardness
Kugeldruckhärteprüfer m ball thrust hardness tester
Kugeldrucklager n ball thrust bearing
Kugeldruckversuch m ball pressure test

Kugeleindruck *m* ball impression, ball indentation
Kugelendmaß *n* spherical end measuring rod
Kugelfallhärte *f* scleroscope hardness
Kugelgelenk *n* ball and socket joint
Kugelgraphitguß *m* nodular iron, spherolitic iron
kugelig *a.* spherical; *(metallo.)* nodular
Kugelkäfig *m* ball cage
Kugellager *n* ball bearing
Kugellängslager *n* ball thrust bearing
Kugellenkkranz *m (auto.)* ball-type turntable
Kugelöler *m* ball oiler
Kugelrohrmühle *f* ball tube grinder
Kugelschmierkopf *m (auto.)* nipple
Kugelschraubstock *m* ball joint vise
kugelstrahlen *v.t.* shot-blast, shot-peen
Kugelventil *n* globe valve
KÜHL ~, (~fläche, ~kammer, ~luft, ~mantel, ~öl, ~pumpe, ~turm) cooling ~
Kühler *m (auto.)* radiator; *(refrigeration)* cooler
Kühlerblende *f (auto.)* radiator cowling
Kühlerhaube *f* radiator cowl, radiator cover
Kühlerjalousie *f (auto.)* radiator shutter
Kühlerschutzgitter *n (auto.)* radiator grille
Kühlerverkleidung *f* radiator cowl
Kühlerverschraubung *f (auto.)* radiator union
Kühlflüssigkeit *f* coolant, cooling agent
Kühlgebläse *n (auto.)* fan
Kühlluftregler *m (auto.)* thermostat
Kühlmaschine *f* refrigerator
Kühlmittel *n* coolant
Kühlschlange *f (auto.)* radiator coil; (Labor:) cooling coil

Kühltruhe *f* food freezer
Kühlwagen *m* refrigerator vehicle
Kühlwasser *n* cooling water
KÜHLWASSER ~, (~leitung, ~mantel, ~pumpe, ~regler, ~rohr, ~umlauf) cooling water ~
Kulisse *f (kinematics)* slotted link
Kulissenrad *n (crank shaper)* bull gear, rocker gear
Kulissenstein *m* [sliding] block
Kulissentrieb *m (shaper)* crank drive gear
Kumaronharz *n* coumarone resin
kümpeln *v.t.* (Bleche:) flange; dish
Kundendienst *m* after-sales service
Kundendienstabteilung *f* service department
Kundengießerei *f* jobbing foundry
Kundenguß *m* jobbings
Kunstbronze *f* art bronze
Kunstguß *m* art castings
Kunstharz *n* synthetic resin
Kunstharzemaillelack *m* synthetic resin enamel
Kunstharzleim *m* synthetic resin glue
Kunstharz-Preßholz *n* synthetic resin-compressed wood
Kunstharzpreßstoff *m* synthetic plastic material
Kunstharzüberzugslack *m* synthetic resin finish
Kunstleder *n* artificial leather
künstliche Beleuchtung, artificial light
Kunstseide *f* artificial silk; rayon
Kunststoff *m* synthetic material, plastics
Kunststoff-Folie *f* plastic sheet, plastic film, plastic foil
Kunststoffindustrie *f* plastics industry
Kunststoff-Kleber *m* adhesive for plastics

Kunststoff-Klebstoff *m* adhesive made from plastics
Kunststoffzahnrad *n* micarta gear
Kupfer *n* copper
KUPFER ~, (~bergwerk, ~draht, ~erz, ~folie, ~hütte, ~kies, ~legierung, ~plattierung, ~überzug, ~verhüttung) copper ~
Kupfergewinnung *f* extraction of copper
Kupferguß *m* cast copper
kupferhaltig *a.* copper-bearing
kupferplattiertes Blech, copper-clad metal
Kupolofen *m* cupola furnace
Kupolofenbetrieb *m* cupola practice
Kupolofenherd *m* cupola hearth
Kupolofenmantel *m* cupola shell
Kuppe *f* (e. Schraube:) point, end
kuppeln *v.t.* couple
Kuppelzapfen *m (rolling mill)* wobbler
Kuppenwerkzeug *n* ending tool
Kupplung *f* (starre:) coupling; (elastische:) clutch
KUPPLUNGS ~, (~bremse, ~büchse, ~flansch, ~gabel, ~getriebe, ~hebel, ~lager, ~lamelle, ~magnet, ~muffe, ~nocken, ~pedal, ~trommel) clutch ~
Kupplungsschalter *m (electr.)* coupled switch; *(gearing)* clutch shifter
Kupplungssteckdose *f* socket coupler
Kupplungsstecker *m (electr.)* coupler plug
Kurbel *f* crank
KURBEL ~, (~anlasser, ~antrieb, ~gehäuse, ~griff, ~hub, ~kasten, ~presse, ~schenkel, ~wange) crank ~
Kurbelgetriebe *n* crank drive [mechanism]
Kurbelinduktor *m* magneto generator
Kurbellager *n (auto.)* crankshaft bearing
kurbeln *v.t.* crank
Kurbelschwinge *f* rocker arm
Kurbelschwingenantrieb *m* crank mechanism
Kurbeltrieb *m* crank mechanism, crank gear; *(slotter)* crank drive
Kurbelwanne *f (auto.)* crankcase sump
Kurbelwelle *f* crankshaft
Kurbelwellenanlasser *m* crank starter
Kurbelwellendrehmaschine *f* crankshaft turning lathe
Kurbelwellenlager *n* crankshaft bearing
Kurbelzapfen *m* crankpin
Kurbelzapfenlager *n* crankpin bearing
Kurbelzapfenschleifmaschine *f* crankpin grinder
Kurbelziehpresse *f* double-action cam drawing press
Kurve *f* (≈ Linienzug:) curve; (≈ Rundung:) radius; (≈ Steuerkurve:) cam
Kurvenautomat *m* cam-operated automatic lathe
Kurvenberechnung *f (mach.)* cam calculation
Kurvenblatt *n (drawing)* graph
Kurvendreheinrichtung *f* cam turning attachment
kurvenförmig *a.* curved, curvilinear
Kurvenfräsen *n* radius milling
kurvengesteuert *a. (mach.)* cam-controlled
Kurvengetriebe *n* cam mechanism
Kurvenhobeleinrichtung *f* radius-planing attachment
Kurvenhub *m* cam throw
Kurvenlineal *n* drawing curve
Kurvenschablone *f* master cam
Kurvenschaltung *f* (e. Maschine:) cam control

Kurvenscheibe *f* plate cam
Kurvenschleifeinrichtung *f* cam-grinding attachment
Kurvensteuerung *f* cam control
Kurventrommel *f* drum-type cam
kürzen *v.t.* shorten
Kurzgewinde *n* short thread
Kurzgewindefräsen *n* plunge thread milling
Kurzgewindeschleifen *n* plunge thread grinding
Kurzhobler *m* cf. Waagerechtstoßmaschine
Kurzkegel *m* steep taper, short taper
Kurzprüfung *f* short-time test
kurzschließen *v.t.* short-circuit
Kurzschluß *m* short circuit
Kurzschlußanker *m* squirrel-cage rotor
Kurzschlußankermotor *m* squirrel-cage motor
Kurzschlußkäfig *m* cage rotor
Kurzschlußmotor *m* squirrel-cage induction motor
Kurzschlußspannung *f* short-circuit voltage
Kurzschlußstrom *m* short-circuit current
Kurzstab *m* short test bar
Kurzversuch *m* short-time test
Kurzwellen *fpl.* short waves
KURZWELLEN ~, (~antenne, ~bereich, ~empfang, ~empfänger, ~sender, ~übertragung) short-wave ~
Kurzzeitmesser *m* microchronometer
Kurzzeittemperung *f* short-cycle annealing
Kurzzeitversuch *m* short-time test
Kutschenlack *m* coach varnish

L

Labyrinthdichtung *f* labyrinth packing
Lack *m* varnish; (≈ Farblack:) lake; (gefärbter:) enamel
Lackanstrich *m* varnish coat
Lackbaum *m* lacquer tree
Lackbenzin *n* paint thinner
Lackdraht *m* enamelled wire
Lackfarbe *f* varnish color
Lackfarbenanstrich *m* varnish color coat
Lackharz *n* varnish gum
lackieren *v.t.* (mit Blanklack:) varnish; (mit Farblack:) lacquer
Lackpapier *n* varnished paper
Lackpapierdraht *m* varnished cambric wire
Lackpflegemittel *n (auto.)* lacquer preservative
Lackspachtel *m* varnish filler
Lacküberzug *m* varnish coat
LADE ~, *(electr.)* (~ gleichrichter, ~kondensator, ~kreis, ~schaltung, ~spannung, ~strom, ~widerstand) charging ~
Ladeanzeigeleuchte *f (auto.)* charging indicator lamp
Ladebrücke *f* tailboard loader
Ladeeinrichtung *f* loading attachment
Ladegerät *n (auto.)* battery charger
laden *v.t.* load; *(electr.)* charge
Lader *m (auto.)* battery charger
Ladestelle *f* (e. Werkzeugmaschine:) loading station
Ladevorrichtung *f (automatic lathe)* feeding attachment, loading attachment, magazine
Ladung *f (electr.)* charge; *(com., mach.)* load

Lage *f* position; location; (≈ Schicht:) layer; situation; condition
Lager *n (mach.)* bearing; (Stütze:) support; (≈ Speicher:) store; *(geol.)* deposit
LAGER ~, *(mach.)* (~bohrung, ~buchse, ~deckel, ~druck, ~einbau, ~gehäuse, ~käfig, ~luft, ~metall) bearing ~
Lagerbestand *m* stock
Lagerbock *m* bearing pedestal
Lagerhalter *m* storekeeper
lagerhaltig *a.* available ex stock
Lagerhaltung *f* storekeeping
Lagerhaus *n* warehouse, magazine
Lagermeister *m* storekeeper
lagern *v.t.* (≈ montieren:) mount; (≈ stützen:) support; (≈ anordnen:) arrange; (≈ speichern:) store; (Rohre:) stockpile, stack
Lagerplatz *m* [storage] yard
Lagerraum *m* storeroom
Lagerschale *f* split bearing, brass
Lagerschild *n* end shield
Lagerschuppen *m* storage shed
Lagerung *f* (≈ Auflager:) bearing; (≈ Befestigung, Anbringung:) mounting; (≈ Abstützung:) support; (≈ Speicherung:) storage
Lagerverwalter *m* storekeeper
Lagerweißmetall *n* babbitt metal
Lagerzapfen *m* journal
Lamelle *f* (e. Kupplung:) disc, plate
Lamellenbremse *f* multiple-disc brake
Lamellenkühler *m* gilled radiator
Lamellenkupplung *f* multiple-disc clutch
Lampe *f* (= Leuchte) lamp
Lampenfassung *f (electr.)* bulb socket

Lampenruß *m* lampblack
Langdrehautomat *m* Swiss bush-type automatic screw machine
Langdrehschlitten *m (lathe)* straight-turning slide
Länge *f* length
längen *v.t.* lengthen, extend, stretch; (Walzgut:) elongate
Längenmaß *n* length dimension
Längenmeßgerät *n* length measuring instrument
Langfräsmaschine *f* manufacturing type milling machine; (große:) planer-type milling machine
Langgewindefräsen *n* traverse thread milling
Langgewindefräsmaschine *f* milling machine for milling long threads
Langhobelfräsmaschine *f* planer-miller
Langhobelmaschine *f* planer, planing machine
Langholz *n* long trunks
Langholzwagen *m* timber transporter, trunk transport car
Langloch *n* slot; oblong hole
langlochen *v.t.* slot; *(woodworking)* dado
Langlochhobel *m* dado plane
längs *adv., prep.* lengthwise; longitudinal; along
Längsanschlag *m (lathe)* length stop
Längsdrehen *n* straight turning
Längsdrehsupport *m (lathe)* saddle
Längsdruck *m* thrust
Längsfalzmaschine *f* side seaming machine
Längskeil *m* ordinary key
Längskopiereinrichtung *f* longitudinal copying attachment
Längskugellager *n* single thrust ball bearing
Längslager *n* thrust bearing
Längsnaht *f (welding)* longitudinal seam
Längsnut *f* longitudinal groove
Längsprobe *f* longitudinal test specimen
Längsschlag *m* (e. Seils:) Lang lay
Längsschleifen *n* traverse grinding
Längsschlitten *m (lathe)* plain turning slide
Längsschnitt *m (drawing)* longitudinal section
Längsspiel *n* end play
Längssupport *m (lathe)* saddle
Langstab *m* long test bar
Längsvorschub *m* (e. Maschinentisches:) longitudinal feed, sliding feed
Langversuch *m* long-time test
Langwelle *f* long wave
LANGWELLEN ~, (~band, ~bereich, ~empfänger, ~sender) long-wave ~
Langzeitversuch *m* long-time test, long-term test
LÄPP ~, (~dorn, ~flüssigkeit, ~maschine, ~mittel, ~öl, ~spindel, ~werkzeug) lapping ~
läppen *v.t.* lap
Läppkäfig *m* retainer, cage
Läppring *m* ring lap
Läppscheibe *f* lapping wheel, abrasive lap, lap
Lasche *f* butt strap, cover plate; *(railway)* fishplate; (e. Kette:) side bar; (e. Feder:) shackle
Laschenbolzen *m* fish-plate bolt
Laschenkette *f* steel side bar chain
Laschennietung *f* butt-joint riveting
Laschenpunktschweißung *f* bridge spot weld

Laschenschiene f fishplate rail
Laschenstoß m (welding) strapped joint
lasieren v.t. glaze
LAST ~, (~bremse, ~druck, ~faktor, ~haken, ~kette, ~rolle, ~verteilung) load ~
Lastdrehzahl f on-load speed
Lastdruckbremse f load reaction brake
Lasthebekette f elevating chain
Lasthebemagnet m lifting magnet
Lastkraftwagen m (US) truck; (UK) lorry
Lastkraftwagenheber m truck jack
Lastschalter m power circuit-breaker
Lastschaltgetriebe n change-underload transmission
Lastseil n hoisting rope
Lastwagen m road transport vehicle
Lastwechsel m (Dauerversuch:) reversal of stress
Lastwiderstand m (electr.) load impedance
Lastzug m truck-trailer combination
Lasur f glaze
Lasuranstrich m glazing coat
Lasurfarbe f transparent color
Latte f batten; slat; lath; (≈ Meßlatte:) rod
Lattenkiste f crate
Lauf m travel, traverse, movement; (Motor:) running; course, way
Laufbahn f runway, track; (Lager:) raceway
Laufbuchse f (e. Zylinders:) liner
Laufeigenschaften fpl. running characteristics
Läufer m (electr.) rotor
LÄUFER ~, (~anlasser, ~kreis, ~spannung, ~strom, ~wicklung) rotor ~
Lauffläche f (e. Rades:) tread
Laufgeräusche n noise under load

Laufgewicht n (e. Waage:) jockey weight, rider
Laufgewichtswaage f weighing lever with movable jockey
Laufkatze f trolley
Laufkatzenfahrbahn f trolley runway
Laufkran m travelling crane
Laufring m (e. Lagers:) raceway
Laufschiene f runway rail
Laufsitz m running fit
Laufsteg m gangway
Laufzeitentzerrer m delay equalizer
Laufzeit f (electron.) transit time
Laufzeitverzerrung f delay distortion
Lauge f caustic solution; alkaline solution; (Seifenlauge:) suds; (Alaunlauge:) alum liquor; (Kalilauge:) caustic potash solution; (Salz:) brine
laugenbeständig a. resistant to caustic solutions; alkali-proof
laugenrißbeständig a. resistant to caustic cracking
Laut m sound; tone
Lautfernsprecher m telephonograph
Lautsprecher m loudspeaker
Lautstärke f sound level; (radio) volume of sound, signal strength
Lautstärkemesser m sound level meter
Lautstärkeregelung f volume adjustment; volume control
Lautverstärker m sound volume amplifier
Lautverzerrung f sound distortion
LD-Stahl m oxygen converter steel
LD-Verfahren n Linz-Donawitz process
Lebensdauer f service life, lifespan
Leckstrom m leakage current
Lederlack m leather varnish
Lederleim m leather glue
Lederriemen m leather belt

Leerhub *m* idling stroke
Leerlauf *m* idle running
LEERLAUF ~, (~drehmoment, ~drehzahl, ~einstellung, ~kennlinie, ~reibung, ~schaltung, ~spannung, ~strom, ~verlust, ~zustand) no-load ~
leerlaufen *v.i.* run idle
Leerlaufhub *m* (*auto.*) idle stroke
Leerlaufstellung *f* neutral position, idle position
Leerweg *m* idle motion, idle movement
Leerwert *m* idle load
Leerzeit *f* lost time, idle time
legieren *v.t.* alloy
Legierung *f* alloy
Legierungsbestandteil *m* alloying constituent
Legierungselement *n* alloying element
Legierungsmittel *n* alloying constiuent
Legung *f* (Kabel, Rohre:) laying
Lehm *m* loam
Lehmformen *n* loam molding
Lehmformerei *f* loam molding shop
Lehmguß *m* loam castings
Lehmstopfen *m* clay plug
Lehrboden *m* (*molding*) bottom board
Lehrdorn *m* screw plug gage
Lehre *f* (*metrol.*) gage; (*UK*) gauge; (als Bohrlehre:) drill jig; boring jig
lehren *v.t.* gage; (*UK*) gauge; caliper; (*UK*) calliper
Lehrenbohren *n* jig boring
Lehrenbohrmaschine *f* jig boring machine
lehrengenau *a.* true to gage
Lehrengenauigkeit *f* accuracy to gage
Lehrenschleifmaschine *f* jig grinder
Lehrgerüst *n* (*civ. eng.*) falsework
Lehrmutter *f* ring gage
Lehrring *m* ring gage

Leichtbau *m* light construction
Leichtbauplatte *f* lightweight building board
Leichtbeton *m* lightweight concrete
Leichtmetall *n* light alloy, light metal
Leichtmetallegierung *f* light metal alloy, light alloy
Leichtprofil *n* light section
Leichtstahlbau *m* light section engineering
Leierbohrer *m* brace drill
Leim *m* glue
leimartig *a.* gluey
leimen *v.t.* glue
Leimfarbe *f* distemper
Leimfolie *f* lime foil, lime film
Leimknecht *m* glue-press, glueing-cramp
Leimspachtel *m* glue filler
Leinenmaßband *n* linen tape
Leinöl *n* linseed oil
Leinölfarbe *f* linseed oil paint
Leinölfirnis *m* varnlsh
Leinölkitt *m* linseed oil putty
Leinwand *f* canvas
Leiste *f* (≈ Stelleiste:) gib; (Holz:) lath, strip; (≈ Anschlagleiste:) fence
Leistung *f* (e. Maschine:) capacity; (e. Motors:) horsepower rating; (≈ Wirkungsgrad:) efficiency; (*electr.*) energy, power
Leistungsabgabe *f* (*electr.*) power output
Leistungsangabe *f* (Motor:) power rating
Leistungsaufnahme *f* (*electr.*) power input
Leistungsbemessung *f* rating
Leistungsbereich *m* (Maschine:) capacity range
Leistungsbeschreibung *f* (*building*) contract specifications
Leistungsentlohnung *f* incentive payment

Leistungsfähigkeit f capacity; efficiency
Leistungsfaktor m *(electr.)* power factor
Leistungsfaktormesser m power-factor meter
Leistungsgradschätzen n *(work study)* performance rating
Leistungslohn m incentive earnings
Leistungslohnsystem n wage incentive plan
Leistungsmesser m *(electr.)* wattmeter
Leistungsmeßgerät n *(electr.)* power meter
Leistungsprüfung f performance test
Leistungsreserve f reserve power
Leistungsschalter m circuit-breaker
Leistungsschild n rating plate
Leistungstransformator m power transformer
Leistungsverbrauch m power consumption
Leistungsverlust m *(electr.)* power loss
Leistungsverstärker m power amplifier
Leistungsverzeichnis n schedule of prices
Leitapparat m *(mach.)* forming attachment
Leitbacke f *(threading)* follower
Leitblech n *(auto.)* baffle plate
leiten v.t. guide; direct; convey; *(electr.)* conduct
leitend a. conductive
Leiter f ladder
Leiter m *(electr.)* conductor
Leiterfahrzeug n mobile turntable ladder
Leitergerüst n ladder scaffold
leitfähig a. conductive
Leitfähigkeit f conductivity
Leitfähigkeitsmesser m conductivity meter
Leitkurve f *(mach.)* lead cam

Leitlineal n *(lathe)* taper bar, guide bar
Leitpatrone f *(threading)* leader
Leitrolle f *(copying)* tracer roller; (Riementrieb:) guide pulley
Leitschiene f guide rail, guide bar; *(copying)* template; *(electr.)* conductor rail; *(railway)* check rail
Leitsilber n *(electr.)* conductive silver
Leitspindel f *(lathe)* leadscrew
Leitspindelmutter f leadscrew nut, split nut
Leit- und Zugspindeldrehmaschine f engine lathe
Leitung f (Kabel, Draht:) line wire, cable; (Rohr:) pipeline; conduit, piping; *(electr.)* (≈ Netzleitung:) mains; (Wärme:) conduction; (e. Betriebes:) management; *(tel.)* line; *(auto.)* pipe, duct
LEITUNGS ~, *(electr.)* (~draht, ~frequenz, ~geräusch, ~kontakt, ~störung, ~verlust, ~verstärker, ~system, ~wähler, ~widerstand) line ~
Leitungsnetz n distributing network
Leitungsrohr n pipe [line]
Leitungsschnur f flexible cord
Leitungswasser n tap water, mains water
Leitvorschub m *(mach.)* sliding feed
Leitwert m *(electr.)* conductance
Leitzahl f index number
Lenkachse f steering axle
Lenkbolzen m *(auto.)* pivot bolt
lenken v.t. *(auto.)* steer, direct, drive
Lenker m *(auto.)* steering wheel
Lenkerarm m *(auto.)* handle bar arm
Lenkgehäuse n *(auto.)* steering-gear housing
Lenkgestänge n *(auto.)* steering linkage
Lenkgetriebe n *(auto.)* steering gear
Lenkhebel m *(auto.)* steering control arm

Lenkkranz m *(auto.)* steering worm sector
Lenkrad n *(auto.)* steering wheel
Lenkradnabe f *(auto.)* steering-wheel hub
Lenkradschaltung f *(auto.)* steering column gear change
Lenkradschloß n *(auto.)* steering-wheel lock
Lenkrohr n *(auto.)* steering-column tube
Lenksäule f *(auto.)* steering column
Lenkschaltung f *(auto.)* steering column gear change
Lenkschloß n *(auto.)* steering column lock
Lenkspurhebel m *(auto.)* drag link
Lenkspurstange f *(auto.)* track rod
Lenkstange f *(motorcycle)* handle-bar
Lenkung f *(auto.)* steering gear, steering mechanism; steering
Lenkwelle f *(auto.)* steering shaft
Lenzpumpe f bilge pump
Leonardschaltung f Ward-Leonard control
Leseband n picking belt
Letternmetall n type metal
Leuchte f lamp
leuchten v.i. light, shine
Leuchtfähigkeit f luminosity
Leuchtfarbe f luminous paint, phosphorescent paint
Leuchtfeuer n beacon
Leuchtgas n illuminating gas
Leuchtkraft f luminous power, luminosity
Leuchtpetroleum n kerosene
Leuchtröhre f luminous discharge lamp, vacuum tube lamp
Leuchtschirm m fluorescent screen
Leuchtstoffflampe f fluorescent lamp
Libelle f (≈ Wasserwaage:) level; (≈ Glasröhrchen:) vial

Libellenwinkelmesser m spirit level clinometer
Licht n light
LICHT ~, (~batterie, ~blitz, ~bündel, ~filter, ~fleck, ~geschwindigkeit, ~intensität, ~modulation, ~punkt, ~relais, ~schirm, ~spur, ~stärke, ~steckdose, ~strahl, ~stromkreis, ~teilchen, ~verteilung, ~weg, ~welle, ~zeichen) light ~
Lichtanlasser m starter-dynamo
Lichtbatteriezünder m *(auto.)* dynamo-battery ignition unit
Lichtbeständigkeit f fastness to light
Lichtbogen m [electric] arc
LICHTBOGEN ~, (~elektrode, ~entladung, ~festigkeit, ~flamme, ~gleichrichter, ~ofen, ~preßschweißung, ~schweißmaschine, ~schweißung) arc ~
Lichtbogenstrahlungsofen m indirect arc furnace
Lichtbogen-Widerstandsofen m resistance arc furnace
lichtdurchlässig a. translucent
Lichtdurchlässigkeitsmesser m diaphanometer, transparency meter
lichtecht a. fast do light, light-proof
lichtelektrische Zelle, photoelectric cell
Lichtleitung f lighting circuit; lighting mains
Lichtmagnetzünder m *(auto.)* dynamo-magneto
Lichtmaschine f *(auto.)* dynamo
Lichtnetz n lighting mains
Lichtnetzantenne f mains antenna; *(UK)* mains aerial
Lichtpause f photo-print
Lichtschalter m installation switch
Lichtstrom m luminous flux

Lichttechnik f lighting engineering
Lichtweite f internal width; inside diameter
Lierung f packing
Lieferkraftwagen m (auto.) pick-up truck, delivery van
Lieferungsabnahmelehre f master gage
Lieferwagen m van, pick-up truck
Lieferwerk n manufacturing plant, suppliers
Liegesitz m (auto.) reclining seat
Limonit m limonite, brown iron ore
Limousine f sedan; (UK) limousine
Lineal n rule, straightedge; (Kegeldrehen) guide bar; (Kopieren:) former; (e. Schieblehre:) beam
Linksdrall m left-hand helix
Linksdrehung f counter-clockwise rotation
Linksgang m left-hand travel
linksgängig a. left-hand, left-handed
Linksgewinde n left-hand thread
Linkslauf m left-hand rotation
Linksschweißung f leftward welding
Linkssteuerung f (auto.) left-hand drive
Linse f lens
Linsenkuppe f rounded end
Linsenschirm m (telev.) lens screen
Linsensenkkopf m raised countersunk head
Lithopon n lithopone
Litze f stranded wire, strand; cord
Lizenzgebühr f licence fee
Loch n hole; borehole, bore; cavity
Lochabstand m hole center distance
Lochband n punched tape
Lochbandsteuerung f punched tape control
Lochblende f diaphragm
Lochbohren n drilling
Lochbohrmaschine n drilling machine

Lochdorn m (rolling) piercer, mandrel, plug
Lochdornstange f (rolling mill) piercer rod
Locheisen n hollow punch
lochen v.t. (kalt:) punch a hole; (= perforieren:) perforate; (warm:) hollow-forge; (Rohre:) pierce; (Holz:) bore
Lochfeile f riffler
Lochfraß m pitting
Lochfraßanfälligkeit f pitting corrosion susceptiblity
Lochkarte f punched card
Lochkartensteuerung f punched card control
Lochkreis m hole circle
Lochlehre f master gage
Lochmutter f ring nut
Lochpfeife f saddler's drive punch
Lochsäge f compass saw
Lochscheibe f (indexing) index plate
Lochschweißung f plug welding
Lochstanze f punching machine, punch press
Lochstein m perforated brick; (e. Stopfenpfanne:) sleeve brick
Lochstempel m punch
Lochstreifen m punched tape
lochstreifengesteuert a. tape-controlled
Lochstreifensteuerung f punched tape control
Lochtaster m inside caliper
Lochteilscheibe f index plate
Loch- und Gesenkplatte f swage block
Lochung f hole-punching; perforation; piercing; hollow forging; boring; s.a. lochen
Lochwalzwerk n piercing mill
Lochzange f punch plier
lockern v.t. loosen; – v.r. slacken
Löffelbagger m shovel excavator

Löffelprobe f *(founding)* cup test
Löffelschaber m spoon scraper
Lohnkosten pl. labor cost
Lohnsatz m labor rate
Lohnwesen n wage accounting
Lokomotivrahmenbohrmaschine f locomotive frame drilling machine
Lokomotivwinde f stone jack
Losboden m (e. Konverters:) detachable bottom
löschen v.t. (Feuer:) extinguish; (Kalk:) slake, slack; (Koks:) water, quench
Löscher m (Feuer:) extinguisher
Löschfunken m quenched spark
löschfunkensender m quenched spark-gap transmitter
Löschfunkenstrecke f quenched spark gap
Löschkondensator m quenching condenser
Löschspannung f *(telev.)* extinction voltage; *(electroerosion)* deionization voltage
Löschturm m *(coking)* quenching tower
Löschung f (Koks:) quenching; (Kalk:) slaking
Löschwagen m *(coking)* quencher car, watering car
lösen v.t. (≈ lockern:) loosen, slacken; *(chem.)* dissolve; *(electr.)* disconnect; release
Losgröße f batch
Loslager n floating bearing
loslassen v.t. release
Löslichkeit f solubility
loslösen v.t. detach
Losrollenlager n loose roller bearing
Losscheibe f (Riementrieb:) loose pulley

losschrauben v.t. unscrew, unbolt
Lösung f solution
lösungsmittel n solvent
Lot n solder; filler; *(math.)* perpendicular; *(tool)* plumb
lötbar a. solderable
Lötbrenner m *(welding)* brazing torch
Lotdraht m wire solder
löten v.t. (weich:) solder; (hart:) braze; (reiblöten:) tin; (feuerlöten:) sweat; (tauchlöten:) dip braze
Lötflußmittel n soldering flux
Lötgebläse n blow torch
Lötkolben m soldering copper
Lötlampe f [soldering] torch
lotrecht a. perpendicular
Lötrohr n blowpipe
Lötsalz n soldering salt
Löt-Schutzgas n brazing atmosphere
Lötspalt m close joint
Lötstelle f joint space
Lötstoß m soldering joint
Lötwasser n soldering fluid
Lötzinn n lead-tin solder
Lücke f gap; space; (Gewinde:) groove; (Sägezähne:) gullet; (Zahnradzähne:) gash
Luft f air; atmosphere; (≈ Spiel:) play, clearance; (Lager:) slackness
LUFT ~, (~druck, ~erhitzer, ~filter, ~härtung, ~kompressor, ~kondensator, ~kühlung, ~pumpe, ~spalt, ~strecke, ~strömung, ~verdichter, ~vorwärmer, ~wirbel, ~zufuhr) air ~
Luftansaugfilter m *(auto.)* air suction filter
luftbereift a. pneumatic-tired
Luftdrehkondensator m variable air condenser
Luftdrossel f air-core choke

Luftdruckmesser m barometer; *(auto.)* tire gage

Luftdruckprüfer m *(auto.)* air pressure gage

lüften v.t. ventilate, air; aerate

Lüfter m *(auto.)* fan, ventilator; (e. Bremse:) lifter

Lüfterhaube f *(auto.)* fan cowling

Lüftermotor m ventilated motor

Luftfeuchtigkeit f humidity

luftgetrocknetes Holz, air-seasoned wood

Lufthammer m *(forge)* pneumatic hammer

Lufthärter m *(met.)* air-hardening steel, self-hardening steel

Lufthärtung f air-hardening

Luftklappe f (Vergaser:) choke

Luftklappenventil n *(auto.)* choke valve

Luftlack m air-drying varnish

Luftleere f vacuum

Luftlinie f linear distance

Luftloch n vent hole

Luftpresse f air-operated press, pneumatic press

Luftpresser m *(auto.)* air brake pressurizer

Luftreifen m pneumatic tire

Luftsauerstoff m atmospheric oxygen

Luftschalter m air-break switch

Luftschlauch m *(auto.)* air tube

Luftschlauchkupplung f *(auto.)* air hose coupling

Luftschraube f propeller

Luftschütz n air-break contactor

Luftspule f air-cored coil

Luftstrom m draft; draught; air current

Lufttransformator m air-core transformer

lufttrocken a. air-dried, air-dry

Lüftung f ventilation, airing, aeration

Lüftungsklappe f *(auto.)* ventilation flap

lüftungstechnische Anlage, ventilating system

Luftwaage f aerometer

Luftwirbel m air vortex

Luftzutritt m admission of air, access of air

Lünette f steadyrest

Lünettenständer m *(boring mill)* end support column, boring stay

Lunker m *(met.)* shrinkhole; blowhole; (im Gußblock:) pipe

Lunkerbildung f shrinking; (im Block:) piping

Lunkern n (Guß:) shrinking; (Block:) piping

Lupe f magnifying glass

Luppe f ball (of iron)

Luppenwalzwerk n puddle mill, muck mill

Lüsterklemme f terminal block for light fixtures

M

Madenschraube *f* hollow set screw
Magazin *n* (*work feeding*) magazine; (*Lager:*) storeroom
Magazinautomat *m* magazine automatic
Magazinverwalter *m* storekeeper
Magazinzuführeinrichtung *f* magazine feeding attachment
Magazinzuführung *f* magazine feed
Magerkalk *m* poor lime
Magerkohle *f* lean coal, non-baking coal
magisches Auge, (*radio*) magic eye
Magnesia *f* magnesia, magnesium oxide
magnesiahaltiges Kalkgestein, magnesian limestone
Magnesitauskleidung *f* magnesite lining
Magnet *m* magnet
MAGNET ~, (~abscheider, ~band, ~bremse, ~feld, ~filter, ~fluß, ~futter, ~kerze, ~kompaß, ~kraft, ~kupplung, ~lautsprecher, ~nadel, ~scheider, ~tonband, ~unterbrecher, ~verstärker, ~waage) magnetic ~
Magnetanker *m* magnet keeper
Magnetantrieb *m* magneto drive
magnetbandgesteuert *p.a.* magnetic tape-controlled
Magnetbremslüfter *m* brake lifting magnet
Magneteisenstein *m* magnetic iron ore, magnetite
magnetisch *a.* magnetic
magnetische Feldstärke, magnetic field strength
magnetischer Pol, magnetic pole
magnetisieren *v.t.* magnetize
Magnetisierung *f* magnetization
Magnetismus *m* magnetism

Magnetjoch *n* magnet yoke
Magnetkern *m* magnet core
Magnetophon *n* magnetic recorder
Magnetpol *m* magnet pole
Magnetregler *m* field regulator
Magnetschalter *m* solenoid switch; (*auto.*) electromagnetic relay
Magnettongerät *n* magnetic recorder
Magnettonverfahren *n* magnetic sound recording system
Magnetventil *n* solenoid valve
Magnetzünder *m* (*auto.*) magneto
Magnetzündung *f* magneto ignition
mahlen *v.t.* mill, (*grob:*) crush; (*fein:*) grind; (*feinst:*) pulverize, powder
Malteserkreuz *n* Geneva stop
Mangan *n* manganese
manganarm *a.* low in manganese
Manganerz *n* manganese ore
manganhaltig *a.* manganiferous
Manganhartstahl *m* austenitic manganese steel
Mangansiliziumstahl *m* silico-manganese steel
Mangel *m* (≈ Fehler:) defect; deficiency
Mannloch *n* manhole
Manometer *n* pressure gage
Manschette *f* gasket, packing; sealing member; (*auto.*) piston cup
Mantel *m* (*techn.*) jacket; shell (*geom.*) surface; (*Kabel:*) sheath
Mantelrohr *n* casing tube
Markenbenzin *n* branded gasoline
Markenschmierstoff *m* branded lubricant
markieren *v.t.* mark
Martinflußstahl *m* open-hearth steel

Martinofen ~ cf. Siemens-Martin~
Martinofenschlacke f open-hearth cinder
Martinroheisen n open-hearth pig
Martinstahl m cf. Siemens-Martin-Stahl
Masche f mesh
Maschendraht m netting wire
Maschenschaltung f (electr.) network circuit
Maschenweite f mesh aperture, mesh
Maschine f machine; (Kraftbrennmaschine:) engine; (electr.) motor
maschinell a. mechanical
maschinelles Schweißen, mechanical welding, machine welding
MASCHINEN ~, (~arbeiter, ~ausfall, ~ausrüstung, ~bauer, ~bett, ~brachzeit, ~gestell, ~leistung, ~leuchte, ~sockel, ~ständer, ~taktierung, ~unterteil) machine ~
Maschinenanlage f mechanical equipment
Maschinenbau m machine-building industry
Maschinenbauingenieur m mechanical engineer
Maschinenbaustahl m machinery steel, engineering steel
Maschinenbauwerkzeug n machinists' tool
Maschinenfabrik f engineering works
Maschinenguß m machinery castings
Maschinenhauptzeit f machining time
Maschinenreibahle f chucking reamer
Maschinensäge f power saw
Maschinensatz m (electr.) generator set
Maschinenschlangenbohrer m machine bit
Maschinenschlosser m fitter

Maschinenschraubstock m drill press vise, toolmakers' vise
Maschinenschweißung f automatic welding
Maschinenwerkzeug n machine-shop tool; metal-cutting tool
Maschinenzeit f (work study) machining time
Maschinenzeitabweichung f (cost accounting) efficiency variance
Maschinist m machinist
Maserfurnier n figured veneer
Maske f (shell molding) half shell mold
Maskenform f shell mold
maskenformen v.t. shell-mold, shell-cast
Maskenformgießerei f shell molding type foundry
Maskenformmaschine f shell molding machine
Maskengußstück n shell-mold casting, shell casting
Maskenkern m (shell molding) shell core
Maß n dimension; size; measure; (≈ Volumen:) volume; (als Meßzeug:) measuring tool; (≈ Lehre:) gage; (UK) gauge; (≈ Maßstab:) rule
Maßabweichung f offsize
Maßanalyse f volumetric analysis
Maßänderung f dimensional change
Maßband n tape rule
Masse f mass; (phys.) substance, matter; (plastics) material; (electr.) earth; (≈ Menge:) quantity
Masseanschluß m (electr.) earth connection
Masseband n earthing strap
Masseform f (founding) dry sand mold
Masseformerei f dry sand molding
Maßeinheit f unit of measure

Maßeinteilung f graduation, division, scale
Massekabel n earthing lead, ground cable
Massekern m *(electr.)* iron dust core
Masseklemme f earth terminal
Massel f *(met.)* pig
Masselbett n pig bed
Masselbrecher m pig breaker
Masselform f pig mold
Masselgießmaschine f pig casting machine
Masselgraben m *(met.)* sow
Massenbeton m mass concrete, bulk concrete
Massendefekt m *(nucl.)* mass defect
Massenerhaltung f *(phys.)* mass conservation
Massenfertigung f long-run production
Massenstahl m rimming steel
Massenteilchen n *(nucl.)* discrete particle
Massenträgheit f inertia
Massenwirkung f mass action
Massenwirkungsgesetz n law of mass-action
Masseschluß m *(electr.)* earthing
Maßfehler m dimensional error
Maßgebung f dimensioning
maßgenau a. true to size
Maßgenauigkeit f dimensional accuracy
maßhaltig a. true to size, accurate to size
mäßig a. moderate
massivprägen v.t. *(cold work)* (Münzen:) coin
Massivprüfung f coining
Maßkontrolle f size control; gaging
maßläppen v.t. form-lamp
maßlich a. dimensional
Maßnahme f measure
maßprägen v.t. size

Maßschleifeinrichtung f sizing attachment
Maßschleifen n size grinding
Maßskizze f dimension sketch
Maßstab m (als Meßwerkzeug:) rule, straightedge; (≈ Skala:) scale; *(fig.)* measure, standard
maßstäblich a. to scale
Maßsteuerung f *(grinding)* sizing
Maßtoleranz f size tolerance
Maß über Eck, size across flats
Maßüberwachung f size control
Maßwalze f *(rolling mill)* sizing roll
Maßzugabe f size allowance
Mast m *(telec.)* pole; mast
Mastic m mastic
Material n material; (für die laufende Fertigung:) stock; *s.a.* Werkstoff
Materialermüdung f fatigue of material
Materialprüfmaschine f material testing machine
Materialprüfung f material testing
Materialzugabe f *(mach.)* machining allowance
Materie f matter, substance
Matrize f [lower] die
Matrizenfräsmaschine f die-sinking machine
Matrizenwalze f lower roll, bottom roll
matt a. dull, mat; unpolished, dead
Matte f mat
Mattenrahmen m matting frame
Mattglanz m mat finish, dull finish
Mattglas n frosted glass
mattieren v.t. tarnish, mat finish; (Holz:) flat down
Mattlack m flat varnish
Mattscheibe f *(photo.)* focusing screen
Mattschweiße f cold shut

mauern *v.t.* lay bricks, mason
Mauersand *m* building sand
Mauerwerk *n* brickwork, masonry
Mauerwerksverband *m* masonry bond
Mauerziegel *m* brick
Maul *n* (e. Schraubenschlüssels:) head, jaw
Maulschlüssel *m* engineer's wrench
Maurer *m* bricklayer, mason
Maurerhammer *m* mason's hammer
Maurerkelle *f* brick trowel
Maurerwerkzeug *n* brickworkers' tool
Mechanik *f* mechanics; mechanism
Mechaniker *m* mechanic
Mechanikerdrehmaschine *f* precision lathe; watchmakers' lathe; bench lathe
mechanisieren *v.t.* mechanize
Mehl *n* (techn.) dust, powder
mehratomig *a.* polyatomic
MEHRFACH ~, (~antenne, ~drehen, ~empfang, ~endschalter, ~ionisation, ~kanal, ~punktschweißung, ~röhre) multiple ~
Mehrfachbetrieb *m* (tel.) multiplex system
Mehrfachbrenner *m* multi-flame burner; multijet blowpipe
Mehrfachdrehkondensator *m* gang capacitor
Mehrfacherdschluß *m* polyphase earth
Mehrfachgesenk *n* multiple die
Mehrfachkabel *n* multicore cable, multiconductor cable
Mehrfachkondensator *m* multiple unit capacitor
Mehrfachleitung *f* multicore line
Mehrfachmeißelhalter *m* combination toolholder
Mehrfachnebenwiderstand *m* (electr.) universal shunt

Mehrfachschalter *m* multiple contact switch
Mehrfachschweißmaschine *f* multiple operator welding machine
Mehrfachstecker *m* multi-contact plug
Mehrfachtarifzähler *m* (electr.) multiple tariff-hour meter, multi-rate meter
Mehrfachtelefonie *f* multiplex telephony
Mehrfachtelegraf *m* multiplex telegraph
Mehrfachumschalter *m* multi-way switch
Mehrfachverkehr *m* multiplex transmission
Mehrfachverstärker *m* (tel., telegr.) multistage amplifier
Mehrfachwerkzeug *n* (plastics) multi-impression mold
Mehrfarbenschreiber *m* multicolor recorder
Mehrflammenbrenner *m* multi-jet blowpipe
Mehrganggetriebe *n* change-speed gear
mehrgängiges Gewinde, multiple thread
mehrgerüstiges Walzwerk, multiple stand rolling mill
mehrgeschossige Rüstung, (building) multistage scaffolding
Mehrgitterröhre *f* multi-grid tube
Mehrkammerofen *m* multi-chamber kiln
Mehrkanaltelegrafie *f* multi-channel telegraphy
Mehrkantdrehmaschine *f* polygonal turning machine
Mehrklanghorn *n* (auto.) multi-tone horn
Mehrkurvenautomat *m* multiple-cam operated automatic turret lathe
Mehrlagenschweißung *f* multi-run weld, multi-layer weld
Mehrlagenspule *f* multi-layer coil
Mehrleiterkabel *n* multi-conductor cable

Mehrlochbohreinrichtung f multi-spindle drilling attachment
Mehrlochdüse f multi-hole nozzle
MEHRPHASEN ~, (~generator, ~gleichrichter, ~motor, ~schaltung, ~strom, ~transformator, ~wicklung) multiphase ~
mehrpolig a. multiple-pole, multipolar
Mehrrillenscheibe f (rope drive) multiple-groove sheave
Mehrscheibenbremse f multiple-disc brake
Mehrscheibenkupplung f multiple-disc clutch
mehrschnittiges Werkzeug, multi-edge cutting tool
Mehrspindelautomat m multiple-spindle automatic machine
Mehrspindelbohrmaschine f multi-spindle drilling machine
Mehrspindelfräsmaschine f multiple-head milling machine
Mehrspindelkopf m (multi-driller) multiple-spindle drill head
Mehrspindelreihenbohrmaschine f multiple-spindle gang drilling machine
Mehrstempelpresse f multiple-die press, multiple plunger press
Mehrstößelräummaschine f multiple-ram broaching machine
Mehrstufengesenk n follow-on die
Mehrstufenpresse f multi-stage press
mehrstufiges Getriebe, multiple-speed gearbox (or gear drive)
mehrtouriger Motor, multiple-speed motor
Mehrwalzen-Walzwerk n cluster mill
Mehrwegebohrmaschine f multiple-way boring machine
Mehrwegemaschine f multi-way machine
Mehrwegeschalter m multiple-way switch
Mehrzweckautomat m multi-purpose automatic
Mehrzweckkraftwagen m multi-purpose vehicle
Mehrzweckmaschine f general-purpose machine
Meiler m (nucl.) atomic pile
Meißel m (metal cutting) cutting tool; (tool) chisel; (in Wortverbindungen der Zerspantechnik nur:) tool
Meißelabhebung f tool lift
Meißelanstellung f tool setting
Meißelauflage f tool rest
Meißeleinsatz m cutter bit
Meißeleinspannung f tool mounting
Meißeleinstellehre f tool setting gage
Meißelhalter m (lathe) toolholder; (shaper) toolhead
Meißelhammer m chipping hammer
Meißelklappe f (shaper, planer) clapper
Meißelklappenhalter m (planer, shaper) clapper box
Meißelschieber m (shaper) toolhead slide
Meißelschleifmaschine f tool and cutter grinder
Meißelschleifvorrichtung f cutter grinding attachment
Meißelvorschub m tool feed
Meister m foreman
Meldeleitung f (tel.) messenger line
Meldeleuchte f indicator lamp
Melder m (Feuer:) alarm box
Meldung f report; (Feuer:) alarm
meliert a. speckled, mixed; (Kohlen:) blended; (Roheisen:) mottled
Membran f diaphragm
Membranpumpe f diaphragm pump

Mengenmesser m volumeter
Mennige f red lead
Merklampe f signal lamp
Merkmal n feature
MESS ~, (~anschlag, ~bereich, ~brücke, ~düse, ~ebene, ~einrichtung, ~fehler, ~funkenstrecke, ~genauigkeit, ~gerät, ~glied, ~instrument, ~latte, ~maschine, ~mittel, ~punkt, ~spannung, ~spule, ~stelle, ~verfahren, ~verstärker, ~vorrichtung, ~werk, ~werkzeug, ~zeug, ~zylinder) measuring ~
Meßautomatik f automatic measuring system
meßbar a. measurable
Meßblock m end block, slip gage
Meßdraht m (threading) thread measuring wire
messen v.t. measure; (mittels Lehre:) gage; caliper
Messer m measuring instrument; (opt., electr.) meter; (metrol.) gage
Messer n knife; blade; (für Kabel:) [sheath] splitting knife
Messerklinge f knife blade
Messerkopf m inserted blade milling cutter
Messerlineal n bevelled steel straightedge
Messerschalter m knife switch
Messerschneidwaren fpl. cutlery
Messerwelle f (woodworking) circular cutter head
Meßgefühl n touch, feel
Meßgleichrichter m meter rectifier
Messing n [yellow] brass
MESSING ~, (~blech, ~draht, ~gießerei, ~guß, ~klemme, ~druckguß) brass ~
Messingpreßguß m brass pressure castings
Messingwaren fpl. brass-ware

Meßkette f surveyor's chain
Meßlänge f (Prüfwesen) gage length
Meßlineal n rule
Meßlupe f magnifying lens
Meßoptik f optical measuring system
Meßscheibe f toolmakers' flat
Meßschiene f graduated steel straightedge
Meßsender m signal generator
Meßstab m (lubrication) dipstick
Meßstift m (e. Meßuhr:) feeler pin
Meßstrecke f gage length
Meßtechnik f measuring practice, metrology; (radio) direction finding
meßtechnisch a. metrological
Meßtisch m dial bench gage; (jig borer) coordinate table
Meßtransformator m instrument transformer
Meßuhr f dial gage, dial test indicator
Meßuhrtiefenlehre f dial depth gage
Messung f measurement; gaging; metering
Meßverfahren n method of measurement
Meßwagen m (railway) recording coach
Meßwandler m instrument transformer
Meßwert m measured value
Meßwesen n metrology
Meßzeigergerät n dial test indicator
Metall n metal
METALL ~, (~arbeiter, ~bearbeitung, ~faden, ~fadenlampe, ~folie, ~gewebe, ~gießen, ~gießerei, ~gußstück, ~karbid, ~kreissäge, ~legierung, ~lichtbogen, ~maßstab, ~physik, ~spritzen, ~spritzpistole, ~spritzverfahren, ~überzug, ~verarbeitung) metal ~
Metalldichtung f metallic packing

metalldrücken v.t. (Bleche:) spin
Metallguß m cast metal; non-ferrous castings
metallhaltig a. metalliferous
Metallhüttenkunde f non-ferrous metallurgy
Metallichtbogenschweißen n unter Schutzgas, gas-shielded metal-arc welding
Metallindustrie f non-ferrous metal industry
Metall-Inertschweißen n inert-gas arc welding
metallisieren v.t. metallize
Metallographie f metallography
Metallprobe f assay
Metallsäge f metal-cutting saw
Metallschlauch m flexible metal hose
Metallschmelze f molten bath
Metallumformung f [plastic] metal deformation
Metallurge m metallurgist
Metallurgie f metallurgy
metallurgisch a. metallurgical
metallverarbeitende Industrie, metalworking industry
Metallwaren fpl. metalware, hardware
Metermaß n meter rule, meter stick
Methylalkohol m methyl alcohol
metrisch a. metric
metrisches Maßsystem, metric system of measurement
Mietwagen m hire-car
Mikroaufnahme f photomicrograph
MIKRO ~, (~endschalter, ~gefüge, ~gestalt, ~härte, ~manometer, ~mechanik, ~physik, ~riß, ~schalter, ~telefon, ~waage, ~welle, ~zoll) micro ~
Mikrofon n microphone

MIKROFON ~, (~kabel, ~kreis, ~sender, ~strom, ~transformator, ~verstärker) microphone ~
Mikrofotografie f photomicrography
Mikrometer n micrometer caliper gage
Mikrometermeßuhr f dial indicator micrometer
Mikrometerokular n micrometer eyepiece
Mikrometerprüflehre f micrometer check gage
Mikrometerschraube f micrometer screw; micrometer caliper
Mikron n micron
Mikroskop n microscope
Mikrotastgerät n comparator gage, comparative measuring instrument
Millimeterpapier n millimeter-ruled paper
Mindestlohnsatz m base wage rate
Mindestmaß n minimum; *(metrol.)* minimum size
Mineralöl n mineral oil, rock oil, petroleum
Mineralölraffinerie f mineral oil refinery
Minette f minette ore
Minimeter n minimeter-type indicator gage
Minuspol m negative pole
Mischbrenner m proportioning mixer
Mischer m mixer; *(met.)* hot mixer
Mischgas n mixed gas
Mischkollergang m pug mill, pan grinder
Mischkristall n solid solution, mixed crystal
Mischleim m mixed glue
Mischpult n *(telev.)* central control desk, mixer; *(US)* master control
Mischröhre f converter valve, modulator, mixer tube
Mischschaufel f *(concrete)* mixing blade
Mischstrom m undulatory current
Mischstufe f *(radio)* mixer stage

Mischtafel f *(telev.)* audio mixer
Mischturm m *(concrete)* concrete mixing tower
Mischung f mixture
Mischventil n mixing valve
Mischwaage f *(met.)* mixing scales
mißgriffsicher a. foolproof
Mitfahrer m *(auto.)* passenger; *(motorcycle)* sidecar rider
Mithörtaste f *(tel.)* listening key
mitlaufende Körnerspitze, live center
mitlaufender Setzstock, follow rest
Mitnahme f (e. Schneidewerkzeuges:) driving
Mitnehmer m driver; (≈ Werkzeuglappen:) tang; (≈ Mitnahmenut:) driving slot; *(conveying)* flight attachment
Mitnehmerkeil m driving key
Mitnehmerkupplung f friction clutch
Mitnehmerlappen m flat tang
Mitnehmerscheibe f driving plate, dog plate
Mitnehmerstift m follower pin
mitschwingen v.i. resonate
Mittel n means; medium; *(chem.)* agent; (≈ Durchschnitt:) average; *(grinding, lapping)* compound
Mittelblech n light plate, medium plate
Mittelblechwalzwerk n light plate rolling mill
Mittelfrequenzmotor m medium-frequency motor
Mittelgerüst n *(rolling mill)* intermediate roll stand
Mittellage f central position
Mittelleiter m *(electr.)* neutral wire
Mittelpunkt m center; *(UK)* centre; midpoint
Mittelschneider m *(tapping)* plug tap

Mittelspannung f *(electr.)* medium voltage
Mittelstellung f central position, neutral position
Mittelwelle f *(radio.)* medium wave
Mittelwert m average, mean value
Mittenabstand m center distance
Mittenentfernung f center distance
Mittenlehre f center gage
mittig a. central; concentric
Mittigkeit f central position; concentricity
Möbelkraftwagen m furniture van
Möbellack m furniture varnish
Möbelschleiflack m furniture rubbing varnish
Möbelüberzugslack m furniture finish
Möbelwagen m removal van
Möbelwagenanhänger m furniture trailer
Modell n model; type; *(founding)* pattern
Modellabhub m (e. Formmaschine:) pattern draw
Modellbauerei f *(founding)* pattern shop
Modellformerei f plate molding shop
Modellfräseinrichtung f pattern milling attachment
Modellherstellung f *(foundry)* pattern-making
Modellplatte f *(foundry)* pattern plate
Modellsand m *(foundry)* facing sand
Modellschreiner m *(foundry)* wood-pattern maker
Modellschreinerei f wood-pattern shop
Modelltischlerei f pattern shop
Modul m module
Modulation f *(telec.)* modulation
MODULATIONS ~, (~drossel, ~frequenz, ~messer, ~messung, ~röhre, ~schaltung, ~spannung, ~spule, ~träger, ~transformator, ~verzerrung) modulation ~

modulieren *v.t.* modulate
Mol *n* gram molecule, mol
Molekül *n* molecule
MOLEKULAR ~, (~bewegung, ~energie, ~gewicht, ~kraft, ~pumpe, ~strahl, ~strom, ~volumen, ~zertrümmerung) molecular ~
Möller *m (met.)* blast-furnace burden
möllern *v.t.* mix the burden, burden
Moment *n (phys.)* moment; momentum, impulse; (≈ Drehmoment:) torque
momentan *a.* momentary, instantaneous
Momentanbeschleunigung *f* instantaneous acceleration
Momentanspannung *f* transient voltage
Momentaufnahme *f* snapshot
Momentschalter *m* quick make and break switch
Momentschraubzwinge *f* quick-action clamp
Momentschweißung *f* shot welding
Montage *f* assembly; erection; mounting; installation
Montageanweisung *f* instruction for assembly; (für Rohrverlegung:) laying instruction
Montageband *n* assembly line
Montagebock *m* repair stand
montagefertig *a.* ready for assembly
Montagegerüst *n* assembling scaffold
Montagehalle *f* assembly shop
Montagekran *m* erection crane
Montageleuchte *f* inspection lamp
Montagemeister *m* erecting supervisor
Montageplan *m* sectional assembly view
Montagetisch *m* assembly bench
Montagewerkzeug *n* assembly tool
Monteur *m* maintenance man; fitter

montieren *v.t.* assemble; install; mount; (Maschinen:) erect
Moped *n* motor-assisted bicycle, autocycle
Morsekegel *m* Morse taper
Morsekegelbohrung *f* Morse taper shank hole
Morsekegelhülse *f* Morse taper sleeve
Morsetaste *f* Morse key
Morsezeichen *n* Morse signal
Mörtel *m* mortar
Mörtelschlamm *m* grout
Mosaikarbeit *f* mosaik work
Mosaikparkettdiele *f* mosaic parquet deal
Motor *m* (Kraftverbrennungsmaschine:) engine; (Elektromotor:) motor
Motorantrieb *m (mach.)* motor drive
Motoraufhängung *f (auto.)* engine mounting
Motorblock *m (auto.)* engine block, cylinder block
Motorbremse *f (auto.)* engine dynamometer
Motorbremslüfter *m* motor-operated brake-lifter
Motordrehzahl *f* motor speed
Motorenöl *n (auto.)* engine oil, motor oil
Motorfahrzeug *n* motor-car, motor vehicle
Motorgehäuse *n* motor casing; (engine): crankcase
Motorgenerator *m* motor generator
Motorhaube *f (US)* hood; *(UK)* bonnet
Motorkonsol *n* motor bracket
Motorkraftstoff *m* motor fuel
Motorlastdrehzahl *f* motor on-speed
Motorleistung *f (auto.)* engine power, engine performance; engine rating; *(mach.)* motor output; (Nennleistung:) motor rating

Motormontagebock n *(auto.)* engine stand
Motorpanne f engine failure
Motorprüfstand m *(auto.)* engine testing stand
Motorrad n motorcycle
Motorradfahrer m motor-cyclist
Motorradscheinwerfer m motorcycle headlamp
Motorroller m motor-scooter
Motorschutzschalter m protective motor switch
Motorwanne f crankcase sump
Motorwippe f hinged motor plate
Muffe f (Kabel:) coupling box; (Rohr:) socket; (≈ Hülse:) sleeve; *(automatic lathe)* (Werkstoffvorschub:) spool
Muffelofen m muffle furnace, retort furnace
Muffenkupplung f sleeve coupling
Muffendruckrohr n socketed pressure pipe
Muffenrohr n socket pipe
Muffenverbindung f spigot and socket joint
Mühle f mill; (Grobmahlung:) crusher; (Feinmahlung:) grinder; (Feinstmahlung:) pulverizer
Mühlsägenteile f mill saw file
Mulde f *(met.)* (zum Chargieren:) box; *(mach.)* *(planer bed)* cavity; *(surface finish)* pit, depression, pocket
Muldenband n trough belt
Muldenkipper m dump truck
Muldenkippwagen m dump-type-hopper truck
Muldenwagen m hopper truck, pan car
Mundstück n (e. Brenners:) tip
Mündung f (e. Konverters:) mouth
Mündungsbär m (e. Konverters:) skull
Münzfernsprecher m coin-box telephone, taxiphone
Münzgold n standard gold
Münzlegierung f coinage alloy
Münzzähler m prepayment meter; slot meter
Muschelkalk m shelly limestone
Muschelsand m shell-sand
Mutter f *(techn.)* nut
Muttergewinde n nut thread, female thread
Muttergewindebohrer m machine nut tap
Muttergewindebohrmaschine f nut tapper
Mutterlauge f mother liquor
Mutternautomat m nut automatic
Mutterneinziehmaschine f nut runner
Mutterschloß n *(lathe)* half-nuts
Mutterzange f nut plier

N

Nabe *f (techn.)* hub
Nabenbohrung *f* hub bore
Nabenfläche *f* boss
nabensenken *v.t.* spot face
Nabensenker *m* spot facer
NACH ~, (~bohren, ~drehen, ~fräsen, ~glühen, ~justieren, ~läppen, ~polieren, ~regeln, ~reiben, ~schärfen, ~schleifen, ~schmieren, ~schweißen, ~spannen, ~stemmen, ~walzen, ~wärmen, ~wuchten, ~zentrieren, ~ziehen) re- ~; (in Verbindung mit Bearbeitungsvorgängen außerdem:) finish ~
Nacharbeit *f* (spanlose:) rework; reworking; (spanende:) re-machining
nacharbeiten *v.t.* (spanlos:) rework; (zerspanend:) re-machine
Nachbearbeitung *f* reworking; re-machining; finishing; *s.a.* Nacharbeit
Nachblasen *n (steelmaking)* after-blow
nachdunkeln *v.t. & v.i. (painting)* darken
nacheilen *v.i.* (Walzen:) lag
Nacheilung *f (electr.)* lag
NACHFORM ~, (~bohren, ~drehen, ~fräsen, ~hobeln, ~räumen, ~schleifen, ~waagerechtstoßen) copy ~
Nachformbohrmaschine *f* copy drilling machine
Nachformdrehmaschine *f* copying lathe
nachformen *v.t.* copy, duplicate; (Gesenke:) die-sink
Nachformfräsmaschine *f* copy-milling machine; profiler; duplicator; (für Gesenke:) die-sinker
Nachformmodell *n* master, template
nachfüllen *v.t.* refill, top up
Nachhärten *n (plastics)* after-hardening
nachlaufen *v.i.* lag
nachlinksschweißen *v.t.* weld back-hand, weld leftward
nachmessen *v.t.* check dimensions
Nachmischer *m (concrete)* agitator
nachprüfen *v.t.* check, control; inspect
Nachprüfung *f* check, control, revision, inspection
nachrechtsschweißen *v.t.* weld rightward
Nachregelung *f* readjustment
Nachricht *f* message, communication; information
NACHRICHTEN ~, (~kabel, ~kanal, ~kreis, ~signal) communication ~
Nachrichtendienst *m (radio)* news service
Nachschlagsicherung *f (power press)* backing, pawl, locking pawl
Nachschliff *m* regrind
nachschneiden *v.t.* finish-cut; (Außengewinde:) cut fuller; (Innengewinde:) retap
Nachschwindung *f (plastics)* after-shrinkage
nachstellbar *a.* adjustable
Nachstellschraube *f* adjusting screw
Nachstellung *f* adjustment; readjustment
Nacht *f* night
NACHT ~, (~arbeit, ~frequenz, ~ruf, ~schicht, ~strom, ~tarif) night ~
Nachteil *m* disadvantage, drawback
Nachwärmofen *m* reheating furnace
Nachzug *m* (Drahtziehen:) follow-up draft

nackter Draht, bare wire
Nadel f needle; (≈ Räumnadel:) broach; (≈ Zeiger:) pointer
Nadelboden m (Konverter:) pinhole plug
Nadelgalvanometer m needle galvanometer
Nadelholz n coniferous wood, pinewood
Nadelkäfig m needle cage
Nadellager n needle [roller] bearing
Nadelventil n needle valve
Nagel m nail
Nagelbohrer m gimlet
Nagelhammer m spike maul
Nagel v.t. nail
Nagelschraube f drive screw
Nagelung f nailing
Nahansicht f close-up view
Nahaufnahme f (telev.) close-up view
Nahempfang m (radio) short-distance reception
Näherung f (math.) approximation
Näherungswert m approximate value
Nahschwund m (radio) local fading
Naht f (Gußnaht:) feather, bin; (Schweißnaht:) seam; weld; (überwalzte:) lap
Nahtaufbau m weld composition
Nahtbreite f weld width
Nahtdicke f (welding) effective throat
nahtlos a. seamless
nahtlos gezogenes Rohr, seamless drawn tube (or tubing)
nahtschweißen v.t. seam-weld; (bei Reihenpunktschweißungen auch:) stitch-weld
Nahtschweißmaschine f seam welding machine
Nahtschweißung f seam welding; stitch welding; s.a. nahtschweißen
Nahtüberhöhung f reinforcement of weld

Nahtwurzel f (welding) weld root
Nahtzone f (welding) zone adjacent to the weld
Nahverkehr m (tel.) local traffic, toll traffic
Napfmanschette f cup seal
napfziehen v.t. (cold work) shallow form, cup
Naphthentrockenstoff m (painting) naphthenate drier
Narbe f (Oberflächenfehler:) pit, scar
Narben m (Leder:) grain side
Nase f (techn.) nose; lug; projection; (e. Keils:) gib-head; (e. Hammerschraube:) nib; (lathe) dog
Nasenkeil m gib-head key
Naßbagger m dredger
Naßdreheinrichtung f wet turning attachment
Naßfräseinrichtung f wet milling attachment
Naßgußform f green sand mold
Naßgußformerei f green sand molding
Naßgußsand m green sand
Naßmahlung f wet grinding
Naßmetallurgie f hydrometallurgy
Naßputztrommel f wet tumbler
Naßschleifeinrichtung f wet grinding attachment
Naßschleifen n wet abrasive cutting
Naßschleifmaschine f wet grinding machine
Naßzerkleinerung f wet crushing
Naßzug m (Draht:) wet draw
Natrium n sodium
Natron n sodium bicarbonate
Natronlauge f caustic soda
Natronwasserglas n soda waterglass
Naturasphalt m natural asphalt
Naturbitumen n asphaltic bitumen

Naturgasfeuerung f natural gas firing
Naturhärte f (e. Stahles:) temper
naturharter Stahl, self-hardening steel
Naturharz n natural resin
Nautik f nautics, navigation
Navigation f (radio) navigation
Nebel m mist; fog
Nebelkammer f [Wilson] cloud chamber
Nebellampe f (auto.) fog lamp
Nebelscheinwerfer m (auto.) long range fog lamp
Nebelschmierung f mist lubrication, fog oiling
Nebenanschluß m (tel.) branch extension
Nebenleistung f additional service
Nebenproduktengewinnung f by-product recovery
Nebenproduktenkokerei f by-product coking practice
Nebenproduktenofen m by-product oven
Nebenschluß m (electr.) shunt
Nebenschlußmotor m shunt-wound motor
Nebenschlußstromerzeuger m shunt generator
Nebenschlußverhalten n shunt characteristic
Nebenschlußwicklung f shunt field winding
Nebenschnittfläche f machined surface
Nebenstelle f (tel.) extension telephone
Nebenstellenanlage f private branch exchange switchboard
Nebenwelle f (auto.) (des Getriebes:) layshaft
Nebenwiderstand m (electr.) shunt [resistance]
Nebenzeit f (mach.) idle time, down-time
neigbar a. inclinable; tiltable

Neigung f incline, inclination; slope; tilt; trend, tendency
Neigungswinkel m (e. Meißels:) back-rake angle; (UK) front rake
NE-Metall n non-ferrous metal
Nenn-Dauerfestigkeit f rated fatigue limit
Nenndrehmoment n rated torque
Nenndrehzahl f rated speed
Nenndurchmesser m nominal diameter
Nenner m (math.) denominator
Nennleistung f rated output; (e. Motors:) rating
Nennmaß n nominal size, design size
Nennspannung f rated voltage
Nennstrom m rated current
Neonlampe f neon lamp
Netz n (Draht:) netting; (Rohr:) system; (electr.) mains, network; distributing mains; (math.) grid system; (railway) network
NETZ ~, (~antenne, ~frequenz, ~spannung, ~transformator) mains~
Netzanschluß m main circuit connection; mains supply
Netzanschlußempfänger m all-mains set
netzanschlußfertig a. ready for connection to the mains
Netzschalter m power supply switch
Netzstrom m line circuit
Netzstruktur f (metalloí.) cellular structure
Netzwerk n (metalloí.) network, mesh work
Netzwerkstruktur f (metalloí.) network structure, cellular structure
Neubau m new building
Neuentwicklung f recent development
Neutron n neutron
NEUTRONEN ~, (~beschuß, ~dichte, ~einfang, ~fluß, ~strahl, ~strahlung, ~zähler) neutron ~

NF-Stufe *f* low-frequency stage
NF-Verstärkung *f* audio-frequency amplification, low-frequency amplification
Nichteisenmetall *n* non-ferrous metal
nichtfluchtend *a.* out-of-line
nichthärtbare Formmasse, thermoplastic material
nichthärtbare Masse, thermoplastic compound
nichthärtbarer Kunststoff, thermoplast, thermoplastics
Nichtleiter *m (electr.)* non-conductor
Nichtmetall *n* metalloid, non-metal
nichtmetallisch *a.* non-metallic
nichtrostend *a.* rustless; stainless
nichttrocknend *a. (painting)* non-drying
Nickel *m* nickel
NICKEL ~, (~erz, ~gewinnung, ~granalien, ~legierung, ~plattierung, ~überzug) nickel ~
nickelhaltig *a.* nickeliferous
nickelplattiertes Blech, nickel-clad sheet
Niederblasen *n* (e. Hochofens:) blow out
Niederdruck *m* low pressure
NIEDERDRUCK ~, (~brenner, ~dampf, ~kessel, ~pumpe, ~turbine) low-pressure ~
niederdrücken *v.t.* depress
Niederfrequenz *f* audio-frequency; low frequency
NIEDERFREQUENZ ~, (~kondensator, ~ofen, ~röhre, ~spannung, ~stufe, ~transformator, ~verstärker, ~widerstand) low-frequency ~
Niedergang *m* (e. Gicht:) descent, fall; (e. Kolbens, Stößels:) down-stroke; (e. Aufzugkübels:) downward travel
Niederhalter *m* (e. Schere:) holding-down; (e. Blechbiegemaschine:) jack
Niederschlag *m* deposit, sediment
niederschlagen *v.t.* settle out, settle down; – *v.i.* deposit
niederschmelzen *v.t.* melt down; (Erze:) smelt down
niederspannen *v.t. (electr.)* step down
Niederspannung *f* low voltage
NIEDERSPANNUNGS ~, (~anlage, ~einrichtung, ~isolator, ~kabel, ~leitung, ~schalter, ~sicherung, ~spule, ~wicklung) low-voltage ~
niedriggekohlter Stahl, soft steel, mild steel, low-carbon steel
niedriglegierter Stahl, alloy-treated steel
niedriglegierter Werkzeugstahl, low alloy tool steel
Niet *m & n* rivet
Nietbarkeit *f* rivetability
nieten *v.t.* rivet
Nietendöpper *m* ~ *cf.* Nietkopfsetzer
Nietenzieher *m* rivet set
Niethammer *m* riveting hammer
Nietkopf *m (US)* rivet point; *(UK)* rivet head
Nietkopfsetzer *m* rivet header, heading set, snap die, rivet set
Nietlochreibahle *f* taper bridge reamer
Nietmaschine *f* riveter, riveting machine
Nietschaft *m* rivet shank
Nietstempel *m* snap head die
Nietung *f* riveting
Nietverbindung *f* riveted joint
Nietwerkzeug *n* rivet set and header
Nietwinde *f* screw dolly
Nietwippe *f* dolly bar
Nippel *m* nipple
NITRIER ~, (~anlage, ~härtung, ~ofen, ~stahl) nitriding ~

nitrieren v.t. *(heat treatment)* nitride
Nitrierschicht f nitrided case
Nitrierung f nitriding, nitrogenization, nitration
Nitroalkydalfarbe f lacquer enamel
Nitrolack m nitrocellulose lacquer
Nitrozellulose f nitrocellulose
Nitrozelluloselack m nitrocellulose lacquer
nivellieren v.t. level
Nivellierinstrument n levelling instrument
Nivellierlatte f levelling staff
Nivellierschraube f levelling screw
Nocken m dog; (≈ Steuerkurve:) cam
NOCKEN ~, (~drehmaschine, ~hub, ~schaltung, ~scheibe, ~schleifmaschine, ~steuerung, ~trommel, ~verriegelung, ~welle) cam ~
Nockenschalter m cam-operated switch
NOCKENWELLEN ~, (~antrieb, ~lager, ~schalten, ~schleifmaschine) camshaft ~
Nockenwellenrad n *(auto.)* timing gear
Noniusablesung f vernier reading
Norm f standard, standard specification
normal a. normal, regular, standard; plain
Normalausführung f standard design, regular design
Normalausrüstung f (e. Maschine:) regular equipment
Normalbeanspruchung f *(mat.test.)* normal stress
Normalfassung f *(electr.)* standard lamp holder
Normalflachlehre f standard flat plug gage
Normalgewinde n standard thread
normalglühen v.t. normalize
Normallehrdorn m standard plug gage
Normallehre f standard gage

Normallehrring m standard ring gage
Normalmotor m standard motor
Normalprofil n regular section, standard section
Normalspannung f *(mat.test.)* normal stress; *(electr.)* normal voltage
Normalspur f *(railw.)* standard gage
Normalstab m standard bar
Normalwinkel m angle to the normal; (tool) standard square
Normalziegel m standard brick
Normblatt n standard specification
normen v.t. standardize
Normprobe f standard test bar
Normvorschrift f standard specification
Nortongetriebe n quick-change gear mechanism
Nortonräderkasten m quick-change gearbox
NOT ~, (~anruf, ~ausgang, ~ausschalter, ~beleuchtung, ~betrieb, ~bremse, ~druckknopf, ~lampe, ~maßnahme, ~reparatur, ~ruf, ~schalter, ~ventil) emergency ~
Notsitz m *(auto.)* folding seat, rumble seat
Notstromaggregat n *(electr.)* emergency generator-set
Nukleon n nucleon
Nukleonenladung f nucleon charge
Null f zero, naught, null
Nullablesung f zero reading
Nullachse f coordinate axis
Nullage f zero position
Nulleinstellung f zero adjustment
Nulleiter m *(electr.)* neutral wire; return wire; neutral conductor
nullen v.t. *(electr.)* neutralize
Nullenzirkel m bow compasses
Nullgetriebe n unmodified gearing

Nullinie f zero line, center line; reference line

Nulljustierung f zero setting, zero adjustment

Nullpunkt m zero point; *(electr.)* neutral point

Nullpunktabweichung f *(metrol.)* zero variation

Nullpunkteinstellung f zero adjustment

Nullpunkterdung f neutral earth

Nullrad n null gear, unmodified gear

Nullspannung f zero potential

Nullspannungsauslösung f no-voltage release

Nullspannungsschalter m disconnect switch; breaker with no-voltage release

Nullstellung f (Hebel:) starting position

Nullstellzähler m reset counter

numerieren v.t. number

numerisch gesteuert, numerically controlled

Nummernschalter m *(tel.)* telephone dial

Nummernschild n *(auto.)* number plate, licence plate

Nummernschild-Leuchte f *(auto.)* number plate lamp

Nummernwahl f *(tel.)* number dialling

Nummernwähler m numerical selector

Nuß f (Werkzeug:) button die

Nußbaumholz n hickory wood

Nut f groove; ($\approx$ Keilnut:) keyway; ($\approx$ Spann-Nut:) flute; ($\approx$ Spann-Nut:) slot; (e. Keilwelle:) splineway; ($\approx$ Schlitz:) slot

nuten v.t. groove; notch; slot; (Schneidwerkzeuge:) flute; (Keilwellen:) spline; (Wellen:) keyway

Nuteneinstechmeißel m groove recessing tool; necking tool

Nutenfräser m shank-type keyway cutter; (für Spann-Nuten:) T-slot cutter; ($\approx$ Schlitzfräser:) slotting cutter

Nutenkreissäge f circular grooving saw

Nutenmeißel m grooving tool, slotting tool

nutenstoßen v.t. slot

Nutenstoßmaschine f keyway slotter

nuten und spunden v.t. *(woodworking)* match

Nutenwelle f ($\approx$ Keilwelle) splineshaft

nutenziehen v.t. groove; keyseat

Nuthobel m match plane; grooving plane

Nutkreissäge f circular slitting saw

Nutmutter f slotted round nut

Nutringmanschette f U-shaped sealing ring

Nut- und Spundhobel m match plane

Nutwelle f splineshaft

Nutzarbeit f useful work

nutzbarer Laderaum, *(auto.)* payload space

Nutzen m use; profit; advantage

Nutzfahrzeug n utility vehicle

Nutzholz n timber

Nutzlast f useful load; *(auto.)* payload

Nutzleistung f effective capacity; (Motor:) useful power, brake horse-power

Nutzungsdauer f service life, useful life

Nutzungsfaktor m *(auto.)* economy tactor, profitability factor

Nutzungsgrad m efficiency

O

Oberbau m superstructure; *(railway)* permanent way

Oberdruckhammer m *(forge)* double-acting hammer

Oberfläche f surface

OBERFLÄCHEN ~, (~behandlung, ~beschaffenheit, ~bild, ~druck, ~entkohlung, ~entladung, ~fehler, ~gestalt, ~güte, ~härte, ~härtung, ~messung, ~rauheit, ~reibung, ~schicht, ~spannung, ~widerstand) surface ~

oberflächengekühlter Motor, surface-cooled motor

Oberflächenverdichter m *(road building)* surface compactor

Oberflächenveredelung f surface treatment

oberflächig a. superficial

Obergesenk n upper die, top die; (als Schmiedegerät:) blacksmiths' top swage

Obergurt m *(building)* top chord

Oberingenieur m chief engineer

oberirdische Leitung, overhead line

Oberkante f top edge

Oberkasten m *(molding)* cope flask

Oberleitung f *(electr.)* overhead line, aerial line

Oberleitungsomnibus m trolleybus

Oberlicht n *(building)* skylight

Oberschieber m *(lathe)* top slide, upper slide

Oberschwingung f harmonic

Oberspannung f *(electr.)* high voltage; *(mat.test.)* maximum stress

Oberstempel m (e. Matrize:) punch

Obersupport m *(lathe)* top slide rest

Oberteil n (e. Reitstockes:) barrel

Oberwelle f harmonic wave

Objektiv n *(opt.)* lens, objective

Obus-Anhänger m trolleybus trailer

Obusaufbau m trolleybus body

Obusfahrgestell n trolleybus chassis

Ocker m ochre

Ofen m (Industrieofen:) furnace; (Kokerei:) oven; (zum Brennen oder Rösten:) kiln

Ofen mit endlosem Band, continuous strand furnace

Ofen mit Förderband, conveyor-type furnace

Ofen mit satzweiser Beschickung, batch-type furnace

OFEN ~, (~auskleidung, ~ausmauerung, ~futter, ~gang, ~mauerwerk, ~reise, ~schacht, ~schlacke, ~wärter, ~zustellung) furnace ~

Ofenansätze mpl. *(steelmaking)* accretions

Ofenbatterie f *(coking)* oven battery

Ofenemaille f baking enamel

ofengeglüht p.p. open annealed

Ofenguß m stove castings

Ofenkopf m *(steelmaking)* port

Ofensau f *(blast furnace)* salamander, furnace sow

Ofentrocknung f oven drying, kiln drying

offener Güterwagen, open top railway car

offenes Lichtbogenschweißen, open arc welding

offene Verlegung, *(electr.)* open installation

öffentliche Ausschreibung, open tendering

Öffnung f opening; (Brennerdüse:) orifice; (≈ Austrittsöffnung:) outlet; (≈ Einlaß:) inlet; (≈ Loch:) hole

Öffnungsweite f (e. Fugennaht:) width of open joint

Ohmscher Widerstand, ohmic resistance

Okular n eyepiece

Öl n oil

ÖL ~, (~ablaßhahn, ~ablaßschraube, ~abscheider, ~abschreckung, ~abstreifer, ~abstrelfring, ~bad, ~behälter, ~brenner, ~dichtung, ~dichtungsring, ~docht, ~druck, ~dunst, ~durchfluß, ~einfüllschraube, ~einfüllstutzen, ~farbe, ~feuerung, ~filter, ~filz, ~förderung, ~füllung, ~härtung, ~heizung, ~kanal, ~kanister, ~kreislauf, ~lack, ~lasur, ~leitung, ~loch, ~manometer, ~meßstab, ~nebel, ~nebelschmierung, ~nippel, ~nute, ~pegel, ~polster, ~pumpe, ~raffinerie, ~rille, ~rückstand, ~schale, ~schlamm, ~schleuder, ~schmierung, ~sieb, ~spachtel, ~spiegel, ~spitze, ~stand, ~standschauglas, ~strom, ~tasche, ~überlaufrohr, ~umlauf, ~verteiler, ~wanne, ~wechsel, ~zerstäuber, ~zufuhr, ~zusatz) oil ~

Ölabziehstein m oilstone

Ölbadschmierung f oil-bath lubrication

Ölberieselung f flood-oiling

Öldruckanzeiger m oil pressure gage

Öldruckbremse f (auto.) oil-hydraulic brake

Öldruckmesser m oil pressure gage

Öldruckschalter m oil pressure switch

Öldunstentlüfter m (auto.) breather

Öler m oiler, oil cup

Ölfangring m (auto.) oil-retaining ring

Ölfilterschlauch m (auto.) flexible oil filter line

Ölgetriebe n oil drive unit, oil hydraulic transmission

Ölgrundierung f oil primer

Ölkabel n oil-filled cable

Ölkännchen n oil squirt

Ölkitt m putty

ölloses Lager, water-lubricated bearing, oilless bearing; grease-packed bearing

Ölrückgewinnungsanlage f (auto.) oil reclamation equipment

Ölsaugpumpe f (auto.) scavenger pump

Ölschalter m oil circuit-breaker

Ölschmierpumpe f (auto.) lubricating oil pump

Ölschütz n oil-filled contactor

Ölsicherung f (electr.) oil-break fuse

Ölspritzschmierung f oil-splash lubrication

Ölstandanzeiger m oil gage

Ölstandsauge n oil level gage

Ölstoßdämpfer m (auto.) hydraulic shock absorber

Öltransformator m oil-filled transformer

Öltropfgefäß n sight-feed lubricator

Ölumlaufschmierung f circulation oiling

Omnibus m omnibus, motorbus, bus

OMNIBUS ~, (~anhänger, ~fahrer, ~fahrgestell, ~haltestelle, ~scheinwerfer) omnibus ~

Optik f optics; (als Einrichtung:) optical equipment

optischer Teilkopf, optical dividing head

optischer Winkelmesser, optical goniometer, optical universal bevel protractor

optisches Tastgerät, *(copying)* optical tracer device
Ordinatenachse *f* axis of coordinates
Ordnungszahl *f* *(math.)* ordinal number; *(nucl.)* atomic number
Organ *n* *(mach.)* member, part, element
Ornamentglas *n* ornamental glass
Ort *m* place; position; spot; *(radio)* station
orten *v.t.* locate; take bearings; *(radar)* find the direction
ortsbeweglich *a.* portable
Ortsbeweglichkeit *f* portability
ortsfest *a.* stationary; fixed
Ortsteilnehmer *m* *(tel.)* local subscriber
ortsveränderlich *a.* portable
Ortsverkehr *m* *(tel., telegr.)* local (telephone or teletype) traffic
Ortung *f* position finding; location; direction finding
Ortungsgerät *n* *(radar)* location finder; *(electr.)* fault localization apparatus; *(radar)* detector

Öse *f* eye; loop
Ösenhaken *m* eye hook
Ösenzange *f* eyelet plier
Ottomotor *m* Otto carburetor engine, gasoline engine
Ovalschleifmaschine *f* oval grinding machine
Ovalstich *m* *(rolling mill)* oval pass, oval groove
Oxid *n* oxide
Oxidationsmittel *n* oxidizing agent
Oxidationsperiode *f* oxidizing period
Oxidationsschlacke *f* oxidizing slag
Oxideinschluß *m* oxide inclusion
Oxidhaut *f* oxide skin, oxidation film
oxidieren *v.t. & v.i.* oxidize
Oxidierung *f* oxidation
oxidisch *a.* oxide
Oxidkeramikplatte *f* oxide-ceramic plate
oxidkeramisches Schneidwerkzeug, oxide-ceramic cutting tool
Oxidschicht *f* oxide film; *(welding)* layer of oxide

P

Paar *n* pair, couple
paaren *v.t. (gears)* mate; pair
Paarigkeit *f (telev.)* pairing
Paarung *f (gearing, telev.)* pairing
Paarungsmaß *n* mating size
Paarverzahnung *f* intermating tooth system
Packung *f (≈ Dichtung:)* packing
paketieren *v.t.* bale, briquet; *(met.)* faggot, pile, bushel
Paketierpresse *f* scrap baling press, cabbaging press
Paketschalter *m (electr.)* multi-section type rotary switch
Palloidverzahnung *f* palloid tooth system
Palmutter *f* patent locknut
Panne *f (auto.)* breakdown
pannensicher *a.* foolproof
Pantograph *m* pantograph
pantographengesteuert *p.a.* pantograph-controlled
Pantographgesenkfräsmaschine *f* pantograph die-sinking machine
Pantographnachformfräsmaschine *f* pantograph miller
Panzer *m* casing; (e. Hochofens:) steel jacket
Panzeraderleitung *f* metal-cased conductor
Panzerblech *n* armor plate
Panzergalvanometer *n* iron-clad galvanometer
Panzerkabel *n* armored cable
panzern *v.t.* armor; *(welding)* surface harden
Panzerplatte *f* armor plate
Panzerplattenhobelmaschine *f* armor plate planer
Panzerplattenwalzwerk *n* armor-plate rolling mill
Panzerschlauch *m* armored hose
Panzerung *f* armoring, metal sheathing
Papier *n* paper; *(recording)* chart
Pappe *f* millboard; cardboard; pasteboard, board
Paraffin *n* paraffin
Paraffinwachs *m* paraffin wax
Parallaxenfehler *m (opt.)* parallax distortion
Parallaxwinkel *m* angle of parallax
Parallele *f* parallel line, parallel
Paralleldrossel *f* shunt inductor
parallelflanschiger Profilträger, I-beam, light beam
Parallelflanschträger *m* parallel-flanged beam
Parallel-Klebschraubstock *m* clamp base vise
Parallelkondensator *m* shunt capacitor
Parallelreißer *m* surface gage
parallelschalten *v.t. (electr.)* connect in parallel
Parallelschaltung *f (electr.)* parallel connection
Parallelschraubstock *m* machinists' vise
Parallelstück *n (tool)* steel parallel
parkerisieren *v.t.* parkerize
Parkettdiele *f* parquet deal
Parkleuchte *f (auto.)* parking light
Parklücke *f (auto.)* parking space
Parkplatz *m* parking place

Parkplatzwächter m (auto.) parking attendant
Parkuhr f parkometer
Parkverbot n no parking
Paßdorn m check plug
passen v.i. fit
Paßfeder f feather key
Paßfehler m form error
Paßfläche f mating surface
Paßflächenkorrosion f fretting corrosion
Paßfuge f fitting joint
paßgerecht a. dimension-fitting
Paßhülse f body-fit sleeve
Paßmaß n fit size
Paßrand m (founding) mold mating face
Paßschraube f fitting bolt, body-fit bolt
Paßsitz m snug fit
Paßstift m set pin, dowel pin
Paßstück n adaptor
Paßteil n fitting member, mating part
Passung f fit; system of fits
Passungsspiel n fitting clearance
Pastenaufkohlen n (heat treatment) paste carburizing
Pastenfarbe f paste color
Patent n patent
PATENT ~, (~amt, ~anmeldung, ~beschreibung, ~schrift, ~verletzung) patent ~
Patentanspruch m patent claim
patentieren v.t. (Draht:) patent
Patentinhaber m patentee
Patrize f upper die, top die; punch
Patrone f (e. Spannzange:) collet; (threading) leader; (e. Schmelzsicherung:) cartridge
Patronenspannfutter n draw-in collet chuck

Pause f (≈ Lichtpause:) print, photostat; (≈ Verweilen:) dwell; (≈ Intervall:) interval
pausenlos a. continuous, uninterrupted
Pausleinwand f tracing cloth
Pauspapier n tracing paper
Pech n pitch
Pechkohle f pitch coal
Pechkoks m pitch coke
Pechschotter m pitch macadam
Pedal n pedal
Pegel m (tel.) level; (civ. eng.) tide gage; (Öl:) level
Pegelmesser m (electr.) transmission level meter
Pegelsteuerung f (tel.) level control
Peilantenne f direction finding aerial
Peilempfänger m directional receiver
peilen v.t. take a bearing; (Echolot:) sound
Peiler m (radio) direction finder
Peilfunk m directional radio
Peilfunkgerät n radio direction finder
Peilgerät n direction finder
Peilsender m directional transmitter
Peilstab m (Öl:) dipstick, level rod
Peilstation f direction finding station
Peilstrahl m signal beam
Peilung f (radio) direction finding
Peilungsrichtung f bearing direction
Peilwinkel m bearing angle
Pendel n (phys.) pendulum; (e. Lampe:) pendant
Pendelachse f pivoted axle; (auto.) swing axle
Pendelbewegung f oscillating motion
Pendelbecherwerk n pivoted bucket conveyor
Pendeldruckknopftafel f pendant control panel

Pendelfräsen *n* cycle milling, reciprocal milling
Pendelfrequenz *f (electr.)* quenching frequency
Pendelfutter *n* floating tool holder
Pendelgleichrichter *m* pendulum rectifier
pendelglühen *v.t.* spheroidize
Pendelhammer *m (mat. test.)* Charpy hammer
Pendelkreissäge *f* swing saw
Pendelkugellager *n* self-aligning ball bearing
Pendellager *n* swivel bearing
pendeln *v.i.* swing, oscillate, reciprocate; (Werkzeug:) float
Pendelreibahle *f* floating reamer
Pendelrückkopplung *f* superregeneration
Pendelschlagwerk *n* pendulum impact machine
Pendelschleifmaschine *f* swing frame grinder
Pendelschnur *f* flexible cord, pendant cord
Pendelschwingachse *f (auto.)* independent axle
Pendeltisch *m* reciprocating table
Pendelverkehr *m* shuttle service
Pendelwarmsäge *f (rolling mill)* hot drop saw
Pendelwerkzeug *n* floating tool
Pendelwinker *m (auto.)* oscillating direction indicator
Pendelwischer *m (auto.)* pendulum-type screen wiper
perforieren *v.t.* perforate
Periode *f (electr.)* cycle
Periodendauer *f (electr.)* period
Periodenzahl *f (electr.)* number of cycles
periodisch *a.* periodic, cyclic

Perle *f (welding)* nugget, globule
Perlit *m (cryst.)* pearlite
Perlitguß *m* pearlitic cast iron
Perlitisieren *n (heat treatment)* isothermal annealing
Permanentrot *n* red toner
Personenkraftwagen *m* passenger car
Personenwagen *m* passenger car
Perspektive *f* perspective
perspektivisch *a.* perspective
Peterwagen *m* radio patrol car
Petroleum *n* (≈ Erdöl:) crude oil, mineral oil, petroleum; (≈ Leuchtpetroleum:) kerosene
Pfaffe *m (forging)* hob, hub
Pfahl *m (building)* pile
Pfahlgründung *f* pile foundation
Pfahlramme *f* pile driver
Pfanne *f* pan; (e. Schneide:) seat; (Gießpfanne:) ladle; *(Späne:)* tray, pan
Pfannenausguß *m* nozzle of a ladle
Pfannenbär *m* ladle skull
Pfannenblech *n* roofing sheet
Pfannenbügel *m* ladle bail
Pfannengabel *f* ladle shank
Pfeiler *m (building)* pier, shaft
Pfeilergründung *f* pier foundation
Pfeilfallversuch *m* falling dart test
Pfeilmotor *m* V-type engine
Pfeilzahn *m* herringbone tooth
Pfeilzahnrad *n* herringbone gear, double-helical gear
Pferdestärke *f* (Motor:) horsepower
Pfette *f (building)* purlin
Pflaster *n* pavement
Pflasterstein *m* paving stone
Pflasterziegel *m* paving brick
Pflugschraube *f* plow bolt
Phase *f (electr.)* phase

PHASEN ~, (~anzeiger, ~ausgleich, ~entzerrer, ~faktor, ~folge, ~gleichgewicht, ~nacheilung, ~regler, ~schwund, ~transformator, ~verschiebung, ~winkel) phase ~

Phasenmesser *m (electr.)* power-factor meter

Phasenschieber *m* phase advancer; phase changer

Phasenschluß *m (electr.)* phase-to-phase short-circuit

Phasenspannung *f* phase-to-neutral voltage

Phenoplast-Preßmasse *f* phenoplastic compression molding material, phenolic molding material

phosphatieren *v.t.* bonderize

Phosphor *m* phosphorus

Phosphorbronze *f* phosphor bronze

Photo ~, *cf.* Foto ~

Physik *f* physics

Pigmentfarbstoff *m* pigment dye

pigmentieren *v.t.* pigment

Pilgerdorn *m (rolling mill)* piercer

Pilgerschrittschweißung *f* step-back welding

Pilgerschrittverfahren *n* reciprocating rolling process

Pilgerschweißung *f* step-back welding

Pilgerwalze *f* pilger roll

Pilzdecke *f (building)* flat-slab floor

Pilzfräser *m* semi-circular milling cutter

Pinole *f (lathe)* center sleeve; *(miller)* quill

Pinsel *m* brush

Pinzette *f* tweezers

PIV-Getriebe *n* PIV-variable speed transmission

plan *a.* plane; flat; level

plan ~, (gekoppelt mit verbalen Zerspanungsbegriffen:) (~abrichten, ~fräsen, ~kopieren, ~läppen, ~schleifen) face ~

Plan *m* plan; project; layout; system; arrangement; program

Plananschlag *m (lathe)* cross stop

planarbeiten *v.t.* face, surface

PLANDREH ~, (~einrichtung, ~hebel, ~maschine, ~meißel, ~schlitten, ~vorrichtung, ~werkzeug) facing ~, surfacing ~

plandrehen *v.t.* face, surface

Plandrehrevolver *m* transverse rotary turret

Plandrehsupport *m* cross slide rest, facing slide rest

Plandrehvorschub *m* cross feed

planen *v.t.* project; plan; *(mach.)* face

Planenstoff *m (auto.)* tilt material

Planetengetriebe *n* planetary gear drive

Planetenrad *n* planet gear, planet wheel

Planfräser *m* face-milling cutter

Planfräsmaschine *f* fixed-bed type miller: *(UK)* bench milling machine

Plangang *m (metal cutting)* cross traverse, transverse feed

Planglas *n* optical flat

planieren *v.t.* level, plane, smooth

Planierraupe *f* caterpillar; bulldozer

Plankosten *pl.* (= Richtkosten) predetermined standard rates

Plankurve *f* face cam

Planrad *n* crown gear, crown wheel, crown rack

Planrevolverdrehmaschine *f* fixed-center turret lathe

Planscheibe *f (lathe)* faceplate; *(vertical turret lathe)* table

Planschleifmaschine *f* face grinder

Planschlitten m *(lathe)* cross slide, facing slide
Planung f planning
Planverzahnung f rack-tooth system
Planvorschub m *(metal cutting)* cross feed
Platine f *(met.)* sheet bar; (Weißblechfabrikation) tin plate bar; *(presswork)* blank; (≈ Ronde:) circular blank
Platinenschere f *(rolling mill)* sheet bar shear
Platinenwalzwerk n sheet bar rolling mill
Plättchen n (Hartmetall:) tip
Platte f plate; disc; *(building)* slab
Plattenbalken m *(concrete)* T-beam
Plattengleichrichter m plate rectifier
Plattenleger m street mason
Plattenspieler m record player
Plattenwender m *(rolling mill)* manipulator
Plattenzink n slab zinc
plattieren v.t. clad
plattiertes Blech, clad sheet, clad plate
Plattierungsmetall n cladding metal
Plattierverfahren n cladding process
Platz m place; yard; *(building)* site
Platzbedarf m space occupied
Pleuel n *(US)* pitman; *(UK)* connecting rod
Pleuelbolzen m *(auto.)* connecting rod bolt
Pleuelbuchse f *(auto.)* connecting rod bush
Pleuellager n *(auto.)* connecting rod bearing, big-end bearing
Pleuelstange f connecting rod
Plombe f [lead] seal
plombieren v.t. seal
Plombierzange f lead sealing plier
Pluspol m positive pole
Plusseite f (e. Rachenlehre:) go-side; (e. Lehrdorns:) not go-side
pneumatisch a. pneumatic

pochen v.t. (Erze:) stamp, pound
Pochmühle f stamp mill
Pochtrübe f stamp pulp
Pochwerk n stamp mill
Podest n platform, stand; *(building)* landing platform
Pol m *(magn., math., electr.)* pole
POL ~, (~block, ~bogen, ~platte, ~prüfer, ~schuh, ~stärker, ~teilung, ~wechsler) pole ~
polarisieren v.t. polarize
Polarität f polarity
Polarkoordinate f polar coordinate
polen v.t. polarize
Polierdrehmaschine f polishing lathe
Polierdrücken n die burnishing
polieren v.t. polish, burnish, smooth; (Walzgut:) planish
Polierfilzscheibe f felt polishing wheel
Poliergerüst n *(rolling mill)* planishing stand
Polierschleifen n flexible grinding
Polierstählen n burnishing
Polierwalzen n roller burnishing
Politur f polish, lustre, bright finish, gloss
Polizei f *(building)* inspectorate
polizeiliche Vorschrift, police regulation
Polster n (≈ Kissen:) cushion; *(auto.)* upholstery
Polsterleder n *(auto.)* upholstery leather
Polstermaterial n upholstery material
polstern v.t. cushion; upholster
Polsterschonbezug m *(auto.)* loose cover
Polsterstoff m *(auto.)* upholstery fabric
Polsterung f upholstery
polumschaltbarer Motor, pole-changing motor
Polumschalter m pole-changing switch
Polumschaltung f pole changing

Polung f polarity
Polwechsel m changing the polarity
Polwender m commutator; pole-changing switch
Polypgreifer m multi-blade grab
Polyvinylchlorid n polyvinyl chloride
Pontonbrücke f floating pontoon bridge
Pore f pore; void
Porenfüller m (woodworking) pore filler
Porigkeit f porosity
porös a. porous
Portalautomat m portal automatic
Portalfräsmaschine f double-column planer-miller
Portalkran m gantry crane
Portlandzement m Portland cement
Portlandzementbeton m Portland cement concrete
Porzellan n porcelain
Porzellankitt m adhesive for porcelain
positionieren v.t. position, locate
Positionsgenauigkeit f positional accuracy
Positionsleuchte f (auto.) side lamp
Potential n potential
Potentialgefälle n potential drop
Potenz f power
potenzieren v.t. raise to a higher power
Poterieguß m sanitary ware
Prägeform f embossing die; (Massivprägung:) coining die
prägen v.t. (cold work) squeeze; stamp; press; (hohl:) emboss; (massiv:) coin; (auf Maß:) size
prägepolieren v.t. press-burnish
Prägestanze f stamping die; embossing die; coining die; s.a. prägen
prägeziehen v.t. (cold work) (Knöpfe, Uhrdeckel:) stamp, press

Prämie f bonus
Präzision f precision
PRÄZISIONS ~, (~bearbeitung, ~bohren, ~drehen, ~drehmaschine, ~einstellung, ~gewinde, ~gerätebau, ~gleitlager, ~instrument, ~maßstab, ~meßmaschine, ~wasserwaage, ~werkzeug) precision ~; s.a. GENAU ~ und GENAUIGKEITS ~
Prellschlag m (forging) percussion; impact
Preßbacke f upsetting die; (Schraubenherstellung) bolt die
Preßblech n sheet suitable for press-work
Preßbolzenschweißung f stud welding
Preßdruck m (plastics) molding pressure
Preßdruckschmierung f pressure lubrication
Presse f power press, mechanical press; (lubrication) gun
Presse mit C-förmigem Gestell, C-frame press
Presse mit Magazinzuführung, magazine feed press
Presse mit Rutschenzuführung, gravity-chute feed press
Presse mit Streifenfördergerät, strip-feed press
Presse mit Walzenvorschubeinrichtung, roll feed press
Presse mit Werkstoffzuführung im Zickzackschritt, zig-zag feed press, stagger-feed press
Presse mit Zangen- oder Greiferzuführung, gripper feed press
pressen v.t. press, compress; (cold work) stamp, squeeze, press; (hot work) diepress, die-form; (Radscheiben:) cone, dish; (≈ hohlprägen:) emboss; (≈

Pressentisch massivprägen:) coin; (≈ strangpressen:) extrude; *(plastics)* compression-mold

Pressentisch *m* (fester:) bed; (verstellbarer:) table; (als Aufspannplatte:) bolster plate

Pressenunterteil *n* press bed

Presserei *f* pressroom; *(plastics)* molding shop

Preßform *f (cold work)* stamping die; *(extruding)* extrusion die; *(plastics)* impression die

Preßformmaschine *f (molding)* squeezer

Preßfuge *f* interference interface

preßgießen *v.t.* die-cast, press-cast

Preßgießen *n* press-cast process

Preßgießmaschine *f* pressure die-casting machine

Preßgrat *m* burr

Preßguß *m* press casting, pressure casting, die casting

Preßgüte *f* stamping quality

Preßling *m* pressed part; stamping; diestamping; cold pressed forging: die-formed part

PRESSLUFT ~, (~anlasser, ~filter, ~futter, ~hammer, ~hebezeug, ~kondensator, ~meißel, ~mengenmesser, ~motor, ~niethammer, ~rüttler, ~spannfutter, ~werkzeug compressed-air ~, air ~

Preßmantelelektrode *f (welding)* extruded electrode

Preßmasse *f (plastics) (obs.)* cf. Formmasse

Preßmatrize *f* (Strangpreßverfahren) extrusion die

Preßmessing *n* hot-pressed brass

Preßnietmaschine *f* compression riveter

Preßöler *m* pressure lubricator

Preßölschmierung *f* force-feed lubrication

Preßpassung *f* interference fit

preßpolieren *v.t.* die-burnish

preßschweißbar *a.* pressure weldable

preßschweißen *v.t.* pressure-weld

Preßschweißen *n* pressure welding

Preßschweißen *f* pressure welding

Preßsitz *m* force fit

Preßspan *m* laminated fibre sheet

Preßspritzen *n (plastics)* cf. Spritzpressen

Preßstempel *m* (e. Strangpresse:) extrusion die; (e. Kunstharzpresse:) impression die; (e. Stanzpresse:) ram, slide

Preßstoff *m (plastics)* molded plastics

Preßstumpfschweißen *n* pressure butt welding, upset butt welding

Preßteil *n* die-formed part, pressed part, pressing, stamping; *(plastics)* compression molded article

Preßtisch *m* (e. Formmaschine:) squeezing table

Preßverfahren *n (plastics)* compression molding process

Preßwerkzeug *n* press tool; compression mold

Preußischblau *n (painting)* Prussian blue

PRIMÄR ~, (~anker, ~batterie, ~element, ~klemme, ~kreis, ~spannung, ~spule, ~strom, ~wicklung primary ~

Primärkosten *pl (cost accounting)* original incurrence

Primärlunker *m* (im Gußblock:) pipe

Primzahl *f* prime number

Prinzipschaltung *f* basic circuit arrangement

Prinzipskizze *f* block diagram

Prisma *n* prism; *(tool)* V-block; (e. Bettführungsbahn:) Vee-way, V-way; V-guideways

Prismenfräser m double-angle milling cutter

Prismenführungsbahn f *(lathe)* V-guideways

Pritschenwagen m plankbed car, platform truck

Probe f *(mat. test.)* test specimen, test piece, (≈ Muster:) sample; (≈ Probestab:) test bar; (≈ Versuch:) trial

Probeentnahme f sampling

Probekörper m test specimen, test piece

Probelöffel m *(founding, steelmaking)* sample spoon

Probestab m test bar

probieren v.t. try; test; *(met.)* assay

Probiergold n standard gold

Probierofen m assay furnace

Probiertiegel m assay crucible

Produktion f production

PRODUKTIONS ~, (~betrieb, ~drehmaschine, ~fräsmaschine, ~kosten, ~leistung, ~maschine, ~planung, ~prämie, ~schleifmaschine, ~serie, ~vorschau, ~werkzeugmaschine) production ~

Profil n profile; cross-section; form

profildrehen v.t. profile-turn

Profilelektrode f contour electrode

profilfräsen v.t. profile-mill, form-mill

Profilfräser m form cutter

Profilfräsmaschine f contour miller

profilgerecht a. profile-true

Profilhobeleisen n molding plane iron

profilieren v.t. profile, contour, shape, form

Profillinie f contour line

Profilloch n formed hole

Profilschere f section shearing machine

profilschleifen v.t. form-grind, contour-grind

Profilschleifmaschine f contour grinder

Profilstahl m section steel, sectional steel

Profilteil n shaped part, shape

Profilverzerrung f profile distortion

Profilwalze f shape roll

Profilwalzen n section rolling

Profilwalzwerk n shape rolling mill, section mill

PROGRAMM ~, (~ablauf, ~einstellung, ~fräsen, ~gesteuert, ~kreis, ~schaltung, ~speicherung, ~stecker, ~tafel, ~wähler) program ~

Programmautomatik f automatic program control

programmieren v.t. program

Programmierer m programmer

Programmplaner m programmer

Programmschalter m timer switch, sequence switch

Programmsteuerung f program control

Programmwalze f sequence control drum

Programmwiederholung f re-run

Projektion f projection

PROJEKTIONS ~, (~ablesung, ~ebene, ~okular, ~optik, ~röhre, ~schirm) projection ~

Projektionsapparat m projector

Projektionslampe f projector lamp

Projektionsrundtisch m projection-type rotary table

Projektionsskaleninstrument n projected-scale instrument

projizieren v.t. *(opt.)* project

Proportionalitätsgrenze f limit of proportionality

Proportionalstab m proportional test bar

PRÜF ~, (~anweisung, ~attest, ~befund, ~belastung, ~bericht, ~einrichtung, ~gerät, ~klemme, ~klinke, ~lampe, ~last, ~maschine, ~spannung, ~stab,

~stand, ~stromkreis, ~vorschrift) test ~, testing ~
prüfen v.t. test; examine; inspect; check; investigate
Prüfendmaß n reference gage block
Prüfer m *(electr.)* detector
Prüffeld n test room, test bay
Prüflehrdorn m precision plug gage
Prüflehre f master gage, check gage
Prüfmaß n inspection gage
Prüfschrank m *(electr.)* test board
Prüfung f test; examination; inspection; check; investigation
Puddelherd m puddling hearth
puddeln v.t. puddle
Puddelofen m puddling furnace
Puddelroheisen n forge pig iron
Puddelstahl m puddle steel
Puddelstahlwalzwerk n muck mill
Puffer m buffer; cushion
PUFFER ~, (~batterie, ~feder, ~generator, ~kontakt, ~lösung, ~schaltung, ~wirkung) buffer ~
Pufferlage f *(welding)* interlayer
pulsieren v.i. pulsate
Pult n control desk, station
Pulveraufkohlen n *(heat treatment)* powder carburizing
Pulverbrennschneiden n powder cutting
pulverig a. pulverulant
Pulvermetallurgie f powder metallurgy

Pulververzinkung f sherardizing
Pumpe f pump
pumpen v.t. pump
PUMPEN ~, (~förderung, ~gehäuse, ~hebel, ~hub, ~kolben, ~kurbel, ~motor, ~stiefel, ~ventil, ~zylinder) pump ~

PUNKT ~, (~belastung, ~berührung, ~entladung, ~gitter, ~ladung, ~masse, ~messung, ~raster, ~wert) point ~
Punktbeleuchtung f spot lighting
Punktelektrode f *(welding)* spot welding electrode
punktschweißen v.t. spot weld
Punktschweißmaschine f spot welding machine
Punktschweißung f spot welding
Punktschweißverbindung f spot weld
Punktstrahllampe f spot lamp
Pupinkabel n coil-loaded cable
putzen v.t. (Gußstücke:) clean, dress, fettle; (Blöcke:) deseam
Putzen m punching
Putzer m *(building)* plasterer
Putzerei f *(founding)* cleaning room
Putzmeißel m chipper
Putzstern m rattler star, milling star
Putztrommel f tumbling barrel, tumbler, rolling barrel
Putzuntergrund m plaster background
Putzwolle f cotton waste

Q

Quaderstein *m* cut stone, ashlar
Quadrat *n* square
quadratisch *a.* square; *(math.)* quadratic
quadratische Gleichung, quadratic equation
quadratischer Mittelwert, root-mean-square [value]
Quadratwurzel *f* square root
Qualität *f* quality
Qualitätsblech *n* high-quality sheet steel
Qualitätsguß *m* high-quality cast iron, high-strength cast iron
Qualitätskontrolle *f* quality control
Qualitätsstahl *m* high-grade steel
Qualitätsüberwachung *f* quality control
Quarz *m* quartz
QUARZ ~, (~filter, ~glas, ~glaslampe, ~gleichrichter, ~kristall, ~lampe, ~oszillator, ~röhre, ~stab, ~uhr) quartz ~
Quecksilber *n* mercury
QUECKSILBER ~, (~dampf, ~faden, ~lampe, ~legierung, ~säule, ~thermometer) mercury ~
Quecksilberdampfgleichrichter *m* mercury vapor rectifier
Quecksilberdampflampe *f* mercury arc lamp
Quelle *f* source
quellen *v.i.* swell, soak
quellfest *a.* swelling-resistant
quer *a.* cross, transverse; radial
Querbalken *m* (building) cross beam; (planer-miller) crossrail
Querbalkenmotor *m* (planer) rail-traverse motor

Querbalkenschieber *m* (planer) saddle, crossrail slide
Querbalkensupport *m* crossrail carriage
Querbalkenverstellmotor *m* rail-elevating motor
Querbeanspruchung *f* transverse stress
Querbiegeversuch *m* transverse bending test
Querbohreinrichtung *f* cross drilling attachment
Querdruck *m* (e. Lagers:) radial load
Querfaltversuch *m* cross folding test
Querfelddynamo *m* self-stabilizing dynamo
Querglied *n* (tel.) shunt element
Querhaupt *n* (planer) cross-beam
querhobeln *v.t.* cross-plane
Querkardanwelle *f* (auto.) transverse cardan-shaft
Querkeil *m* cotter
Querkondensator *m* shunt capacitor
Querlager *n* radial bearing
Querleitwert *m* transverse conductance
Querprobe *f* transverse test specimen
Querriegel *m* latch
Querrippe *f* cross brace
Querriß *m* transverse crack
quersägen *v.t.* cross-cut
Querschlitten *m* (lathe) cross slide; (grinder, miller, shaper) saddle
querschmieden *v.t.* forge to the grain
Querschnitt *m* cross-section
Querschnittfläche *f* cross-sectional area
Querschnittsverminderung *f* (rolling) draft

Querspule f (tel.) leak coil
Querstrahler m broadside aerial
Querstraße f cross road
Querstrom m transverse current
Quertisch m (grinder) saddle
Querträger m cross girder, cross beam
Quervergrößerung f (televl.) lateral magnifying power
querverstellbar a. transversely adjustable

Quervorschub m (metal cutting) cross feed
Querzugversuch m tensile test across the grain
Quetschgrenze f (Prüfwesen) compressive yield point
Quetschkondensator m compression capacitor
Quetschnaht f (welding) mash weld
Quickton m quick clay

R

Rabitzwand f expanded lathing wall
Rachenlehre f snap gage
Rad n (Laufrad:) wheel; (Zahnrad:) gear
Radabzieher m wheel puller
Radantenne f cartwheel antenna (or aerial)
Radar n radar; (UK) radiolocation, radar
RADAR ~, (~anlage, ~ausrüstung, ~bild, ~echo, ~gerät, ~messung, ~ortungsgerät, ~schirm, ~störung, ~strahl) radar ~
radargesteuert p.a. radar-controlled
Radarkuppel f radome
Radarmessung f radar ranging
Radarzeichen n radar trace
Radbolzen m (auto.) hub stud
Radbremse f (auto.) hub brake
Radeffekt m (radar) spoking
Räderblock m (gearing) gear train, gear cluster, gear block
Räderfräsmaschine f gear milling machine
Rädergetriebe n gear train
Räderkasten m (lathe) apron
Räderplatte f (lathe) apron wall
Räderpumpe f gear pump
Räderspindelkasten m geared headstock
Räderübersetzung f gear transmission ratio
Räderuntersetzung f gear reduction
Rädervorgelege n back gears
Räderwechselgetriebe n change gear drive
Räderziehpresse f reducing press
Radfahrer m cyclist, bicyclist
Radfahrweg m cycle track, bicycle path

Radfelge f wheel rim
Radialbohrmaschine f radial drilling machine
Radialkugellager n radial ball bearing
Radialspannung f (electr.) radial potential
Radialspiel n radial play
Radialturbine f radial flow turbine
Radiaxlager n single-row grooved type ball bearing
Radienfräskopf m radius milling head
Radienlehre f radius gage
Radienschleifvorrichtung f radius grinding attachment
Radio n (≈ Funk:) radio, wireless; (≈ Rundfunk:) broadcasting
RADIO ~, (~apparat, ~bake, ~empfänger, ~frequenz, ~gerät, ~kanal, ~ortung, ~ortungsgerät, ~peilung, -röhre, ~schaltung, ~sender, ~sendung, ~störung, ~technik, ~techniker, ~übertragung, ~welle) radio ~
radioaktiv a. radioactive
radioaktiver Abfall, hot waste
radioaktiver Niederschlag, radioactive deposit
radioaktive Verseuchung, radioactive contamination
Radioaktivität f broadcasting station
Radiotelegrafie f radio telegraphy, wireless telegraphy
Radkappe f wheel cap
Radkörper m (gearing) gear body
Radkranz m (gearing) gear rim; (e. Rades:) wheel rim
Radmutterkurbel f speeder

Radmutternschlüssel m *(auto.)* wheel nut spanner
Radmutersteckschlüssel m rim wrench
Radnabe f wheel hub, boss
Radnabenabzieher m *(auto.)* wheel hub puller
Radnabendeckel m *(auto.)* wheel hub cap
Radnabenkappe f *(motorcycle)* wheel hub clevis
Radreifenbohrbank f tire boring mill
Radreifenkarusselldrehmaschine f vertical tire boring mill
Radsatzdrehmaschine f car wheel lathe
Radscheibe f wheel center, center web
Radschlepper m industrial wheel tractor; wheeled tractor
Radspeiche f wheel spoke
Radspur f *(auto.)* track
Radstand m wheel base
Radsturz m *(auto.)* camber
Radvorspur f *(auto.)* toe-in
Radzierkappe f *(auto.)* ornamental hub cap
Radzierring m *(auto.)* ornamental wheel ring
Raffinat n refined product
Raffinatblei n refined lead
Raffinationsanlage f refining plant, refinery
Raffinationsofen m refining furnace
Raffinatkupfer n refined copper
Raffinatzink n refined spelter
Raffinerie f refinery
raffinieren v.t. refine
Rahmenantenne f frame antenna
Rahmenblechschere f guillotine plate shear
Rahmenempfänger m loop receiver
Rahmenwasserwaage f box-type spirit level
Rakete f rocket
RAKETEN ~, (~antrieb, ~batterie, ~düse, ~raumschiff, ~vortrieb) rocket ~
Ramme f *(civ. eng.)* pile driver
rammen v.t. *(building)* drive
Rammgerüst n piling frame
Rammhaube f *(building)* pile helmet, driving cap
Rammpfahl m bearing pile
Rampe f (e. Brücke:) approach
Rampenbeleuchtung f stage lighting
Rand m edge; border; margin; periphery; (e. Walze:) collar
Randblase f *(surface finish)* subcutaneous blowhole
Rändel n knurl
Rändelkopf m (e. Schraube:) knurled head
Rändelmutter f knurled nut
rändeln v.t. knurl
Rändelstoßmaschine f knurl shaper
Rändelwerkzeug n knurling tool
Randentkohlung f surface decarburization
Randgebiet n edge zone
Randnaht f *(welding)* edge weld
Randschicht f rim zone, outer zone, surface layer
Randschichthärten n surface layer hardening
Randschweißung f *(welding)* edge welding
Randzone f *(founding)* shell; rim zone, rim
Rangieranlage f marshalling yard
Rangierbahnhof m marshalling yard
Rangierbewegung f shunting movement
rangieren v.t. shunt

Rangiergleis *n* classification yard line
Rangierheber *m* garage jack, floor jack
Rangierlokomotive *f* shunting locomotive (*or* engine)
Rangierstellwerk *n* shunting tower
Raseneisenerz *n* bog iron ore
Raspe *f* rasp
Rast *f* (*met.*) bosh
Raste *f* (e. Rastenscheibe:) notch, slot
Rastenscheibe *f* notched disc
Raster *m* (*televi., opt.*) raster, grating, screen
Rastwinkel *m* (*met.*) bosh angle
Ratsche *f* ratchet
Rattermarke *f* chatter mark
rattern *v.i.* chatter
Rauch *m* smoke
Rauchabzugskanal *m* flue
Rauchgas *n* flue gas, stack gas
Rauchkanal *m* waste-gas flue
Rauheit *f* roughness
Rauhlgkeit *f* roughness
Rauhputz *m* rough cast plastering
Raum *m* room, space; compartment, chamber, ($\approx$ Spiel:) clearance
Räumasche *f* (Zinkgewinnung:) retort residue
Raumbedarf *m* (e. Maschine:) floor space occupied; (*shipping*) cubic contents
Raumbeleuchtung *f* room lighting
Raumbeständigkeit *f* constancy of volume
Räumdorn *m* push broach
räumen *v.t.* (*mach.*) broach; (Zinkdestillation:) remove retort residues
Räumer *m* (*auto.*) bulldozer
Raumfahrzeug *n* space vehicle
Raumformfräsen *n* three-dimensional tracer milling

Raumgeometrie *f* space geometry, solid geometry
Raumgewicht *n* weight by volume; ($\approx$ Wichte:) volumetric weight; specific gravity
Raumgitter *n* (*cryst.*) space-lattice
Räumhub *m* broaching stroke
Rauminhalt *m* volume; cubic capacity, volume capacity
Raumladung *f* space charge
Raumladungsgitter *n* space charge grid
Räumlänge *f* length of cut, length of run
Räumleistung *f* broaching capacity
räumlich *a.* spatial, cubic; three-dimensional
Raumluftschiffahrt *f* space navigation, astronautics
Räummaschine *f* broaching machine
Raummaß *n* volumetric measure, cubic measure
Raummenge *f* volume
Räumnadel *f* broach
Räumnadelhalter *m* broach holder
Räumnadelziehmaschine *f* pull-type broaching machine
Raumpeilung *f* three-dimensional direction finding
Raumschiff *n* space ship
Räumschlitten *m* broach-handling slide
Räumstoßschlitten *m* broach-handling slide
Räumstoßschlitten *m* broach push slide
Raumstrahlung *f* space radiation
Räumvorrichtung *f* broaching fixture
Räumwerkzeug *n* broaching tool
Räumzahn *m* broach tooth
Raumzeit *f* space-time
raumzentriert *p.a.* (*cryst.*) body-centered

Räumzeug *n* external broach, surface broach
Räumzeugschleifmaschine *f* broach grinding machine
Räumzeugschlitten *m* broach slide, broach head
Räumziehkopf *m* broach pull head
Raupe *f (welding)* bead, run
Raupenblech *n* padded plate
Raupenfahrzeug *n* track-type vehicle, crawler-type vehicle, tracked vehicle
Raupenkette *f* track
Raupenschlepper *m* crawler-type tractor
Rauschen *n (radio)* noise; (im Mikrofon:) hissing; (e. Lichtbogens:) frying
Rauschpegel *m (electron.)* noise level
Rauschsender *m (radar)* noise transmitter
Rauschspannung *f (electron.)* noise potential
Rauschstörung *f* noise jamming
Raute *f (rolling mill)* diamond; *(geom.)* rhombus, lozenge
rautenförmig *a.* diamond-shaped; rhombic
reagieren *v.i.* react
Reaktion *f* reaction
reaktionsfähig *a.* reactive
reaktionsträge *a.* inert, inactive
Reaktionsträgheit *f* inertness, inactivity
Reaktionsturbine *f* reaction turbine
Reaktor *m* reactor
Reaktorleistung *f* reactor output
Reaktorsteuerung *f* reactor control
Rebschwarz *n* vegetable black
Rechenfehler *m* computational mistake, miscalculation
Rechengerät *n* calculator
Rechengröße *f* calculation factor

Rechenmaschine *f* calculating machine, computer
Rechenschieber *m* slide rule
Rechentafel *f* nomogram
rechnen *v.i.* calculate, compute, figure, reckon
Rechner *m* computer
Rechteck *n* rectangle
rechteckig *a.* rectangular
rechts *adv.* right, right-hand, right-handed, rightward
RECHTS ~, (~drall, ~drehung, ~flanke, ~gang, ~gewinde, ~meißel, ~schneide, ~schneidend, ~spirale, ~steuerung, ~verkehr) right-hand ~
rechtsgängig *a.* right-handed, right-hand, clockwise
Rechtslauf *m* clockwise rotation
rechtsläufig *a.* right-hand, clockwise
Rechtslenker *m* car with right-hand steering
Rechtsschweißung *f* rightward welding
rechtwinklig *a.* right-angled; (≈ rechteckig:) rectangular; square
rechtwinkliges Dreieck, right-angled triangle
reckaltern *v.t.* strain-age, strain-age-harden
Reckalterung *f* strain age hardening
recken *v.t.* stretch, strain, elongate, extend; *(hot work)* hammer-forge
Reckhammer *m* forging hammer
reckschmieden *v.t.* hammer-forge
Reckspannung *f* strain
Reduktionsschlacke *f (met.)* reducing slag; deoxidizing slag
reduzieren *v.t. (cold work)* (Kartuschhülsen:) reduce
Reduziergetriebe *n* speed reduction gear

Reduzierstück n reducer
Reduzierventil n reducing valve
Reflexschaltung f reflex circuit
Regal n shelving rack
Regel f rule, regulation
Regelanlasser m rheostatic starter
regelbar a. adjustable; controllable; *(speeds)* variable
regelbares Getriebe, variable speed gear drive
Regelbarkeit f adjustability, controllability; variability
Regelbauart f standard design
Regelgerät n control gear
Regelgetriebe n variable speed gear drive
Regelkreis m *(electr.)* control circuit; *(electron.)* input signal; *(programming)* feedback loop
Regelmotor m variable speed motor
regeln v.t. regulate; control; vary
Regelschalter m regulating switch
Regelscheibe f *(grinding)* regulating wheel, control wheel
Regelung f adjustment; regulation; *(traffic)* control
Regelungstechnik f automatic control technology
Regelventil n regulating valve
REGENERATIV ~, (~feuerung, ~flammofen, ~gasfeuerung, ~kammer, ~koksofen) regenerative ~
Regenfallrohr n rainpipe
regenfest a. rainproof
Regionalsender m regional broadcasting station
Registervergaser m *(auto.)* two-phase carburetter (*or* carburetor)
REGISTRIER ~, (~ballon, ~blatt, ~galvanometer, ~gerät, ~papier, ~streifen, ~trommel, ~vorrichtung) recording ~
registrieren v.t. record, register
Regler m controller; regulator; governor; (Wärme:) thermostat
Reglerwelle f *(auto.)* governor shaft
regulierbar a. adjustable, controllable
Regulierwiderstand m regulating resistance
REIB ~, (~antrieb, ~belag, ~kupplung, ~kupplungsgetriebe, ~säge, ~scheibe, ~spindelpresse, ~trieb) friction ~
Reibahle f reamer
reiben v.t. *(mach.)* ream
Reibesitz m transmission fit
reiblöten v.t. tin
Reibradgetriebe n friction gearing
Reibtrennsäge f circular friction saw
Reibtriebspindelpresse f friction screw press
Reibtriebspindelziehpresse f friction-screw-driven drawing press
Reibung f friction
REIBUNGS ~, (~drehmoment, ~elektrizität, ~kraft, ~kupplung, ~schluß, ~verlust, ~wärme) frictional ~
Reibungsarbeit f friction energy
Reibungsbelag m friction lining
Reibungsbremse f friction brake
Reibungsschweißen n friction welding
Reibungswiderstand m (e. Motors:) engine drag
Reichweite f range, reach
Reifen m *(US)* tire; *(UK)* tyre
Reifendecke f *(auto.)* outer cover
Reifendruck m *(auto.)* inflation pressure
Reifendruckmesser m tire gage
Reifenfüllflasche f tire inflation cylinder
Reifenheber m tire remover, rim tool

Reifenluftdruckprüfer *m* tire gage
Reifenmontierhebel *m* tire lever
Reifenpanne *f (auto.)* flat, puncture, blowout
Reifenprofil *n (auto.)* tread pattern
Reifenwalzwerk *n* tire mill
Reihe *f* row; *(speeds)* range; (≈ Reihenfolge:) sequence; *(math.)* progression, order; series
REIHEN ~, *(electr.)* (~kondensator, ~kreis, ~schaltung, ~spannung, ~widerstand) series ~
Reihenbohrmaschine *f* gang drilling machine
Reihenmotor *m (auto.)* in-line engine; *(electr.)* series-wound motor
Reihenparallelschaltung *f* series-parallel connection
Reihenpunktschweißung *f* straight-line spot welding, single-row spot welding
Reihenschlußerregung *f* series excitation
Reihenschlußmotor *m* series-wound motor
Reihenschlußstromerzeuger *m* series generator
Reihenschlußverhalten *n* series characteristics
Reinigung *f* cleaning, cleansing, washing; *(chemische R.:)* dry cleaning; *(Abwasser:)* purification
Reise *f* (e. Hochofens:) campaign
Reißbrett *n* drawing board
reißen *v.t.* rupture, break, tear; – *v.i.* rupture, fracture, disrupt, crack, burst, break, spring; (Holz:) check
Reißfeder *f* drawing pen
Reißfestigkeit *f* tensile strength; breaking strength
Reißlast *f* breaking load

Reißnadel *f* pocket scriber
Reißschiene *f* T-square
Reißspan *m* tear chip, fragmental chip, discontinuous chip
Reißstock *m* universal surface gage
Reißversuch *m* (Draht:) hanging test
Reißzeug *n* drawing instruments *pl.*
Reißzirkel *m* bow compass
Reitnagel *m* center sleeve
Reitstock *m (lathe)* tailstock
REITSTOCK ~, (~führung, ~körner, ~kurbel, ~oberteil, ~pinole, ~planscheibe, ~spindel, ~spitze, ~untersatz) tailstock ~
Reitstockausladung *f* tailstock overhang
Rekuperativofen *m* recuperative furnace
Relais *n* relay
RELAIS ~, (~klemme, ~kontakt, ~magnet, ~sender, ~spule, ~station, ~steuerung, ~übertragung, ~wicklung) relay ~
relative Dielektrizitätskonstante, relative permittivity
Relaxationsgenerator *m* relaxation generator
Relaxationskriechen *n* relaxation creep
Reliefschweißung *f* projection welding
Rennmotor *m* racing-car engine
Rennsport *m* motor racing
Rennstahl *m* bloomery steel
Rennstrecke *f* racing course
Rennverfahren *n* direct process (for the production of wrought iron)
Rennwagen *m* racing car
Rentabilität *f* profitability
Reparatur *f* repair
Reparaturbetrieb *m* repair shop
Reparaturbock *m* repair stand

Reparaturkostenstelle f maintenance cost center
Repulsionsmotor m repulsion motor
Reserve f reserve
Reservebatterie f spare battery
Reservekanister m *(auto.)* jerrycan
Reserverad n spare wheel
Reservereifen m *(auto.)* spare tire
Reservetank m reserve tank
Resonanz f resonance
RESONANZ ~, (~bereich, ~einfang, ~energie, ~feld, ~frequenz, ~kreis, ~leitwert, ~potential, ~relais, ~schaltung, ~schwingung, ~spannung, ~strahlung, ~transformator, ~verstärker, ~widerstand) resonance ~, resonant ~
Rest m rest, remainder; (≈ Rückstand:) residue
REST ~, (~entladung, ~feuchtigkeit, ~ladung, ~magnetismus, ~spannung, ~strom) residual ~
Retortenofen m retort furnace
Retortenverkokung f retort coking
Reversierblechwalzwerk n reversing plate mill
Reversierduowalzwerk n two-high reversing mill
Reversiermotor m reversible motor
Reversierstraße f reversing mill train
Reversierstreckgerüst n reversing stand of rolls for roughing
Reversierwalzgerüst n reversing rolling stand
Reversierwalzwerk n reversing mill
Revision f revision
Revisionslehre f reference gage
Revolver m *(turret lathe)* turret head, turret
Revolver ~, (~automat, ~bohrmaschine, ~drehmaschine, ~kopierdrehmaschine, ~nachdrehmaschine, ~schlitten) turret ~
Revolverkopf m turret head, turret
REVOLVERKOPF ~, (~bohrmaschine, ~schaltachse, ~schaltung, ~schlitten, ~verriegelung, ~zapfen) turret ~
Revolverpresse f dial feed press
Revolverteller m turret-type table
Revolvertisch m rotary table
Rheostat m rheostat
Richtamboß m straightening anvil
Richtantenne f directional antenna (or aerial)
richten v.t. *(cold work)* straighten; (Stangen, Draht:) roller-level; (Bleche:) stretcher-level; (Wellen, Rohre:) crossroll, reel; *(radio)* direct, guide, beam
Richter m *(electr.)* converter
Richtfunk m directional radio
Richtkosten pl. (= Plankosten) predetermined standard rates
Richtlineal n straightedge
Richtmaschine f straightening machine; (Bleche:) stretcher leveller
Richtpresse f straightening press; (für Rohre:) reeling machine
Richtschiene f parallel
Richtsender m directional transmitter
Richtstrahler m directional aerial
Richtstrahlung f directive radiation
Richtung f direction
Richtungsanzeiger m *(auto.)* direction indicator
Richtungsschalter m directional switch
Richtungsschild n *(traffic)* advance sign
Richtwalze f straightening roll
Richtwert m reference value; approximate value

Riefe f *(metal cutting)* score
riefig a. scored
Riegel m latch-bolt, locking bolt
Riemen m belt
RIEMEN ~, (~antrieb, ~fallhammer, ~fett, ~gabel, ~leitrolle, ~rutsch, ~scheibe, ~schlupf, ~spanner, ~spannrolle, ~spannung, ~trieb, ~verbinder, ~verdeck, ~zug) belt ~
Riemenschloß n round belt coupling
Riementrum n strand of a belt
Riemenvorgelege n countershaft
Riemenvorspannung f initial belt tension
Riffelblech n checkered sheet
Riffelblechwalzgerüst n checker mill
riffeln v.t. serrate; groove
Riffelverzahnung f serration
Riffelwalze f fluted roller
Rille f groove; (Schleifscheibe:) rib
Rillenkugellager n deep groove ball bearing
Rillenscheibe f grooved pulley
Rillenschiene f grooved rail
Rillenschwelle f *(US)* grooved tie; *(UK)* grooved sleeper
Ring m ring; (Draht:) coil; (e. Walze:) collar
Ringabstreifvorrichtung f coil stripper
Ringabwickelvorrichtung f coil holder
Ringbrenner m annular burner
Ringbuckel m *(welding)* annular projection
Ringglühofen m coil annealing furnace
Ringlager n cf. Radiallager
Ringleitung f *(blast furnace)* bustle pipe
Ringschlüssel m box wrench
Ringschmierlager n ring-oiling bearing
Ringschneide f (e. Schraube:) cup point
Ringschraube f lifting eye bolt

Ringsicherung f (für Kolbenbolzen:) circlip
Ringwaage f ring balance
Rinne f trough, channel, gutter
Rippenblech n ribbed pattern floor plate
Rippenkühler m *(auto.)* gilled radiator
Rippenrohr n *(auto.)* finned tube
Riß m *(drawing)* sectional drawing; (als Fehler:) crack
Rißfestigkeit f *(concrete)* resistance to cracking
Ritz m scratch; crack
Ritzel n pinion, driving gear
Ritzelwelle f pinion shaft
Ritzhärte f scratch hardness
Ritzhärteprüfer m scratch hardness tester
Rockwellhärteprüfung f Rockwell hardness test
Rohbau m *(building)* carcase
Rohblock m *(met.)* ingot
Rohdichte f gross density
Roheisen n pig iron; pig
Roheisenerzeugung f pig-iron production
Roheisen-Erz-Verfahren n pig iron-ore process
Roheisenmischer m hot-metal mixer
Roheisenpfanne f hot-metal ladle
Roheisenschrottverfahren n pig iron-scrap process
Rohgas n crude gas
Rohglas n rough cast glass
Rohhaut f rawhide
Rohkupfer n crude copper
Rohleinöl n raw linseed oil
Rohling m blank
Rohpetroleum n crude oil, petroleum
Rohr n (Gas-, Wasserleitungsrohr:) pipe; (aus Stahl, NE-Metallen, Glas, Zement:) tube

ROHR ~, (~anschluß, ~bogen, ~bruch, ~flansch, ~formstück, ~gewinde, ~kluppe, ~krümmer, ~legung, ~leitung, ~muffe, ~schelle, ~schlosser, ~schneider, ~schraubenschlüssel, ~schraubstock, ~verbindung, ~verlegung, ~zange) pipe ~
Rohrdraht m (electr.) conduit wire, leading-in wire
Röhre f (electr.) (US) tube; (UK) valve
RÖHREN ~, (~electr.) (~antenne, ~empfang, ~empfänger, ~fassung, ~frequenz, ~fuß, ~generator, ~gleichrichter, ~lampe, ~rauschen, ~sicherung, ~spannung, ~strom, ~ventil, ~verstärker, ~widerstand) (US) tube~; (UK) valve~
Rohrendenverjüngen n swaging
röhrengesteuert p.p. valve-controlled; electronic
Röhrenkühler m tubular radiator
Röhrenlibelle f tubular spirit level
Röhrenstreifen m (rolling) skelp
Röhrenwalzen n tube rolling
Röhrenwalzwerk n tube rolling mill
Rohrgerüst n tubular scaffold
Rohrknie n elbow
Rohrleitungsbau m pipeline construction
Rohrleitungsgraben m pipeline trench
Rohrmast m tubular mast, tubular pole
Rohrmühle f tube mill
Rohrniet m tubular rivet
Rohrschweißung f tube welding
Rohrverschraubung f screw union; screwed pipe joint
Rohrwelle f tubular shaft
rohrziehen v.t. (kalt:) sink; (warm:) hot draw
Rolladen m roller shutter

Rollbahn f (Lager:) raceway
Rollbahnwaage f roller-type conveyor balance
Rollbandmaß n tape-rule
Rolle f (Kettentrieb:) pulley, sheave
Rolleinrichtung f (Gewinde:) flat die rolling equipment (or attachment)
rollen v.t. (Gewinde:) roll
Rollenelektrode f (welding) roller-type electrode
Rollenherdofen m roller-hearth furnace
Rollenkäfig m roller cage
Rollenkettengetriebe n roller chain drive
Rollenkettenrad n roller chain sprocket
Rollenkranzlager n roller thrust ring bearing
Rollenlagerfett n roller bearing grease
Rollenlehre f roller screw gage
Rollenlünette f roller steady
Rollennaht f (welding) roller seam weld
Rollenpunktschweißmaschine f roller spot welding machine
Rollenrichtmaschine f roller leveller
Rollenrichtpresse f roller levelling machine
Rollenschrittverfahren n (welding) step-by-step seam welding, intermittent seam welding
Rollensetzstock m roller steady
Rollenspurlager n roller supported step bearing
Rollenstromabnehmer m trolley wheel
Rollfilm m roll film
Rollgabelschlüssel m adjustable nut wrench
Rollgang m (rolling mill) table roller
Rollgangwaage f roller bed balance
Rollgesenk n fuller
Rollherdofen m roller hearth furnace

rollieren *v.t.* (Oberflächen:) tumble
Rollkufe *f* rocker
Rollmaß *n* measuring tape
Rollofen *m* roll-over type furnace
Rollprofilieren *n* (Schleifscheiben:) roller crushing
Rollreibung *f* rolling friction
Rolltreppe *f* escalator
rommeln *v.t.* (Guß:) tumble, barrel
Ronde *f* circular blank
Röntgenaufnahme *f* radiograph
Röntgenbild *n* radiograph, X-ray photograph, roentgenogram
Röntgendurchleuchtung *f* radioscopy
Röntgeneinrichtung *f* X-ray outfit
Röntgenogramm *n* radiograph
Röntgenographie *f* X-ray photography
Röntgenologe *m* radiologist
Röntgenologie *f* radiology
Röntgenphotographie *f* radio-photography
Röntgenprüfung *f* X-ray examination
Röntgenröhre *f* X-ray tube, roentgen tube
Röntgenstrahl *m* X-ray
Röntgenstrahlung *f* X-ray radiation
Röntgenuntersuchung *f* X-ray examination, radiographic inspection
rosa *a.* pink
Rost *m* (Eisenrost:) rust; (Feuerrost:) grate; (Siebanlage:) grid
Rostanfressung *f* corrosion
Rostangriff *m* corrosive action, corrosive attack
Rostbeschickungsanlage *f* grate stoker
rostbeständig *a.* rust resisting
Rostbeständigkeit *f* resistance to corrosion, rust-resisting property
rosten *v.i.* rust, corrode, oxidize, become rusty

rösten *v.t.* roast; sinter; calcine
Rösterz *n* calcined ore
Rostfeuerung *f* grate firing
Rostfraß *m* corrosion
rostfreier Stahl, stainless steel
Rostnarbe *f* corrosion pit
Röstofen *m* drying kiln
Röstreaktionsverfahren *n* roasting reaction method
Röstreduktionsverfahren *n* roasting reduction method
Rostschutz *m* rust protection
Rostschutzfarbe *f* rust protection paint
Rostschutzmittel *n* rust preventative, rust preventive
rostsicher *a.* corrosion-resistant
rostsicherer Stahl, stainless steel
Rostung *f* rusting, corrosion
Röstung *f* roasting, calcination
Röstverfahren *n* roasting process, calcining method
Rotation *f* rotation, revolution
Rotbruch *m* red shortness, red brittleness
rotbrüchig *a.* red short
Rotbuche *f* beech
Rotglut *f* red heat, redness
Rotgluthärte *f* red-hardness
Rotgluthitze *f* red heat
Rotguß *m* red brass
Rotorwerkzeug *n* flexible drive tool
Rotwarmhärte *f* red-hardness
Rückblickspiegel *m* (auto.) driving mirror
Rücken *m* back; (e. Werkzeugschneide:) land
Rückenlehne *f* (auto.) seat back
Rückenwinkel *m* (e. Schneidzahnes:) clearance angle
Rückfahrleuchte *f* (auto.) reversing lamp
Rückfenster *n* (auto.) rear window

Rückfluß m return flow
ruckfrei a. smooth
Rückgang m return motion, return travel; (Walzgut:) return pass
rückgewinnen v.t. recover, reclaim
Rückgewinnung f recovery, reclamation
Rückhub m return stroke, backstroke
rückknallen v.i. (welding) backfire
rückkohlen v.t. recarburize
Rückkohlung f (met.) recarburization
rückkoppeln v.t. couple back, feed back
Rückkopplung f feedback, reaction
RÜCKKOPPLUNGS ~, (~kondensator, ~kreis, ~regelung, ~schaltung, ~spule, ~transformator, ~verstärker) reaction ~, feedback ~
Rücklauf m return movement, return; (≈ Rückhub:) return stroke; (von Flüssigkeiten:) recirculation; (telev.) kickback, flyback
Rücklaufimpuls m (telev.) flyback pulse
Rücklaufkupplung f reversing clutch
Rücklaufzeit f (telev.) retrace period
Rückleiter m (electr.) return conductor
Rückleitung f (electr.) return wire, return line; (Rohr:) return pipe
Rücklicht n (auto.) tail lamp
Rückprallhärteprüfung f rebound hardness test
Rückruf m (tel.) recall
Ruckschaltwerk n (Sperrgetriebe:) ratchet and pawl mechanism
Rückschlag m (welding) flashback; (auto.) backfiring
Rückschlagsicherung f (mach.) backlash eliminator
Rückschlagventil n non-return valve, check valve
rückseitig a. rear; – adv. rearward

Rücksitz m (auto.) back seat
Rückspiegel m driving mirror
Rückspiegelleuchte f (auto.) illuminated driving mirror
Rücksprung m rebound
Rücksprunghärte f rebound hardness
Rückstand m residue
Rückstoß m repulsion, rebound; (electron., nucl.) recoil
RÜCKSTOSS ~, (~atom, ~bahn, ~elektron, ~feder, ~potential) recoil ~
rückstrahlen v.i. reflect
Rückstrahler m (auto.) reflector
Rückstrahlung f reflection
Rückstrahlvermögen n reflecting power
Rückstrom m (electr.) return current; (Flüssigkeiten:) return flow, reflux
Rücktrittbremse f (bicycle) coaster brake
Ruckvorschub m (lathe) intermittent feed
Rückwand f rear wall
rückwärts adv. backward
rückwärtsfahren v.i. reverse
Rückwärtsgang m (auto.) reverse gear
Rückwärtsgeschwindigkeit f reverse speed
Rückwärtskipper m (US) end dump truck; (UK) end-tipping lorry
Rückwärtsschweißung f leftward welding, backhand welding
ruckweise a. intermittent
Rückwirkung f (electr.) reaction
Ruf m (tel.) call; ringing
Rufnummer f (tel.) subscriber's number
Rufzeichen n (tel.) ringing tone
Ruhekontakt m (electr.) break contact
ruhende Reibung, static friction
Ruhestellung f position at rest; rest position
Ruhezustand m state of rest

ruhig *a.* silent, quiet; smooth
ruhiger Lauf, (e. Maschine:) smooth running; (e. Getriebes:) silent running
Rührwerk *n* stirrer, agitator
rund *a.* round; circular; cylindrical
Rundblock *m* round ingot
Rundbuckel *m (welding)* circular projection
runderneuern *v.t.* (Autoreifen:) retread, recap (tires)
Rundfahrtwagen *m* sight-seeing car
Rundfräsen *n* circular milling
Rundfunk *m* broadcasting; (≈ Funk) radio
RUNDFUNK ~, (~ansager, ~empfänger, ~gerät, ~sender, ~spule, ~station, ~technik, ~übertragung, ~verstärker, ~welle) broadcasting ~, radio ~
Rundgesenk *n* blacksmiths' swage
Rundgewinde *n* round thread
Rundheit *f* roundness; cylindricity
Rundheitsprüfung *f* concentricity test
Rundhobeleinrichtung *f* radius planing attachment
Rundhobeln *n* radial planing
Rundholz *n* log
Rundknüppel *m (met.)* round billet
Rundkopiereinrichtung *f* circular copying attachment
Rundkopieren *n* circular profiling
Rundläppen *n* cylindrical lapping
Rundlauf *m* concentric running; concentricity
Rundlaufabweichung *f (gears)* radial run-out
rundlaufen *v.i.* rotate, revolve; run true, run concentric
Rundlauffräsen *n* rotary-table milling
Rundlauffräsmaschine *f* rotary miller
Rundlaufprüfung *f* concentricity test

Rundnaht *f* circumferential seam
Rundnahtschweißung *f* circular seam welding
Rundofen *m* mit Drehherd, circular furnace with rotating hearth
Rundpassung *f* cylindrical fit
Rundpassungslehre *f* cylindrical limit gage
Rundräummaschine *f* rotary broaching machine
Rundräumverfahren *n* rotary broaching method
Rundsäulenbohrmaschine round column drill
rundschleifen *v.t.* grind cylindrical
Rundschleifen *n* plain grinding, cylindrical grinding
Rundschleifmaschine *f* plain grinder, cylindrical grinder
Rundschweißnaht *f* circumferential seam
Rundschweißung *f* edge welding
Rundsichtverglasung *f (auto.)* panorama windscreen (*or* windshield)
Rundskala *f* dial
Rundstab *m* round bar, rod; – *pl.* rounds
Rundstahl *m* round bar steel
Rundstange *f* rod
Rundteiltisch *m* rotary indexing table
Rundtisch *m* circular table; rotary table
Rundtischfräsmaschine *f* rotary-table miller
Rundtischschaltmaschine *f* rotary indexing-table machine
Rundtischschleifmaschine *f* rotary table grinder
Rundung *f* roundness; radius
Runzellack *m* wrinkle finish
Ruß *m* soot, lampblack
rüsten *v.t. (mach.)* set up

Rüstung *f (building)* scaffolding
Rüstzeit *f* setting-up time, setup time
Rutsch *m* slip, slippage
Rutsche *f* chute
rutschen *v.i.* slip
rutschfest *a.* slip-proof
Rutschgefahr *f (auto.)* skidding risk
Rutschkupplung *f* friction safety clutch
Rutschpflaster *n* slippery set paving
rutschsicheres Reifenprofil, non-skid tread
Rüttelbeton *m* vibrated concrete
Rüttelformmaschine *f* jar-ram molding machine
Rüttelgerät *n (concrete)* vibrating equipment
rütteln *v.t.* shake; (Formsand:) jar, jolt
Rüttelpreßformmaschine *f* jar squeezer
Rüttelstampfer *m* vibrating tamper
Rütteltisch *m (concrete)* shaking table; *(molding)* jarring table, jolting table
Rüttelverfahren *n (molding)* jar ramming method
Rüttelwendeformmaschine *f* turnover-table jolter
Rüttler *m (concrete)* vibrator; *(molding)* jolter

S

sachkundig *a.* skilled, competent, expert
Sachverständiger *m* expert
Sackloch *n* blind hole, bottom hole
Säge *f* saw
Sägeautomat *m* automatic sawing machine
Sägeblatt *n* sawblade; (e. Bandsäge:) band; (e. Kreissäge:) web
Sägebügel *m* saw frame
Sägegatter *n* saw frame; (Maschine:) saw mill
Sägemaschine *f* machine saw, power saw
Sägemehl *m* saw dust
sägen *v.t.* saw; rip; cross-cut
Sägengewinde *n* buttress thread
Sägewerk *n* saw mill
Sägezahngenerator *m* sawtooth generator
Salmiak *m* sal ammoniac
Salmiakgeist *m* liquid ammonia
Salpeter *m* saltpeter
Salzbadeinsatzhärtung *f* salt bath case hardening
Salzbadhärtung *f* salt bath hardening
Salzbadnitrieren *n* salt bath nitriding
Salzbadofen *m* salt bath furnace
Salzsäure *f* hydrochloric acid
Sammelanschluß *m (tel.)* collective number
Sammelbehälter *m* collecting tank, reservoir
Sammelbunker *m* storage bin
Sammelkondensator *m* reservoir capacitor
Sammelleitung *f* collecting main
Sammelnummer *f (tel.)* collective number
Sammelschiene *f (electr.)* bus-bar, omnibus bar
Sammler *m (electr.)* storage battery
SAND ~, (~aufbereitung, ~aufbereitungsanlage, ~bestrahlung, ~blasdüse, ~form, ~grube, ~kern, ~mischer, ~papier, ~schleuderformmaschine, ~schüttelsieb) sand ~
Sandguß *m* (Prozeß:) sand mold casting; (Produkt:) sand castings
Sandkasten *m (shell molding)* dump box, investment bin
Sandpapier *n* abrasive paper, sandpaper
Sandputzmörtel *m* sanded plaster
sandschleifen *v.t.* sand
Sandstein *m* sandstone
Sandstrahlbläserei *f* sand-blasting
sandstrahlen *v.t.* sandblast
Sandstrahlgebläse *n* sandblast unit
Sandstrahlgebläsedüse *f* sandblast nozzle
Sandstrahlreinigung *f* sandblast cleaning
Sandstreifenbildung *f (concrete)* sand streaking
Sanitätsguß *m* castings for sanitary appliances
Sanitätskraftwagen *m* ambulance car
Sattdampf *m* saturated steam
Sattel *m (motorcycle)* saddle
Sattelanhänger *m* semi-trailer
Satteldach *n* saddle roof
Sattelrevolverdrehmaschine *f* ram-type turret lathe; *(UK)* capstan lathe
Sattelschlepper *m* truck-tractor
Sattelzugmaschine *f* tractor

Sättigung f saturation; *(metallo.)* concentration

Satz m sediment, deposit; set; gang

Satzbohrer m serial hand tap

Satzfräser m gang milling cutter

Satzkoks m coke per charge

Sau f *(met.)* furnace sow

Sauerstoff m oxygen

Sauerstoffanreicherung f oxygenization

Sauerstoffaufblaskonverter m top-blowing oxygen converter

Sauerstoffaufblasverfahren n top-blowing oxygen process

Sauerstoffblasstahlwerk n oxygen converter steel plant

₂₁₇**Sauerstoffbohren** n oxygen cylinder

Sauerstoffhobeln n oxygen deseaming

Sauerstoffkernlanze f *(steelmaking)* oxygen core lance

Sauerstoffkonverterstahl m oxygen-converter steel

Sauerstoffkonverterverfahren n oxygen steelmaking process

Sauerstofflanze f *(steelmaking)* oxygen lance

Sauerstoffpulverlanze f oxygen powder lance

Sauerstoffschneiden n oxy-cutting

Sauerstoffschweißung f oxy-hydrogen welding

Sauerstoffstrahl m oxygen jet

Sauerstoffstrahlrohr n oxygen lance

SAUG ~, (~gebläse, ~höhe, ~kolben, ~korb, ~pumpe, ~rohr, ~stutzen) suction ~

saugfähig a. absorptive

Sauggeräuschdämpfer m *(auto.)* intake muffler

Saugheber m siphon

Saugleitung f suction line, suction pipe; *(auto.)* intake manifold

Saugluftbremsung f *(auto.)* pneumatic vacuum braking

Saugmassel f *(founding)* sinking head, feeder

Saugmotor m *(auto.)* aspirating engine

Saugtrichter m (im Gußblock:) pipe

Saugventil n poppet valve, suction valve

Saugzuggebläse n induced draft fan

Säule f column, post, pillar; *(electr.)* pile

Säulenbohrmaschine f round-column drilling machine

Säulen-Friktionsspindelpresse f column screw press

Saum m edge, border; (e. Spektrums:) fringe

säumen v.t. edge, border; (Bleche:) square

Saumschere f trimming shear

Säure f acid

säurebeständig a. acid-proof, acid-resisting

säurefest a. acid-resistant

Säuregehalt m acid content

saures Futter, acid lining

saures Verfahren, *(met.)* acid process

schaben v.t. (von Hand:) scrape; (maschinell:) shave

Schaber m scraper

Schabfräser m shaving cutter

Schabhobel m spoke shave

Schablone f *(copying)* former plate; *(mach.)* template, templet; *(molding)* strickle: *(painting)* stencil; (≈Steuerkurve:) cam

Schablonendrehmaschine f copying lathe

Schablonendrehvorrichtung f form turning attachment
Schablonenformerei f template molding
Schablonenfräsmaschine f profile milling machine
schablonengesteuert a. cam-controlled
schablonieren v.t. *(molding)* strickle; *(painting)* stencil
Schabloniervorrichtung f *(molding)* sweeping device
Schabotte f *(forge)* anvil-block
Schabrad n *(gear cutting)* rotary shave cutter
Schacht m *(mining)* shaft; (e. Ofens:) stack; (Kanalisation:) manhole; (Kabel:) trench
Schachtbrunnen m excavated well
Schachtfutter n (e. Hochofens:) stack lining
Schachtmauerwerk n stack brickwork
Schachtofen m shaft furnace
Schachtpanzer m stack casing
Schaden m damage; disadvantage
Schadenersatzanspruch m claim for indemnification
schadhaft a. defective
schädlich a. harmful
Schaft m shank; (Schraube:) body; (Ventil:) stem
Schaftfräser m shank-type cutter; end milling cutter
Schaftmeißel m shank-type tool
Schaftritzel n shaft pinion
Schaftschraube f grub screw threaded part way
Schaftwerkzeug n shank tool
Schake f (e. Kette:) link
Schäkel n shackle
Schalbrett n *(scaffolding)* shuttering board

Schale f dish, cup; (Öl, Späne:) pan, tray; *(electron.)* shell; (Schmiedefehler:) scab
schälen v.t. *(lathe)* pre-turn; *(milling)* slab; (Gewinde:) whirl; (Schneckenräder:) skive
Schalengußverfahren n shell molding
Schalenhartguß m chill casting
Schalenkupplung f clamp coupling
schälfräsen v.t. slab-mill
Schälfräser m slab milling cutter
Schall m sound
schalldämpfend a. sound-absorbing
Schalldämpfer m *(auto.)* exhaust silencer
Schallfrequenz f audio-frequency
Schallgrenze f *(aeron.)* sonic barrier
Schallmesser m phonometer
Schallortung f sound ranging
Schallplatte f phonograph record, disc
Schallplattenaufnahme f disc recording
schallschluckend p.a. sound-absorbing
Schallschwingung f sound vibration
Schallsignal n acoustic signal
Schallwelle f sound wave
Schälmaschine f *(threading)* peeling machine
Schälmeißel m *(lathe)* skin turning tool, (Schneckenräder:) skiving tool
Schaltachse f (e. Revolverkopfes:) spindle
Schaltbild n wiring diagram, circuit diagram, connection diagram
Schaltbrett n switchboard; switch panel; *(auto.)* instrument panel
Schaltbrettleuchte f *(auto.)* switchboard lamp
Schalteinrichtung f *(electr.)* switchgear
schalten v.t. *(auto.)* change gears, shift gears; *(electr.)* switch; *(gears)* control;

(Hebel:) shift; (Kupplung:) clutch; (Maschine:) operate; (Motor:) start; stop; (Revolverkopf:) index; (Support, Tisch:) traverse

Schalter m *(electr.)* switch; circuit-breaker; (≈ Schaltorgan:) control [member]

Schaltgehäuse n *(auto.)* gearshift housing

Schaltgerät n *(electr.)* switchgear

Schaltgestänge n *(auto.)* gearshift linkage

Schaltgetriebe n control mechanism; *(auto.)* gearbox; *(gearing)* speed change gears; *(lathe)* feedbox; (Motor:) control gear

Schalthäufigkeit f *(electr.)* switching frequency

Schalthebel m control lever; *(auto.)* gearshift lever, change-speed lever

Schaltkasten m *(electr.)* switchbox

Schaltklaue f (Kupplung:) clutch dog

Schaltklinke f (e. Sperrades:) pawl

Schaltkontakt m switch contact

Schaltkupplung f cut-off coupling, clutch

Schaltleistung f *(electr.)* circuit breaking capacity

Schaltmagnet m solenoid

Schaltorgan n control member; *pl.* control gear

Schaltplan m wiring diagram

Schaltpult n control station

Schaltsäule f pillar-type switchgear

Schaltschema n wiring diagram, circuit diagram

Schaltschlüssel m *(auto.)* ignition key

Schaltschrank m switchgear cabinet

Schaltschütz n contactor

Schaltstange f *(auto.)* gear-shift rod

Schaltstation f control cubicle

Schaltstück n contact piece

Schalttafel f *(auto.)* instrument panel, dashboard; *(electr.)* switchboard

Schalttafelinstrument n switchboard instrument

Schaltung f control; connection; wiring; circuit; *(auto.)* gear changing; (Revolverkopf:) indexing [mechanism]

Schaltvermögen n *(electr.)* breaking capacity

Schaltvorrichtung f *(electr.)* switching device

Schaltwelle f *(auto.) (obs.)* s. Hauptwelle

Schaltzeit f *(electr.)* switching time; *(televl.)* picture feed interval

Schalung f *(building)* formwork, shuttering; boarding

Schalungsgerüst n *(building)* framework for shuttering

Schalungsrüttler m *(concrete)* form vibrator

Schamotte f refractory clay, fireclay

Schamottemörtel m fireclay mortar

Schamottestein m fire-brick, fireclay brick

Schärfe f sharpness; *(opt., televl.)* definition

Scharfeinstellung f sharp adjustment; *(opt., photo.)* focusing; *(radio)* sharp tuning

schärfen *v.t.* sharpen

Scharnier n hinge

Scharrierhammer m bush hammer

Schaubild n diagram

Schaufel f shovel; (Kohlen:) scoop; *(conveying)* loading bucket

Schaufelradbagger m rotary bucket excavator

Schauglas n sight-glass

schaukeln *v.i.* rock; swing

Schaukelofen m tilting furnace, tipping furnace

Schaumbeton *m* foam concrete
schäumen *v.i.* effervesce; (Schlacke:) foam
Schaumfeuerlöscher *m* foam-type fire extinguisher
Schaumgummi *n* foam rubber
Schaumschlacke pumicestone slag, foamed slag
Schaumschlackenbeton *m* foamed slag concrete
Schaumstoff *m* expanded material, foamed material, foamed plastics; *(colloq.)* foam rubber; *(allg.)* plastic foams
Schaumton *m* foam clay
Scheibe *f* disc, disk; plate; (Glas:) pane; (≈ Meßscheibe:) dial; (Radscheibe:) wheel web; (Riemenscheibe:) pulley; (Seilscheibe:) sheave; (Schleifscheibe:) wheel; (Unterlegscheibe:) washer; (Antriebsscheibe:) pulley; (Läppscheibe:) lapping wheel
Scheibenbremse *f (auto.)* disc brake
Scheibendichtung *f* sheet gasket
Scheibenentfroster *f (auto.)* windscreen defroster
Scheibenfeder *f* Woodruff key
Scheibenfräser *m* side milling cutter
Scheibenkupplung *f* flange coupling
Scheibenrad *n (auto.)* disc wheel
Scheibenwalze *f* center disc roll
Scheibenwascher *m (auto.)* windscreen washer
Scheibenwischer *m (auto.)* windscreen wiper
Scheider *m* separator
Scheinleistung *f (electr.)* apparent power
Scheinleistungsmesser *m* volt-ampere meter
Scheinleitwert *m* admittance

Scheinverbrauchszähler *m* volt-ampere-hour meter
Scheinwerfer *m (auto.)* headlamp
Scheinwiderstand *m* impedance
Scheitel *m (geom.)* vertex; (Kurve:) peak
Scheitelpunkt *m* apex
Scheitelspannungsmesser *m* peak voltmeter
Schellack *m* shellac
Schelle *f* (Röhre:) clip, (Kabel:) clamp
Schellhammer *m* set hammer, snap tool
Schenkel *m* (Kurbelwelle:) web; (Magnet:) limb; (Taster, Winkelstahl, Zirkel:) leg; (Winkel:) side
SCHER ~, (~beanspruchung, ~festigkeit, ~kraft, ~messer, ~modul, ~spannung, ~stift) shear ~
Schere *f* shear; (für Wechselräder:) quadrant; *(miller)* overarm braces
scheren *v.t.* shear, cut
Scherenheber *m (auto.)* articulated jack
Scherenwagenheber *m* compound leverage floor jack
Scherspan *m* continuous chip
scheuern *v.t.* clean; scour; (Guß:) tumble
Scheuertrommel *f* tumbling barrel; (Drahtfabrikation:) scouring barrel
Scheuerwirkung *f* abrasive action
Schicht *f* (≈ Lage:) layer; (≈ Deckschicht:) coat; (Film:) film; (Arbeit:) shift; *(building)* course
Schichtpreßstoff *m* molded laminated plastics; laminated plastic material
Schichtpreßstoff-Erzeugnis *n* laminated product
SCHIEBE ~, (~dach, ~fenster, ~keil, ~rad, ~rädergetriebe, ~sitz, ~tisch, ~tür) sliding ~
Schiebebühne *f* transfer table

Schiebemaßstab *m* two-way rule
Schieber *m* (als Meßzeug:) slide rule; (≈ Schlitten:) slide; (≈ Ventil:) slide valve
Schiebezahnrad *n* slip gear
Schieblehre *f* slide caliper rule
schief *a.* oblique; (≈ geneigt:) inclined; (≈ windschief:) skew
Schiefer *m* slate
Schieferbruch *m* fibrous fracture
Schieferdach *n* slate roof
Schieferkalk *m* sparry limestone
Schieferkreide *f* graphitic clay
Schieferöl *n* shale oil
Schiene *f* (electr.) busbar; (metrol.) rule, blade; (railway) rail
SCHIENEN ~, (~bohrmaschine, ~fuß, ~profil, ~steg, ~stoß, ~strang, ~strom, ~stuhl, ~verlaschung, ~walzwerk) rail ~
Schienenbiegemaschine *f* rail cambering machine
Schienenbiege- und Richtmaschine *f* rail bending and straightening machine
Schienenbus *m* rail diesel car
Schienenfahrzeug *n* rail-borne vehicle
Schienenfahrzeugbau *m* rail-borne vehicle construction
Schienenherzstück *n* built-up frog
Schienenkreuzungsstück *n* built-up crossing
Schienennagel *m* rail spike, dog spike
Schienenschwelle *f* tie; sleeper
Schienentrio *n* three-high rail mill
Schiffbau *m* shipbuilding
Schiffbau-Wulstwinkel *m* shipbuilding bulb angle
Schiffsblech *n* shipbuilding plate
Schiffskessel *m* marine boiler
Schiffsrumpffarbe *f* hull paint

Schiffsschraube *f* ship propeller
Schild *n* (für Drehzahlen:) index plate
Schindel *f* (building) shingle
Schirm *m* (magn.) shield; (radio, telev.) screen
Schirmbild *n* (radar) image, pattern
Schirmgitter *n* (radio) screen grid
Schlacke *f* slag, cinder, clinker; dross
SCHLACKEN ~, (~abstichrinne, ~beton, ~bims, ~einschluß, ~faser, ~grube, ~halde, ~pfanne, ~sand, ~stein, ~stichloch, ~wagen, ~wolle, ~zement, ~ziegel) slag ~
Schlackenfrischreaktion *f* (Martinverfahren:) lime boil
schlaff *a.* (Riemen, Seil:) slack
Schlag *m* blow, impact; shock; (e. Seils:) spin, lay
Schlagarbeit *f* (Prüfwesen) energy of blow, impact energy
schlagartig *a.* sudden, abrupt
Schlagbär *m* striking hammer, tup, impact anvil
Schlagbeanspruchung *f* impact stress
Schlagbiegeversuch *m* impact bending test
Schlagdauerversuch *m* endurance impact test
Schlagdruckversuch *m* impact compression test
schlagen *v.t.* strike; – *v.i.* (Kurbel:) knock; (Welle:) whip, chatter; (Schwungrad:) run out of truth
Schlagfestigkeit *f* impact strength
Schlagfolgezeit *f* (forging) sequence period of blows
Schlagfräser *m* fly-cutter
Schlaghärteprüfer *m* impact hardness tester

Schlagkreuzmühle f beater mill
Schlagloch n *(road building)* pothole
Schlaglot n brazing solder
Schlagmühle f beater mill
Schlagnietmaschine f percussion riveter
Schlagpressen n *(plastics)* impact molding
Schlagpreßschweißen n press welding
Schlagschotter m broken stone
Schlagschraube f drive screw
Schlagschweißung f percussion welding
Schlagversuch m impact test; falling weight test, drop test
Schlagwerk n impact testing machine
schlagwettergeschützt a. flame-proof
Schlagzahnfräsen n fly-cutting
Schlagzerreißversuch m impact tension test
Schlamm m mud, sludge; (Aufbereitung:) pulp; (Elektrolysenbetrieb:) slime
Schlämme f slurry
schlämmen v.t. (Erze:) wash, slime; *(chem.)* elutriate, levigate
Schlammventil n mud valve
Schlange f (Kühl-, Heizschlange:) coil
Schlangenbohrer m auger bit
Schlauch m hose
Schlauchkupplung f hose coupling
Schlauchtülle f tail end
schlauchloser Reifen, *(auto.)* tubeless tire
Schlaufe f loop
Schleichgang m inching
Schleichvorschub m inching feed
Schleifapparat m grinding attachment
Schleifautomat m automatic grinder
Schleifband n abrasive belt
schleifbar a. grindable

Schleifdorn m wheelhead
Schleifdorn m grinding arbor
Schleifdraht m *(electr.)* slide wire
Schleifdrahtkompensator m *(electr.)* slidewire potentiometer
Schleifdrahtmeßbrücke f slide-wire bridge
Schleifdruck m wheel pressure
Schleife f *(rolling mill)* loop
Schleifeinrichtung f grinding attachment
schleifen v.t. grind; (mit Schmirgel:) sand; (≈ schärfen:) sharpen; (≈ wetzen:) whet; (Ventilkegel:) reface; – v.i. drag, rub, bear
Schleifgrat m burr
Schleifkontakt m sliding contact
Schleifkopf m wheelhead
Schleiflack m flatting varnish
Schleiflehre f grinding gage
Schleifleinen n abrasive cloth
Schleifleistung f (e. Maschine:) grinding capacity; (e. Scheibe:) cutting action
Schleifleitung f sliding contact line
Schleifmaschine f grinder, grinding machine
Schleifmittel n abrasive
Schleifmittelpaste f abrasive compound
Schleifmotor m wheel motor
Schleiföl n cutting oil
Schleifpapier n abrasive paper
Schleifpaste f grinding compound
Schleifpolieren n flexible grinding
Schleifring m *(electr.)* slip ring
SCHLEIFRING ~, (~anker, ~läufer, ~spannung, ~strom) slip ring ~
Schleifringläufermotor m slip-ring induction motor
Schleifriß m check mark
Schleifscheibe f grinding wheel

Schleifscheibenschutzhaube f wheel guard
Schleifschlamm m sludge, swarf
Schleifschlitten m wheel carriage
Schleifspäne mpl. swarf
Schleifspindelkopf m wheelhead
Schleifspuren fpl. grit marks
Schleifstaub m abrasive grit
Schleifstein m grindstone
Schleifstift m (tool) pencil wheel
Schleifsupport m wheelhead carriage
Schleiftasse f cup wheel
Schleiftopf m cup wheel
Schleifvorrichtung f grinding attachment
Schleifwirkung f abrasive action
Schlepper m tractor
Schleppertriebwerk n tractor transmission
Schleppschrapper m power drag scraper
Schleppschweißung f touch welding
Schleppseil n (auto.) tow rope
Schlepptau n hauling rope
Schlepptisch m drag table
Schleppwalze f (road building) tractor-drawn roller; (rolling mill) dummy roll, drag roll, idle roll
Schleuderbeton m centrifugally cast concrete, spun concrete
Schleuderbetonrohr n spun concrete pipe
Schleuderformmaschine f (foundry) slinger
Schleudergießmaschine f centrifugal casting machine
Schleudergießverfahren n centrifugal casting process
Schleuderguß m centrifugal casting
Schleudergußrohr n centrifugally cast pipe, spun pipe
Schleudergußstück n centrifugal casting

Schleudergußverfahren n centrifugal casting process
schleudern v.t. centrifuge; – v.i. (auto.) skid, swerve
Schleuderpumpe f centrifugal pump
Schleuderschmierung f centrifugal lubrication
Schleudertauchschmierung f forced flood lubrication
Schlichte f (molding) blackwash
schlichten v.t. (spanlos:) smooth, dress; (zerspanend:) finish, finish-machine
Schlichthobel m smooth plane
Schließkopf m (e. Niets:) closing head, closing end
Schließkopfgesenk n snap head die
Schliff m grinding; grinding operation; (Oberfläche:) grinding finish; (e. Werkzeuges:) grind
Schlinge f (e. Seils:) loop
Schlingkette f sling chain
Schlitten m (e. Werkzeugmaschine:) slide; (lathe) saddle, carriage; (miller, planer) saddle; (saw) carriage; (≈ Räumschlitten:) ram; (e. Bohrmaschine:) boring slide
Schlittenführung f carriage guideways; saddle slideways; ramways; s.a. Schlitten
Schlittenrevolverdrehmaschine f saddle-type turret lathe; (UK) combination turret lathe
Schlittenvorschub m (lathe) carriage feed; (drill) headstock feed; (miller) suddle feed
Schlitz m slot; slit
schlitzen v.t. groove; slot; split
Schlitzfräser m slitting cutter, slotting cutter

Schlitzmutter f slotted round nut
Schlitzsäge f slotting saw
Schloß n lock; (Vorhängeschluß:) padlock; (e. Seils:) joint; (e. Leitspindel:) halfnuts; *(auto.)* door lock
Schlosser m locksmith; (Rohr:) fitter
Schlosserhammer m engineers' hammer
Schloßkasten m *(mach.)* apron
Schloßmutter f *(lathe)* split nut, half-nuts pl.
Schloßschraube f square neck carriage bolt
Schlupf m slippage, slip
schlupffrei a. free from slip
Schlupfregler m slip regulator
Schlüssel m (≈ Rohrschlüssel:) wrench spanner; (Kosten:) ratio system
Schlüsselschraube f lag screw
Schlüsselweite f width across flats
Schlußleuchte f *(auto.)* tail lamp
Schlußstein m *(building)* keystone
Schmalbandstraße f narrow strip mill
Schmalspurgleis n narrow-gage track
schmelzbar a. fusible
SCHMELZ ~, *(electr.)* (~draht, ~einsatz, ~leiter, ~patrone, ~sicherung) fuse ~
Schmelzdrahtsicherung f wire fuse
Schmelze f melt; heat; (Birnenbetrieb:) blow
schmelzen v.t. & v.i. melt; (Erz:) smelt
Schmelzer m melter; (Birnenbetrieb:) blower
Schmelzfluß m flux
schmelzflüssiger Stahl, molten metal
Schmelzgang m *(met.)* heat
Schmelzgut n melting charge
Schmelzhütte f smelting plant
Schmelzpunkt m melting point; fusing point

schmelzschweißbar a. fusion weldable
Schmelzschweißung f fusion welding
Schmelzschweißverfahren n fusion welding process
Schmelztiegel m crucible
Schmelzung f melt, heat; fusion
Schmied m blacksmith, smith
schmiedbar a. forgeable
Schmiedbarkeit f forgeability
Schmiede f (Betrieb:) forge; (handwerklich:) smithy
SCHMIEDE ~, (~gesenk, ~matrize, ~ofen, ~presse, ~riß, ~spannung) forging ~
Schmiedeamboß m anvil block
Schmiedeeisen n cf. Schmiedestahl
Schmiedegrat m flash
Schmiedehammer m forging hammer; blacksmiths' hand hammer
Schmiedemessing n hot-working brass
schmieden v.t. (industriell:) forge; (im Gesenk:) drop forge; (von Hand:) smith
Schmiederohling m rough forging
Schmiedestahl m forging grade steel
Schmiedestück n forging
Schmiedeteil n forging
Schmiedewerkstatt f forge; smith's shop, smithy
Schmiedezunder m hammer scale
Schmiege f carpenters' bevel
Schmierapparat m lubricator
schmieren v.t. lubricate; (Öl:) oil; (Fett:) grease; – v.i. load, stick
Schmierloch n oil hole
Schmierlötung f plumbers' wiping
Schmiernippel m lubricating nipple
Schmiernut f oil groove
Schmieröl n lubricating oil

Schmierpresse f grease gun
Schmierpumpe f lubricating pump
Schmierschraube f oil plug
Schmierseife f soft soap
Schmierstoff m lubricant
Schmiertechnik f lubrication technology
Schmierung f lubrication; (Fett:) greasing; (Öl:) oiling
SCHMIRGEL ~, (~leinen, ~papier, ~pulver, ~scheibe, ~staub) emery ~
schmirgelartig a. abrasive
schmirgeln v.t. sand
Schmirgelschleifmaschine f sander
Schmutz m dirt
Schmutzfang m (mach.) splash guard
Schmutzfänger m (auto.) mud flap
schmutzig a. dirty
Schnapper m (rolling mill) catcher
Schnappschalter m quick-break switch
Schnauze f (e. Pfanne:) lip, nozzle
Schnecke f (gearing) worm
Schneckenförderer m worm conveyor, screw conveyor
Schneckenfräsautomat m automatic worm hobbing machine
Schneckengetriebe n worm gearing, worm transmission, worm gear mechanism
Schneckenrad n worm gear, worm wheel
Schneckenradantrieb m worm wheel drive
Schneckenradgetriebe n worm gearing
Schneckenschleifmaschine f worm thread grinder
Schneckenantrieb m worm-gear drive
Schneckenwälzfräser m worm generating hob
Schneckenzahnstange f worm rack

Schneekette f snow chain; tire chain, nonskid chain
schneidbrennen v.t. flame-cut
Schneidbrenner m cutting torch
Schneiddruck m (e. Meißels:) tool thrust
Schneide f cutting edge; (e. Fräsers:) tooth
Schneideinrichtung f (Gewinde:) screw-cutting attachment, thread-cutting attachment; (als Gewindebohreinrichtung:) tapping attachment
Schneideisenhalter m die stock
Schneideisenkopf m die head
schneiden v.t. cut; shear; (Gewinde:) thread; – v.r. intersect
Schneidenwinkel m (e. Meißels:) cutting tool angle
Schneidfähigkeit f cutting quality
Schneidfase f (e. Spiralbohrers:) margin, heel
Schneidhaltigkeit f edge-holding quality
Schneidkante f cutting edge
Schneidkantenrücken m land
Schneidkluppe f screw plate stock, die stock
Schneidlade f miter box
Schneidlegierung f cutting alloy
Schneidlippenfase f (Spiralbohrer:) heel
Schneidmaschine f (Gewinde:) screw-threading machine; tapping machine
Schneidöl n cutting oil
Schneidplatte f (Hartmetall:) carbide tip
Schneidrad n (shaper) shaper cutter; (gear hobber) pinion-type cutter
Schneidrücken m (e. Werkzeuges:) land
Schneidschraube f self-tapping screw
Schneidspäne mpl. cuttings

Schneidwerkzeug *n* cutting tool
Schnellamt *n (tel.)* quick-service toll exchange
Schnellarbeitsstahl *m* high-speed steel
Schnellaufdrehmaschine *f* high-speed lathe
Schnellgang *m (auto.)* overdrive, top gear; (e. Maschinentisches:) rapid traverse; (eines Getriebes:) high-speed motion
Schnellschalter *m* quick-break switch
Schnellschaltgetriebe *n* quick-change gear
Schnellschnittstahl *m* high-speed steel
Schnellspannfutter *n* quick-action collet chuck
Schnellspannschlüssel *m* monkey wrench
Schnellstahl *m* high-speed steel
Schnellverstellung *f* (Schlitten, Tisch:) power rapid traverse
Schnellwaage *f* express scales
Schnellwechselgetriebe *n* quick-change gear drive
Schnellweg *m (traffic)* expressway
Schnitt *m* cutting, cut; *(tool)* cutting die; blanking die; *(drawing, geom.)* section
Schnittbau *m* die-making
Schnittbild *n* sectional view
Schnittdruck *m* tool thrust
Schnittfläche *f* (e. Werkstückes:) work surface; (≈ Hauptschnittfläche:) shoulder of the cut; *(drawing)* sectional area
Schnitthaltigkeit *f* edge-holding quality
Schnittholz *n* timber
Schnitthub *m* cutting stroke
Schnittkraft *f* cutting thrust
Schnittleistung *f* cutting power

Schnittplatte *f* cutting die
Schnittpresse *f* cutting press, punching press
Schnittpunkt *m* point of intersection
Schnittstanze *f* punching press
Schnittstempel *m* blanking, punching and cutting die
Schnittwinkel *m* (e. Meißels:) true cutting angle
Schnittzeichnung *f* sectional drawing
schnitzen *v.t.* carve, cut
Schnur *f (electr.)* cord, flexible lead
Schnurrolle *f* cord pulley
Schnurschalter *m* pendant switch
schonen *v.t.* spare; protect; preserve
Schonganggetriebe *n (auto.)* overdrive
Schonung *f* care
schopfen *v.t. (rolling)* crop
Schöpfprobe *f (met.)* ladle sample
Schopfsäge *f (met.)* crop end saw
Schopfschere *f* end shears
Schornstein *m* chimney, stack
Schornsteineinfassung *f* chimney weathering
Schotter *m* crushed stone, coarse gravel; (für Gleisanlagen:) ballast
Schotterbeton *m* ballast concrete
Schotterbrecher *m* coarse crusher
schraffieren *v.t.* cross-hatch
schräg *a.* oblique; inclined; angular; (von Kanten:) bevelled
Schrägaufzug *m* inclined hoist
Schrägmaß *n* (Meßzeug:) steel protractor
Schrägrampe *f* sloping bench
Schrägrollenlager *n* angular contact bearing
Schrägrost *m* inclined grate
schrägschleifen *v.t. (woodworking)* bevel

Schrägschnitt *m* diagonal cutting; diagonal cut

schrägstehend *a.* inclined

schrägstellbar *a.* inclinable; tiltable, angular adjustable

Schrägungswinkel *m* (e. Gesenkschräge:) draft angle

schrägverstellbar *a.* angular adjustable, inclinable; tiltable

Schrägverzahnung *f* helical tooth system

Schrägwalze *f* cross roll

Schrägwalzverfahren *n* roll piercing process

Schrägwalzwerk *n* rotary piercing mill

Schrägwinkel *m* (Meßzeug) [universal] bevel

Schrägzahn *m* helical tooth

Schrägzahnkegelrad *n* helical tooth bevel gear

Schrägzahnrad *n* helical-toothed gear, helical gear; (US auch:) spiral gear

Schrank *m* cabinet; (Säge:) set; (für Werkzeuge:) chest; (Getriebe:) compartment

Schranke *f (railway)* gate

schränken *v.t.* (Sägezähne:) set

Schränkung *f* set

Schränkzange *f* saw set plier

Schrapper *m* scraper

Schrapperkübel *m* scraper bucket

Schrapperwinde *f* scraper hoist

Schraubautomat *m (electr.)* screw-in circuit-breaker

Schraube *f* (≈ Kopfanziehschraube:) screw; (≈ Durchsteckschraube:) bolt; (e. Schraublehre:) spindle-screw

SCHRAUBEN ~, (~anzieher, ~bock, ~gewinde, ~klaue, ~nut, ~pumpe, ~schlitz, ~schlitzfräser, ~schlitzsäge, ~schneidmaschine, ~schneidwerkzeug, ~spindel, ~stahl, ~verbindung, ~winde, ~zieher) screw ~

Schraubenautomat *m* automatic screw machine

Schraubenbolzenschneider *m* bolt clipper

Schraubendrehmaschine *f* screw-cutting lathe

Schraubenfeder *f* helical spring, coil spring

Schraubenhersteller *m* bolt maker, bolt manufacturer

Schraubenlinie *f* helix; – *pl.* helices

Schraubenrad *n*, **Schrägzahnrad**, *n*, helical gear; (US auch:) spiral gear

Schraubenrädergetriebe *n* helical gear transmission

Schraubenschaft *m* screw body, bolt shank

Schraubenschlüssel *m* wrench; *(UK)* spanner

Schraubfassung *f* screw base

Schraubgetriebe *n* skew gear transmission

Schraubknecht *m* cabinet clamp

Schraubkontakt *m* screwed contact

Schraublehre *f* micrometer caliper

Schraublehrenstichmaß *n* inside micrometer

Schraubrad *n* skew gear

Schraubrohrmuffe *f* screwed pipe joint

Schraubsicherung *f* screw-in type fuse

Schraubstock *m* vise; *(UK)* vice

Schraubtiefenlehre *f* micrometer depth gage

Schraubzwinge *f* C-clamp

Schreiber *m* (= registrierendes Meßgerät) recording instrument

Schreibtelegraph *m* teleprinter
Schreiner *m* joiner; (≈ Zimmermann:) carpenter
Schreinerwinkel *m* carpenters' try square
Schrittmacherofen *m* walking-beam type furnace
schrittschalten *v.t.* inch, jog
Schrittschaltwerk *n* (*electr.*) step-by-step switch; (*mach.*) inching control mechanism
Schrittwähler *m* (*tel.*) step-by-step selector
schrittweise *adv.* gradually, step by step
Schrotkugeln *fpl.* shot
Schrotsäge *f* crosscut saw
Schrott *m* scrap; waste
SCHROTT ~, (~händler, ~lagerplatz, ~paketierpresse, ~platz, ~schmelze, ~zerkleinerungsanlage) scrap~
schrumpfen *v.i.* shrink; contract
Schrumpfpassung *f* shrink fit
Schrumpfriß *m* shrinkage crack; contraction crack
Schrumpfung *f* contraction; shrinkage
Schrumpfungsriß *m* (*founding*) check-crack
schrupp ~, (~bohren, ~drehen, ~fräsen, ~läppen, ~räumen, ~schleifen) rough ~
SCHRUPP ~, (~fräser, ~geschwindigkeit, ~leistung, ~maschine, ~meißel, ~raumnadel, ~reibahle, ~reihe, ~scheibe, ~schleifmaschine, ~schnitt, ~span, ~tiefe, ~vorschub, ~wälzfräser, ~werkzeug) roughing ~
Schruppautomat *m* automatic roughing lathe
schruppen *v.t.* rough, rough-machine
Schrupphobel *m* jack plane

Schub *m* shear; (axialer:) thrust
Schubfestigkeit *f* shearing strength
Schubkarren *m* wheelbarrow
Schublehre *f* slide caliper rule
Schubmodul *m* modulus of shear
Schubspannung *f* shearing strain, shear stress
Schuh *m* (Kabel:) socket
Schukosteckdose *f* socket with protective plug reception
Schukostecker *m* shrouded contact plug
Schulter *f* (*mach.*) shoulder; collar
Schulterkugellager *n* magnet-type ball bearing
schuppig *a.* flaky, flaked, scaly
Schüreisen *n* poker
Schürloch *n* poke hole, stoke hole
Schurre *f* chute
Schuß *m* (Gewebe:) weft
Schüttbeton *m* cast concrete
Schüttdichte *f* apparent density, bulk density
Schüttelrinne *f* shaking trough, reciprocating trough
Schüttelsieb *n* shaking screen, shaking sifter, riddle
Schüttgewicht *n* bulkweight
Schüttgut *n* bulk material
Schütttrichter *m* feed hopper
Schutz *m* (Späne:) guard
Schütz *n* (*electr.*) contactor; (*hydr. eng.*) sluice gate
Schutzanzug *m* (*welding*) protective clothing
Schutzblech *n* guard
Schutzbrille *f* safety goggles
schützen *v.t.* protect, guard
Schützensteuerung *f* contactor control
Schutzerdung *f* protective earthing

Schutzgas n *(heat treatment)* controlled atmosphere
Schutzgas-Lichtbogenschweißung f protective gas arc welding
Schutzgasschweißung f protective gas welding, shielded arc welding
Schutzgitter n fence; *(auto.)* radiator grill
Schutzhaube f protection hood, guard; *(welding)* helmet
Schutzkappe f *(welding)* helmet
Schutzkontakt m protective contact
Schutzkontaktsteckdose f socket with protective plug reception
Schutzkontaktstecker m shrouded contact plug
Schutzmarke f trademark
Schutzmaßnahme f preventive measure, protective measure
Schutzschalter m protective switch
Schutzschicht f protective cover, protective coat, safety film; *(welding)* protective layer
Schutzschiene f *(welding)* shielding plate
Schutzsicherung f protected fuse
Schutzüberzug m preservative coating, protective coating
schwabbeln v.t. buff
Schwachgas n lean gas, weak gas
Schwachstrom m weak current, low-voltage current
Schwachstromkabel n communication cable
Schwachstromsteckdose f low-voltage socket
Schwachstromstecker m low-voltage plug
Schwachstromtechnik f communication engineering; *(US)* signal engineering
schwachwandig a. thin-walled

Schwalbenschwanzführung f dovetail guide
Schwallwasserschutz m hose-proof enclosure
schwanken v.i. vary, fluctuate
Schwankung f variation; fluctuation; *(electr.)* undulation
Schwarzblech n black sheet, black plate
Schwarzbruch m black brittleness, black shortness
schwarzbrüchig a. black-short
Schwärze f *(molding)* black wash
Schwarzfärben n black-finishing
schwarzglühen v.t. black anneal
Schwarzguß m all-black malleable cast iron
Schwarzkalk m magnesian limestone
Schwarzkerntemperguß m black heart malleable
Schwarzsender m non-licensed transmitter
Schwärzungsmesser m densitometer
Schwebe f suspension
Schwebefrequenz f beat frequency
Schwebung f *(wave mech.)* beat
Schwebungsempfänger m beat receiver
Schwefel m sulphur
Schwefelaufnahme f (im Schmelzbad:) sulphur pickup
Schwefelentfernung f desulphurizing
schwefelhaltig a. sulphur-bearing
Schwefelkies m iron pyrites
Schwefelsäure f sulphuric acid
Schwefelwasserstoff m hydrogen sulphide
Schwefelzink n zinc sulphide
Schwefelzink n zinc sulphide
Schwefelzinkweiß n lithopone
schweflige Säure, sulphurous acid

schweifsägen v.t. curve
SCHWEISS ~, (~aggregat, ~anlage, ~brenner, ~brille, ~draht, ~elektrode, ~fehler, ~gut, ~güte, ~lage, ~naht, ~perle, ~schlacke, ~stoß, ~temperatur, ~transformator, ~umspanner, ~vorrichtung, ~wärme) welding ~
Schweißautomat m automatic welding machine
schweißbar a. weldable
Schweißbarkeit f weldability, welding property
Schweißdraht m welding wire, filler rod, welding rod
Schweiße f weld
schweißen v.t. weld
Schweißer m welder
Schweißerei f welding shop
Schweißgang m welding pass
Schweißhitze f (met.) wash heat
Schweißkohle f carbon electrode
Schweißlöten n braze-welding
Schweißmaschine f welder
Schweißnahtrissigkeit f susceptibility of welding seams to cracking
Schweißpaste f welding flux
Schweißraupe f bead, welding run
Schweißrißempfindlichkeit f sensitivity to weld cracking
Schweißschirm m handshield
Schweißspannung f (electr.) welding arc voltage
Schweißstab m welding rod, filler rod, welding electrode
Schweißstahl m wrought iron; (Stahl in Schweißgüte:) steel suitable for welding
Schweißstelle f point of weld
Schweißüberhöhung f weld reinforcement
Schweißumformer m arc welding converter
Schweißumspanner m welding transformer
Schweißung f welding; weld
Schweißung von oben, downhand weld
Schweißverbindung f welded joint
Schweißwurzel f root of the weld
Schweißzange f electrode holder
Schwelanlage f low-temperature carbonizing plant
Schwellbelastung f pulsating stress
Schwelle f (building) sill; (railway) sleeper, tie; (fig.) threshold
Schwellenverlegemaschine f (US) tie laying machine; (UK) sleeper laying machine
Schwellenwert m threshold value
Schwellfestigkeit f pulsating fatigue strength
Schwelofen m low-temperature carbonizing furnace
schwenkbar a. swivelling; swinging; (≈ kippbar:) tiltable
Schwenkbarkeit f swivelling feature
Schwenkbereich m (e. Auslegers:) (US) boom swing; (UK) jib swing
Schwenkbohrmaschine f radial drilling machine
schwenken v.t. swing; swivel; tilt
Schwenkhebel m pivoted lever, tumbler lever
Schwenkkonsole f swivel knee
Schwenkkran m jig crane, swing crane
Schwenklager n pivot bearing
Schwenkmeißelhalter m swivel tool block
Schwenkmöglichkeit f swivelling feature
Schwenkpunkt m pivot
Schwenkrad n tumbler gear

Schwenkradgetriebe n quick-change gear mechanism
Schwenkschalter m pendant switch
Schwenktisch m swivel table
Schwenkung f swivel
Schwerbeton m heavy concrete
Schwerebeschleunigung f acceleration due to gravity
Schwergewichtsachse f center-of-gravity axis
Schwerkraft f gravity
Schwerlasttransportanhänger m heavy duty trailer
Schwermetall n heavy metal
Schweröl n heavy oil
Schwerwerkzeugmaschine f heavy-duty machine tool
schwimmender Estrich, floating mastic
Schwimmer m (Vergaser:) float
Schwimmerventil n (auto.) float valve
Schwimmkran m floating crane
schwinden v.i. shrink; contract; (radio) fade
Schwindhohlraum m shrinkhole
Schwindmaß n shrink rule
Schwindmaßzugabe f shrinkage allowance
Schwindung f shrinkage; contraction
Schwindungsriß m shrinkage crack
Schwingachse f floating axle, swinging axle
Schwingdurchmesser m (lathe) swing
Schwinge f rocker arm; (≈ Schwenkhebel:) tumbler lever
Schwinghebel m (slotter) rocking lever
Schwingkreis m (radio) oscillating circuit
Schwingkurbel f crank arm
Schwingkurbelgetriebe n crank mechanism

Schwingsieb n shaking screen
Schwingtisch m (concrete) vibrating table
Schwingung f (freie:) swing, oscillation; (konstanter Frequenz:) vibration
Schwingungsbeanspruchung f cyclic stress
schwingungsdämpfend a. vibration damping
Schwingungsdämpfer m vibration damper; (auto.) shock absorber
Schwingungserreger m oscillator, exciter
Schwingungserregung f vibration excitation
Schwingungsfestigkeit f dynamic strength
Schwingungsmesser m vibrometer
Schwingungsprüfmaschine f vibratory testing machine
Schwingungsschreiber m vibrograph
schwingungstechnisch a. dynamic
Schwingungsversuch m repeated stress test
Schwingungsweite f amplitude of oscillation
schwitzen v.i. (road building) bleed
Schwitzwasser n condensation water
Schwund m contraction; (radio) fading control, volume control
Schwungkraftanlasser m (auto.) inertia starter
Schwungmagnetzünder m flywheel magneto
Schwungmoment n moment of inertia
Schwungrad n flywheel
Schwungradanlasser m flywheel starter
Sechseck n hexagon
sechseckig a. hexagonal
Sechselektrodenröhre f hexode

Sechskantholzschraube f *(UK)* hexagon head coach screw

Sechskantrevolverkopf m hexagon turret head

Sechskantschraube f hexagon head screw

Sechskantsteckschlüssel m hexagon socket wrench

Sechspolröhre f hexode

Sechsspindelautomat m six-spindle automatic machine

sechsspurige Autostraße, six-lane highway

Sechszylindermotor m six-cylinder engine

Seegerring m *(auto.)* circlip

Seegerzange f piston pin retention pliers

Seekabel n submarine cable

Seele f (Elektrode, Kabel, Seil:) core; Spiralbohrer:) center web

Segeltuch n canvas

Segeltuchverdeck n *(auto.)* canvas top

Segment n segment

Segmentkreissäge f inserted tooth metal slitting saw

Segmentlenkung f *(auto.)* worm and sector steering

Segmentsägeblatt n segmental saw blade

Sehfeld n field of view

Sehne f *(techn.)* chord

Sehnenmaß n chordal measure

Seifenlösung f soap solution

Seifenzinn n stream tin

Seigerung f *(met.)* liquation, segregation

Seigerungsstreifen m *(metallo.)* ghost lines

Seil n rope

SEIl ~, (~führungsrolle, ~kloben, ~rolle, ~scheibe, ~schlag, ~schlinge, ~schloß, ~schlupf, ~spannung, ~trommel, ~winde, ~zug) rope ~

Seilkausche f thimble

Seilstrang m strand of rope

Seiltrieb m rope drive

Seilzugwaage f tackle line balance

Seitenansicht f side view

Seitenband n *(radio)* side band

Seitenentleerer m side discharge car

Seitenkipper m side-dump car, side-dump truck

Seitenmeißel m side cutting tool

Seitenriß m side elevation

Seitenschneider m side cutting plier

Seitenstrahler m *(auto.)* side reflector

Seitensupport m *(planer)* sidehead; *(grinder)* side toolbox; *(lathe)* side carriage

Seitenumkehr f *(tel.)* lateral inversion

seitenverkehrt a. inverse

seitlich a. lateral; adv. sideways

seitlich blasender Konverter, sideblown converter

SEKUNDÄR ~, (~elektron, element, ~emission, ~spannung, ~strahlung, ~strom, ~wicklung) secondary

selbstansaugend a. self-priming

selbstansaugender Motor naturally aspirated engine

Selbstanschluß m *(tel.)* automatic telephone

Selbstauslöser m *(electr.)* automatic release

Selbstentlader m self-discharging truck, dump truck, automatic tipper, self-unloader

Selbstentzündungsmotor m compression-ignition engine

selbsterregte Schwingung, *(metal cutting)* self-induced vibration
Selbstgang *m* (e. Tisches:) power feed
selbstgehendes Erz, self-fluxing ore
Selbsthärter *m* self-hardening steel
selbsthemmend *a.* self-locking; irreversible
Selbstkosten *pl.* prime cost
Selbstschalter *m* automatic circuit breaker
selbstschließend *a.* self-locking
selbstschmierend *a.* self-lubricating
selbsttätig *a.* automatic
selbsttätiger Eilgang, power rapid traverse
selbsttätiger Vorschub, power feed
Selbstwähler *m* (tel.) automatic telephone
Selbstwählferndienst *m* long-distance dialing
selbstzentrierend *a.* self-centering
Selbstzündung *f (auto.)* self-ignition; (Diesel:) compression ignition
Selengleichrichter *m* selenium rectifier
Selenzelle *f* selenium cell
SENDE ~, (~band, ~bereich, ~frequenz, ~gerät, ~kanal, ~kreis, ~leistung, ~reichweite, ~relais, ~röhre, ~schaltung, ~station, ~verstärker, ~welle) transmission ~, transmitting ~
senden *v.t. (radio)* transmit
Sender *m* wireless station, broadcasting station, transmitter
Senderaum *m* studio
Sendung *f (radio)* transmission, broadcast
senken *v.t.* lower; (Walzen:) screw down; (Schraubenköpfe:) countersink
Senkmutter *f* countersunk nut
SENKRECHT ~, (~abhebung, ~außenräummaschine, ~bewegung, ~bohrmaschine, ~dreh- und bohrwerk, ~feinbohrwerk, ~führung, ~innenräummaschine, ~konsolfräsmaschine, ~läppmaschine, ~räummaschine, ~revolverdrehmaschine, ~schnitt, ~spindel, ~verstellung, ~vorschub, ~zahnradstoßmaschine, ~zustellung) vertical ~
Senkrechtfräsmaschine *f* vertical spindle miller
Senkrechtschleifmaschine *f* vertical spindle grinder
Senkrechtstoßmaschine *f* slotting machine, slotter
senkrecht verstellen, adjust vertically
Senkschraube *f* countersunk bolt; countersunk screw
Serie *f* series; (≈ Partie:) batch, lot; (Produktion:) run
Serienbau *m* series-production
Serienbohren *n* repetition drilling
Serienschalter *m* multi-circuit switch
Serienschaltung *f* series connection
Servobremse *f* servo brake, power brake
Servolenkung *f (auto.)* servo-assisted steering mechanism
Servomotor *m* servo-motor
Servosteuerung *f* servo-control
Setzbett *n* settling tank
Setzkopf *m* (e. Niets:) die head, set head
Setzmaschine *f* jigging machine
Setzpacklagestein *m* *(road building)* blockstone, base stone
Setzriß *m* (e. Bodenbelages:) settlement crack
Setzstock *m* steadyrest; *(boring mill)* boring stay
Setzstockbacke *f* steady jaw, steady-rest shoe
Setzstocklager *n* stay bearing

Setzstockständer

Setzstockständer m *(boring mill)* end support column
Sicherheit f safety; reliability; dependability
SICHERHEITS ~, (~auslöser, ~faktor, ~glas, ~gurt, ~koeffizient, ~kupplung, ~kurve, ~maßnahme, ~organ, ~schalter, ~schalthebel, ~schloß, ~ventil, ~vorschub) safety ~
Sicherheitszündschlüssel m ignition safety key
sichern v.f. secure; lock; safeguard
Sicherung f safety feature; *(electr.)* fuse
SICHERUNGS ~, *(electr.)* (~dose, ~kasten, ~patrone, ~schalter, ~tafel) fuse ~
Sicherungsautomat m automatic cut-out
Sicherungsblech n tab washer
Sicherungsstift m locking pin
sichtbar a. visible, visual
Sichtbehinderung f visual obstruction
Sichtbeton m exposed concrete
Sichtfeld n field of view
Sichtkontrolle f visual control
Sichtverhältnisse npl. visibility
Sichtweite f sight distance, visual range
Sicke f bead
sicken v.t. bead
Sickenmaschine f beading machine
sickern v.i. seep, percolate, trickle; leak
Sickerwasser n seepage water
Sieb n strainer; sieve; screen; *(radio)* filter
SIEB ~, *(radio)* (~drossel, ~kondensator, ~korb, ~kreis, ~schaltung, ~spule) filter ~
sieben v.t. screen; sieve, riddle; *(radio)* filter
Siebtrommel f drum screen, rotary screen
Siebung f screening; *(radio)* filtering

Sintereisen

Siedehitze f boiling heat
Siedekurve f boiling-point curve
Siederohr n boiler tube
Siegellack m sealing wax
Siemens-Martin-Flußstahl, m open-hearth steel, Siemens-Martin-steel
Siemens-Martin-Ofen m *(US)* open-hearth furnace; *(UK)* Siemens-Martin furnace
Siemens-Martin-Roheisen n open-hearth pig [iron]
Siemens-Martin-Stahl m open-hearth steel
Siemens-Martin-Verfahren n open-hearth process
Signalhorn n horn
Signalleuchte f indicator lamp
Signalsummer m buzzer
Signalwesen n *(railway)* signalling
Sikkativ n siccative, dryer
Silberelektrolyse f electrolytic silver refining
Silberfolie f silver foil, silver leaf
Silbergewinnung f silver extraction
Silbersand m sea sand, beach sand
Silberscheideanstalt f silver refinery
Silberstahl m *(US)* Stub's steel; *(UK)* silver steel
Silikastein m silica brick
Silikatschlacke f silicate slag
Silizium n silicon
Silo m silo, storage bin
Simmerring m oil-seal ring
Simshobel m rebate plane
sinnfällig a. logical; coordinate; sensible
Sinter m scale, cinder, sinter; agglomerate
Sinteranlage f sintering plant, agglomerating plant
Sintereisen n sintered-powder metal

Sinterhartmetall *n* sintered carbide metal
Sinterkarbid *n* cemented carbide
Sintermetall *n* sintered-powder metal
sintern *v.t.* sinter, frit; vitrify; – *v.i.* clinker, bake, cake, sinter
Sinterröstung *f* blast roasting, sinter roasting
Sinterschlacke *f* clinker
Sinterwagen *m (rolling mill)* roll scale car
Sitz *m* seat; (Passungen:) fit
Sitzfläche *f* (e. Ventils:) seat face
Sitzversteller *m (auto.)* seat adjuster
Skala *f* scale; graduation
Skalascheibe *f* graduated dial
Skalenmeßgerät *n* direct-reading instrument
Skalenring *m* graduated collar
Skalenscheibe *f* graduated dial
Skalenstrich *m* graduation line
skalieren *v.t.* graduate
Skelettbau *m* skeleton structure
Skizze *f* sketch; draft
skizzieren *v.t.* draft; plot
SM-Flußstahl *m* O.H. ingot steel
Sockel *m (mach.)* base; (≈ Ständer:) pedestal; (≈ Winkeltisch:) knee; (≈ Fassung:) mount; (e. Kontaktanschlusses:) socket
Sockelschalter *m* socket switch
Soda *n* soda
Sodalösung *f* soda solution
Soffitte *f* tubular lamp
Soffittenlampe *f (auto.)* festoon lamp
Sohle *f* sole, floor level, base, bottom, level, floor
Sohlplatte *f* baseplate, bottom plate, bed plate
SOLL ~, (~durchmesser, ~frequenz, ~länge, ~maß, ~wert) nominal ~, theoretical ~, rated ~
Sollkosten *pl. (cost accounting)* standard practice budget
Soll-Leistung *f (mach.)* rated output; *(work study)* planned performance
Sollzeit *f (time study)* allocated time
Sonderzubehör *n* (e. Maschine:) extra equipment
Sonnenblende *f (auto.)* anti-dazzle vizor
Sonnenenergie *f* solar energy
Sonntagseisen *n* week-end metal
Sorgfalt *f* care
sortieren *v.t.* assort, sort, grade, classify
Sortierwaage *f* sizing balance
Soziussattel *m* pillion seat
Spachtel *m (painting)* filler, surfacer
Spachtelmasse *f* mastic, putty compound
Spalt *m* gap
Spaltaxt *f* single bit axe
Spaltbandsäge *f* ripping band saw
Spaltbrüchigkeit *f (cryst.)* cleavage brittleness
spalten *v.t.* split
Spaltfilter *m (lubrication)* edge filter
Spaltkreissäge *f* circular rip saw
Spaltlöten *n* close joint brazing
Spaltprodukt *n (nucl.)* fission product
Spaltsäge *f* framed ripping saw
Spaltung *f* splitting; *(nucl.)* fission
Span *m (mach.)* chip; cut; *(woodworking)* shavings pl.
 einen ~ nehmen, take a cut
Spanabfluß *m (mach.)* escape of chips, discharge of chips
Spanabfuhr *f* removal of chips
spanabhebende Bearbeitung, cutting, machining
Spanabnahme *f* metal removal

Spanauffangraum *m* chip compartment
Spanbildung *f* chip formation
Spanbrecher *m* chip breaker
Spanbrust *f* rake
SPÄNE ~, (~abfuhr, ~abscheider, ~auffangraum, ~fall, ~fangschale, ~fluß, ~förderer, ~pfanne, ~raum, ~rutsche, ~schale, ~schutz, ~verstopfung, ~wanne) chip ~
spanen *v.t.* machine, cut
spanende Werkzeugmaschine, metal-cutting machine tool
Spanentfernung *f* chip removal
Spanfläche *f* (e. Drehmeißels:) true rake, cutting face
Spanleistung *f* (e. Maschine:) cutting capacity, metal removing capacity
spanlose Bearbeitung, non-chip forming
spanlose Formgebung, non-cutting shaping, shape flowing
Spannbacke *f* (e. Futters:) gripping jaw
Spannbeton *m* pre-stressed concrete
Spannbolzen *m* tension bolt
Spanndorn *m* draw-in arbor
Spanndraht *m* span wire
spannen *v.t.* clamp; (Werkstücke:) load; (in e. Futter:) chuck; (e. Feder:) tension; (Schrauben, Riemen, Kette:) tighten
Spanner *m* clamping fixture; (Kette:) tightener
Spannflansch *m* mounting flange
Spannfutter *n* chuck
Spannhülse *f* (Lager:) adaptor sleeve
Spannhülsenlager *n* adapter bearing
Spannklaue *f* *(lathe)* [lathe] dog; (Spannpratze:) clamping jaw
Spannkluppe *f* *(woodworking)* dog vise
Spannmuffe *f* clamping sleeve

Spannmutter *f* locknut
Spann-Nut *f* T-slot
Spannpatrone *f* draw-in collet
Spannplatte *f* (Motor:) hinged baseplate; (Presse:) bolster plate
Spannrolle *f* (Riementrieb:) idler pulley, tightener pulley
Spannsäge *f* web saw
Spannschiene *f* (Motor:) slide rail
Spannschloß *n* *(tool)* turnbuckle; *(concrete)* locking device
Spannschraube *f* holding-down bolt
Spannung *f* ($\approx$ Beanspruchung:) stress; strain; (Dampf:) pressure, tension; (Kette, Riemen, Seil:) tension; pull; *(electr.)* voltage
spannungführend *a.* *(electr.)* live
Spannungsabfall *m* *(electr.)* voltage drop
Spannungsanzeiger *m* voltage indicator
Spannungs-Dehnungs-Diagramm *n* stress-strain diagram
Spannungsfeld *n* *(electr.)* electric field
Spannungsfestigkeit *f* dielectric strength, disruptive strength
spannungsfrei *a.* *(electr.)* dead
spannungsfrei glühen *v.t.* stress-relieve
Spannungsfreiglühen *n* stress-relieving anneal
Spannungsgefälle *n* potential drop
Spannungskorrosion *f* stress corrosion
spannungslos *a.* *(electr.)* dead
Spannungsmesser *m* *(electr.)* voltmeter; *(mat. test.)* strain gage; *(road building)* strain meter
Spannungsoptik *f* photoelasticity
Spannungsprüfer *m* *(electr.)* potential tester
Spannungsregler *m* voltage regulator, potential regulator

Spannungsreihe f contact series
Spannungsriß m tension crack
Spannungsteiler m *(electr.)* voltage divider
Spannungsunterschied m *(electr.)* potential difference
Spannungswandler m potential transformer, voltage transformer
Span-Nut f flute
Spannvorrichtung f *(mach.)* workholding fixture
Spannweite f span
Spannzangeneinsatz m draw-in collet
Spannzangenfutter n draw-in collet chuck
Spannzeug n clamping tool, clamping fixture
Spantiefe f depth of cut
Spanung f metal cutting
Spanungsgröße f metal-cutting element
Spanwinkel m rake angle
Spanzustellung f *(grinding)* infeed adjustment
Sparbeize f inhibitor
Sparcatron-Verfahren n spark cutting process
Sparren m *(building)* rafter
Spateisenstein m spathic iron ore
Spaten m spade
Speichenrad n spoke wheel
Speicher m storeroom; accumulator; *(programming)* store; magazine
Speicherung f storage
Speiseleitung f *(electr.)* feeder line
speisen v.t. feed; supply
Speisewasser n feed water
Spektralanalyse f spectrum analysis
Sperrad n ratchet wheel
Sperrbeton m water-repellent concrete
sperren v.t. arrest; stop; lock; block

Sperrflüssigkeit f sealing liquid
Sperrfurnier n plywood veneer
Sperrgetriebe n ratchet gearing
Sperrholzplatte f plywood panel, plywood sheet
Sperriegel m locking bolt
sperrig a. bulky
Sperrklinke f pawl
Sperrklinkenrad n ratched wheel
Sperrschichtwirkung f *(lubrication)* blocking effect
Sperrschwinger m *(telev.)* blocking oscillator
Sperrstörung f *(telev.)* barrage jamming
Sperrung f *(mech.)* locking mechanism; trip gear; *(radar)* blocking
Sperrwirkung f locking action
Sperrzahn m pawl
Spezialnutmotor m double-squirrel cage motor
Spezialprofil n special section
spezifischer Durchgangswiderstand, volume resistivity
spezifizieren v.t. specify; detail
Sphäroguß m nodular iron, spherolitic cast iron, spheroidal graphite cast iron
Spiegel m *(forging)* wad; *(opt.)* reflector; (e. Flüssigkeit:) level; (S. M. – Ofen:) bulkhead
Spiegelbild n reverse image, mirror image
Spiegeleisen n spiegel, specular pig iron
Spiegelfrequenz f image frequency
Spiegelinstrument n mirror instrument
spiegelverkehrt a. mirror-inverted
Spiegelwirkung f *(nucl.)* mirror effect
Spiel n play, clearance; (≈ Totgang:) backlash
spielfrei a. free from clearance; (Gewinde:) free from backlash

Spielpassung *f* clearance fit
Spielraum *m* clearance
Spielschwankung *f* permissible variation
Spießkantkaliber *n* diamond pass
Spin *m (nucl.)* spin
SPIN ~, *(nucl.)* (~bahn, ~drehimpuls, ~elektron, ~moment, ~verdopplung, ~verteilung) spin ~
Spindel *f* spindle; screw spindle, screw
SPINDEL ~, (~bremse, ~bund, ~drehzahl, ~ende, ~hülse, ~lager, ~nase, ~öl, ~pinole, ~rad, ~trommel, ~umdrehung, ~vorschub) spindle ~
Spindelgetriebe *n* geared spindle drive
Spindelkasten *m* headstock
Spindelkopf *m (lathe)* spindle nose; *(grinder)* wheelhead
Spindelpresse *f* screw press
Spindelschlagpresse *f* (≈ Friktionsprägepresse) percussion press
Spindelstock *m* spindle head, head; *(driller)* drill head; *(grinder)* wheelhead
Spindelstockgetriebe *n* headstock gearing
SPIRAL ~, (~antenne, ~bahn, ~bohrer, ~feder, ~kegelrad, ~nut, ~schleifen, ~verzahnung, ~zahn) spiral ~
Spiralbewehrung *f* helical reinforcement
Spiralbohrerschleiflehre *f* drill point gage
Spiralbohrerschleifmaschine *f* twist drill grinder
Spiralbohrerspitzenschleifmaschine *f* drill point grinder
Spirale *f* spiral; (Drall:) twist; (schraubenförmige:) helix
spiralig *a.* spiral; curly; helical
Spiritus *m* alcohol
Spiritusbeize *f* spirit stain
Spirituslack *m* spirit varnish

spitz *a.* pointed; sharp; (Winkel:) acute
Spitze *f* point; (e. Diamanten:) nib; (e. Drehmeißels:) nose; (e. Gewindezahnes:) tip, crest; (e. Kurve:) peak; (e. Schraube:) cone point; (≈ Körnerspitze:) center; *(geom.)* apex
Spitzenabstand *m (lathe)* distance between centers
Spitzenbelastung *f (electr.)* peak load
Spitzendrehmaschine *f* center lathe; *(UK)* centre lathe
Spitzenhöhe *f (lathe)* height of centers
spitzenlose Schleifmaschine, centerless grinder
spitzenlos schleifen, centerless grind
Spitzenschleifen *n* center grinding
Spitzenspannung *f* peak voltage
Spitzenspiel *n* (e. Gewindes:) crest clearance
Spitzenstrom *m* peak current
Spitzenverkehr *m (auto.)* rush-hour traffic
Spitzenweite *f (lathe)* distance between centers
Spitzenwinkel *m* included angle of point; (e. Drehmeißels:) nose angle; (e. Spiralbohrers:) lip angle
Spitzgewinde *n* sharp thread, angular thread, V-thread
Spitzhacke *f* pick
Spitzkerb *m* V-notch
Spitzsenker *m* countersink
spitzwinklig *a.* acute-angled
Spitzzirkel *m* dividers *pl.*
spleißen *v.t.* (e. Seil:) splice
Splint *m* [split] cotter pin; *(shop term)* cotter
Splintholz *n* sapwood
Splitt *m* broken stone, chipping
Splitter *m* (Diamanten:) chip

Splittstreuer m *(road building)* chip spreader, chipper
Sportkabriolett n convertible coupé
Sportwagen m sports car
Sportzweisitzer m roadster
Sprechfunk m radiotelephony
Sprechgerät n *(radio)* telephone
Sprechkreis m *(tel.)* speaking circuit
Sprechverkehr m *(tel.)* telephone traffic
Spreizband n (e. Bremse:) expanding band, friction band
Spreize f *(building)* prop
Spreizreibahle f solid expansion reamer
Spreizringkupplung f expanding-ring clutch
sprengen v.t. burst, blow up, blast; *(concrete)* blast, shoot
Sprengkapsel f blasting cap
Sprengkraft f explosive force
Sprengladung f bursting charge
Sprengmeister m blaster
Sprengniet m explosive rivet
Sprengpulver n gunpowder
Sprengring m *(auto.)* retaining ring, circlip
Sprengringlehre f retainer ring gage
Sprengstoff m explosive
Sprengzünder m detonating fuze, blasting fuze
Spritlack m spirit varnish
Spritzalitieren n alumetizing, aluminum-coating by spraying
Spritzbewurf m *(building)* sprayed roughcast
Spritzblech n splash guard
Spritzdraht m spray wire
Spritzdüse f spray nozzle
Spritze f (Öl:) gun; squirt; injector
spritzen v.t. *(painting)* spray; *(plastics)* inject

Spritzer m spatter; (durch Mattschweiße:) scabs pl.
spritzgießen v.t. *(plastics)* injection mold
Spritzgußform f *(plastics)* injection mold
Spritzgußmaschine injection molding machine
Spritzgußmasse f *(plastics)* injection molding compound
Spritzgußstoff m injection molded material
Spritzgußtechnik f injection molding technique
Spritzgußteil n *(plastics)* injection molded article, injection molded part; injection molding
Spritzgußverfahren n injection molding process
Spritzgußwerkzeug n *(plastics)* injection mold
Spritzlack m spraying varnish
Spritzmaschine f *(concrete)* gun
Spritzmetallisierung f metal spraying
Spritzpistole f spray gun
Spritzpresse f injection molding press
Spritzpressen n *(plastics)* transfer molding
Spritzpreßform f transfer mold
Spritzpreßwerkzeug n *(plastics)* transfer mold
Spritzschmierung f splash lubrication
Spritzvergaser m atomizing carburetor
Spritzverlust m *(welding)* spatter loss
Spritzversteller m injection timing device
Spritzwasserschutz m splash-proof enclosure
spröde a. brittle
Sprödigkeit f brittleness
Spruch m *(radio)* message
sprühen v.t. spray; – v.i. scatter
Sprühentladung f corona discharge

Sprühstrom m corona discharge current
Sprung m jump; (≈ Riß:) crack
Sprungschaltung f jump feed
Sprungvorschub m (e. Maschinentisches:) jump feed, skip feed
Sprungzeit f (electron.) transition time
Spule f (electr.) coil, solenoid; (mach.) reel
spülen v.t. wash, rinse, clean; (Motor:) scavenge
Spulenwicklung f coil winding
Spulenwiderstand m coil resistance
Spülluft f (auto.) scavenging air
Spülöl n wash oil
Spülschlitz m (auto.) scavenging port
Spundbohle f sheet pile
spunden v.t. (woodworking) groove and tongue
Spundhobel m match plane
Spundung f (building) sheeting
Spundwand f steel sheet iling
Spundwandramme f sheeting driver
Spundwandrammung f sheet-pile driving
Spundwandschloß n (civ.eng.) interlock
Spundwandstahl m steel sheet piling
Spur f (chem.) trace: (nucl., radar, traffic) track
spuren v.i. (auto.) keep in track
Spurhaltigkeit f track rod arm
Spurkranz m wheel-flange
Spurmesser m (auto.) track gage
Spurnagel m track spike
Spurstange f (auto.) track rod
Spurweite f (auto.) wheel track; (railway) track gage
Spurzapfen m pivot-journal
Stab m bar; (≈ Stange:) rod
Stabanker m (electr.) bar-wound armature
Stabantenne f rod antenna (or aerial)

Stabbatterie f torch battery, dry pile
Stabbewehrung f (concrete) bar reinforcement
Stabfeder f (auto.) torsion bar spring
stabil a. stable, strong, sturdy, rigid; heavy; steady
stabilisieren v.t. (met.) stabilize
Stabilisierung f stabilization
Stabilität f stability, sturdiness, rigidity
Stablampe f electric torch
Stabmagnet m bar magnet
Stabmaterial n bar steel, bar stock
Stabstahl m bar steel
Stabstahlwalzwerk n bar mill
Stabwicklung f bar winding
Stacheldraht m barbed wire
Stadium n stage
Städtebau m municipal engineering
staffeln v.t. stagger
Stahl m steel
  beruhigter ~, killed steel
  halbberuhigter ~, semi-killed steel
  niedriglegierter ~, alloy-treated steel
  unberuhigter ~, rimming steel, rimmed steel
STAHL ~, (~bandmaß, ~bau, ~blech, ~draht, ~drahtbürste, ~erzeugung, ~feder, ~gießer, ~gießpfanne, ~güte, ~hochbau, ~lineal, ~maßstab, ~muffe, ~rohr, ~schrott, ~skelettbau, ~spundbohle, ~werk, ~winkel, ~wolle) steel ~
Stahlanker m (e. Presse:) tie rod
Stahlbandbewehrung f steel band armoring
Stahlbeton m reinforced concrete
Stahlbetonbalken m reinforced concrete beam
Stahlbetonbau m reinforced concrete construction (or structure)

Stahlbetonbohle f reinforced concrete sheet pile
Stahlbetondecke f reinforced concrete floor
Stahlbetonrippendecke f reinforced concrete ribbed floor
Stahlbetonskelett n reinforced concrete carcase
Stähleschleifmaschine f tool and cutter grinder
Stahlformgießerei f steel-casting foundry
Stahlformgußstück n steel casting
Stahlgießerei f steel-casting foundry, steel foundry, steel works
Stahlgliedermaßstab m folding steel rule
Stahlguß m cast steel; steel castings
Stahlkies m steel shot
Stahlkonstruktion f structural steelwork
Stahlmatte f reinforcing steel mat, fabric reinforcement
Stahlrohrgerüst n tubular steel scaffold(ing)
Stahlsand m steel grit, steel shot
Stahlsandhinterschüttung f (shell molding) shot backup
Stahlschweißkonstruktion f welded steel construction
Stahlwaren pl. steel-ware
Stahlwerker m steel-worker, steelmaker
Stammholz n trunk wood
Stampfasphalt m compressed asphalt
Stampfer m (concrete) tamper
Stampfmasse f stamping mass
Stand m position; station; (mat.test.) bench, bay, floor; (Öl:) level; (Messe:) stand; (Wasser:) level
Standanzeiger m (auto.) level gauge (or gage)

Standardkostenverlauf m standard cost flow
Standardverrechnungsanteil m standard cost charging rate
Standardvorgabe f standard allowance
Ständer m (mach.) stand, pedestal; column, housing, standard; (shaper) frame; (work feeding) supporting stand; (e. Motors:) stator; (electr.) frame
Ständerbohrmaschine f upright drilling machine, [box] column drilling machine
Ständererregung f (electr.) stator excitation
Ständerführung f column-ways
Ständerfuß m (e. Maschine:) column base
Ständerschleifmaschine f column grinder
Ständerschlitten m (boring mill) spindle slide; (miller) spindle carriage
Ständerspule f (electr.) stator coil
Ständerstrom m stator current
Ständerstromkreis m (electr.) stator circuit
Ständerwicklung f stator winding
Standfestigkeit f stability; (mat.test.) creeping strength
Standöl n stand oil
Standlicht n (auto.) parking light
Standmesser m level gage
Standort m place, location; (an der Maschine:) operator's position
Standschauglas n level gage
Standschraublehre f bench micrometer
Standzeit f (e. Werkzeuges:) [tool] life
Stange f rod; (≈ Stab:) bar
Stangenabschnitt m (forging) slug
Stangenanschlag m (automatic) bar stop
Stangenarbeit f (mach.) bar work

Stangenautomat *m* automatic bar machine

Stangendraht *m* wire rod

Stangenzirkel *m* universal divider; beam trammel

Stanzautomat *m* automatic punching machine

Stanze *f* punching machine; *(tool)* punching die

stanzen *v.t.* punch, stamp; blank

Stanzen *n* metal stamping; (≈ Ausschneiden:) blank cutting, blanking; (≈ Lochstanzen:) punching; (≈ Viellochen:) perforating

Stanzerei *f* punching department

Stanzloch *n* punched hole

Stanzmatrize *f* punching die, stamping die; blanking die

Stanzpresse *f* punching press, power punch

Stanzwerkzeug *n* punching tool

Stapelwagen *m* tructier, stacking truck, fork lift

Stapler *m* cf. Stapelwagen

stark *a.* strong; heavy; thick; powerful

Stärke *f (mech.)* strength; ruggedness; stability; *(electr., opt., magn.)* intensity; (Blech:) gage; (e. Lösung:) concentration; (e. Motors:) power; *(chem.)* starch; (≈ Dicke:) thickness

Stärkekleister *m* starch paste

Starkstrom *m* heavy current, power current

Starkstromanlage *f* power installation; power plant

Starkstromkabel *n* power cable

Starkstromkondensator *m* heavy current condenser

Starkstromleitung *f* power line

Starkstromnetz *n* power mains

Starkstromtechnik *f* heavy current engineering

Starktonhorn *n (auto.)* loud-tone horn

Starkverzinkung *f* heavy zinc coating

starkwandig *a.* thick-walled

starr *a.* rigid; fixed; stationary

Starrfräsmaschine *f* rigid milling machine

Starrheit *f* rigidity, stability

Starrhobelmaschine *f* rigid planing machine

Starrschmiere *f* grease

Starter *m (auto.)* starter

Starterklappe *f (auto.)* choke

Starterknopf *m* starter knob

Starterkranz *m (auto.)* starter gear ring

Starterwelle *f (motorcycle)* kick-starter shaft

Statik *f* statics

Statiker *m* structural engineer

Statistik *f* statistics

Stativ *n* stand; tripod

Stator *m (electr.)* stator

STATOR ~, (~anlasser, ~gehäuse, ~spule, ~strom, ~wicklung) stator ~

Staub *m* dust; powder

Staubabsaugung *f* dust exhaust, dust collection

Staubabsaugungsanlage *f* dust collection plant

Staubabscheidung *f* dust separation

Staubfänger *m* dust arrester, dust collector

Staubfeuerung *f* coal dust firing

staubgeschwängert *a.* dust-laden

Staubkalk *m (steelmaking)* powdered lime

Staubkohle *f* dust-coal, powdered coal

staubkohlengefeuert *a.* pulverized coal fired

Staubsack *m (blast furnace)* dust catcher

Staubsauger *m* vacuum cleaner
Staubverzinkung *f* dry galvanizing, sherardizing
Stauchaltern *n cf.* Reckaltern
stauchen *v.t. (cold work)* upset; *(hot work)* pressure-forge; (Bolzenköpfe:) head; *(rolling)* roll on edge; – *v.r.* (Walzgut:) buckle
Stauchgerüst *n* edging stand
Stauchkaliber *n* edging pass
Stauchmaschine *f* upsetting machine, upsetter
Stauchmatrize *f* upsetting die
Stauchpresse *f* upsetting press, upsetter
Stauchschweißung *f* upset welding
Stauchstempel *m* upsetting die; (Köpfe:) heading die, header
Stauchstich *m (rolling)* edging pass
stauchverschränken *v.t.* joggle
Staudruck *m* back pressure
Stauferfett *n* cup grease
Stechmeißel *m (metal cutting)* cutoff tool; (Handwerkzeug:) firmer chisel
Steckachse *f (auto.)* linchpin
Steckachsnabe *f (auto.)* linchpin hub
Steckdose *f* plug socket, plug box; wall socket
Stecker *m (electr.)* plug
STECKER ~, (~anschluß, ~feld, ~schalter, ~sicherung, ~stift) plug ~
Steckkontakt *m* plug connection
Steckrad *n* pick-off gear
Steckschlüssel *m* socket wrench; *(UK)* box spanner
Steg *m* (Kette:) side bar; (Kurbeltrieb:) fixed member; (Spiralbohrer:) web (Schiene:) stem; (Profilstahl:) web
Stegabstand *m* (e. Fugennaht:) width of gap

Stegflanke *f* flank of gap
Steghöhe *f (welding)* depth of abutting gap faces
Stegkette *f* stud-link chain
Steg-Längskante *f (welding)* longitudinal edge of gap
Stehbolzen *m* staybolt
Stehbolzenbohrmaschine *f* staybolt drilling machine
Stehlager *n* pillow block
Stehlampe *f* portable standard, floor standard
Stehsockel *m (civ.eng.)* vertical skirting
steif *a.* stiff, rigid; *(concrete)* consistent
Steife *f* stiffness, rigidity; consistency
Steifigkeit *f cf.* Steife
steigend gießen, bottom cast, cast up-hill, bottom pour
Steiger *m (founding)* riser
Steigfähigkeit *f (auto.)* hill-climbing ability
Steigkanal *m* uptake
Stelgleitung *f* ascension pipe, uptake
Steigrohr *n* standpipe; *(founding)* down sprue
Steigstrom-Vergaser *m (auto.)* updraught carburetter *(or* carburetor)
Steigung *f* rise; gradient; (Feder:) pitch; (Gewinde:) lead; (Kegel:) taper; *(road building)* upgrade
Steigungslehre *f* screw pitch gage
Steigungswinkel *m (threading)* lead angle; *(gearing)* helix angle
steil *a.* steep; (Dach:) high-pitch; (Gewinde:) coarse
Steilgewinde *n* coarse screw thread
Steilheit *f (electr.)* transconductance
Steilwert *m (electron.)* slope
Stein *m* stone; (Fels:) rock; (künstlicher:) brick; (≈ Ablagerung:) scale

Steinbrecher m stone crusher
Steinbruch m stone quarry
Steinbrucharbeiter m quarryman
Steinfaser f rock wool
Steingut n earthenware
Steinkohle f pitcoal, mineral coal, bituminous coal
Steinkohlenbergwerk n colliery, coal-mine
Steinkohlengas n coal gas
Steinkohlenteer m coal tar
Steinkohlenteerpech n coal-tar pitch
Steinschlag m broken stone, broken bricks; (im Gebirge:) rock fall
Steinzeug n stoneware
Steinzeugrohr n stoneware pipe
Stellbogen m (Zirkel:) quadrant
Stelle f place; spot; point; (math.) digit; (building) site; (≈ Lage:) position; (cost acc.) item
Stelleisen n quadrant
Stelleisenrad n quadrant gear
Stelleiste f gib
Stellkeil m adjusting wedge
Stellmacher m wheelwright
Stellmacherei f cartwright's shop
Stellmotor m servo-motor
Stellmutter f adjusting nut
Stellring m set collar
Stellschmiege f sliding T-bevel
Stellschraube f set screw, adjusting screw
Stellung f position; location; situation
Stellwinkel m (Meßzeug:) carpenters' bevel
Stemmarbeit f (building) cutting-away work; (mittels Stemmeißel:) calking work
Stemmaschine f (woodworking) mortising machine

Stemmeisen n calking chisel; (woodworking) mortise chisel
Stemmeißel m (woodworking) mortise chisel
stemmen v.t. (= verstemmen) calk; (UK) caulk; (Holz:) mortise
Stemmhammer m calking hammer
stemmlochen v.t. (woodworking) mortise
Stemmnaht f (welding) calk weld, calking seam
Stemmverbindung f calked joint
Stempel m (tool) punch, upper die; (mining) prop; (coke oven) ram
Stempeleinschlag m (surface finish) die mark
Stempelfräseinrichtung f punch milling attachment
Stempelrichtpresse f gag press
steppen v.t. (Matten:) stitch
Steppnahtschweißen n stitch welding
Steppung f (von Matten:) stitching
Sterndreieckschalter m star-delta switch
Sterngriff m star knop
Sternrad n capstan wheel
Sternrevolverdrehmaschine f hexagon turret lathe
Sternrevolverkopf m hexagon-type turret head
Sternschaltung f star connection
Sternspannung f star voltage
Stetigförderer m continuous conveyor
Steuer n (auto.) steering wheel
STEUER ~, (~befehl, ~einrichtung, ~elektrode, ~energie, ~frequenz, ~gerät, ~gitter, ~gitterröhre, ~gitterspannung, ~hebel, ~impuls, ~kabel, ~kreis, ~leitung, ~pult, ~schalter, ~schaltung, ~schütz, ~spannung, ~stromkreis,

~tafel, ~transformator, ~ventil) control ~

Steuerkanzel f pulpit
Steuerkommando n command signal, impulse
Steuerkurve f cam
Steuermotor m servo motor
steuern v.t. control; regulate; operate
Steuernocken m cam
Steuerrad n (auto.) steering wheel
Steuerschieber m control valve; (lathe) control slide
Steuerstromkreis m control circuit
Steuerung f control; control mechanism; control system; (auto.) steering, drive; (radio) navigation; (acous.magn.) control
Steuerwalze f (electr.) drum controller
Steuerwelle f camshaft
Stich m (rolling mill) pass; (Schmiedefehler:) cold shut; ($\approx$ Lunker:) shrinkhole
Stichel m (engraving) single-lip cutter
Stichelhaus n plain toolpost
Stichloch n (met.) taphole, notch, hole
Stichlochstopfmaschine f blast furnace gun
Stichmaß n inside micrometer
Stichprobe f random test
Stift m pin; ($\approx$ Drahtstift:) nail, brad; ($\approx$ Fixierstift:) locating plug; ($\approx$ Kerbstift:) taper pin; ($\approx$ Paßstift:) dowel pin; ($\approx$ Scherstift:) shear pin; ($\approx$ Taststift:) tracer pin
Stiftbolzen m stud
Stiftschlüssel m socket screw wrench
Stiftschraube f stud screw, stud
Stillegung f (e. Betriebes:) shut-down
Stirnansicht f front view
Stirnbrett n facade board

stirnen v.t. (mach.) face-mill, end-mill
Stirnfläche f end face
Stirnflächenabstand m (e. Stirnnaht:) distance between abutting faces
Stirnfräser m shell end mill
Stirnfugennaht f bevelled edge weld
Stirnkehlnaht f edge fillet weld
Stirnkipper m end-dump truck
Stirnlängskante f longitudinal edge of fusion faces
Stirnmodul m (gearing) real module
Stirnnaht f edge weld
Stirnrad n spur gear
Stirnradgetriebe n spur gearing
Stirnradwälzfräser m spur gear hob
Stirnradwendegetriebe n spur gear reversing mechanism
Stirnschnitt m (drawing) transverse section
Stirnseite f end face, front face
Stirnsenker m counterbore
Stirnstoßkante f (e. Stirnnaht:) abutting edge of end
Stirnwand f front wall
Stocheisen n poker bar, poker
stochen v.t. poke, stir, stoke
Stöckel m (e. Ambosses:) stake
Stockpunkt m (Öl:) setting point
Stockung f (traffic) jam
Stoff m (phys.) matter; (chem.) substance; ($\approx$ Werkstoff:) material; (textile) fabric, cloth
Stoffscheibe f (polishing) cloth wheel
Stoffteilchen n particle
Stollen m (Kabel:) tunnel; (e. Schneidwerkzeuges:) row of cutting teeth
Stopfbuchse f stuffing box
STOPFBUCHSEN ~, (~brille, ~dichtung, ~packung) stuffing box ~

Stopfdämmung f loose-fill insulation
Stopfen m stopper, plug; (e. Lochdorns im Hohlwalzverfahren:) piercer, mandrel
Stopfenstange f stopper rod
Stopfenwalzwerk n plug mill
Stopfer m (railway) tamper
Stopfmaschine f (railway) tamping machine
Stopleuchte f (auto.) stop light
Stoplichtschalter m (auto.) stop light switch
Stopuhr f stop watch
Stöpsel m stopper; (electr.) plug
Stöpselschnur f plug-ended cord
Storchschnabel m pantograph
stören v.t. & v.i. disturb; interfere; (radio) jam
Störimpuls m inference pulse
Störpegel m noise level
Störquelle f source of trouble
Störschutz m noise suppression
Störschutzkondensator m interference suppression capacitor, anti-interference capacitor
Störsender m jamming transmitter
Störspannung f disturbing voltage
Störstrom m disturbing current
Störung f disturbance; trouble; interference
störungsfrei a. trouble-free
Störungsmeßgerät n interference measuring instrument
Stoß m (mech.) shock, impact, blow, push, bump; (electr.) surge, impulse, pulse; (mach.) butt-joint, joint; (phys.) impulse, pulse; (≈ Fuge, Verbindung:) joint; (civ. eng.) thrust; (≈ Ruck:) jolt; jerk
Stoßbank f push bench

Stoßbeanspruchung f impact stress
Stoßbelastung f (welding) pulsating load
Stoßdämpfer m shock absorber, dashpot
Stößel m (power press) slide, ram; (Ventil:) tappet
STÖSSEL ~, (~antrieb, ~führung, ~gleitbahn, ~hub, ~kopf, ~schlitten, ~support) ram ~
Stößelrücklauf m return stroke of the ram
stoßen v.t. push; (≈ fügen:) join; (metal cutting) slot; shape; (Nuten:) keyway; – v.i. strike (against)
Stoßentladung f impulse discharge
Stoßerregung f shock excitation
Stoßfänger m shock absorber; (auto.) bumper, fender
Stoßfläche f (welding) abutting face
stoßfrei a. shockless
Stoßfuge f joint
Stoßläppen n ultrasonic grinding
Stoßmaschine f (≈ Senkrechtstoßmaschine:) slotter, slotting machine; (≈ Waagerechtstoßmaschine:) shaper, shaping machine
Stoßmeißel m slotting tool
Stoßmischer m (concrete) batch mixer
Stoßnaht f butt joint
Stoßofen m pusher-type furnace
Stoßrad n (gear cutting) pinion-type cutter
Stoßräumen n push broaching
Stoßräummaschine f push-type broaching machine
Stoßspannung f transient voltage
Stoßstange f (auto.) bumper; (slotter) slotting bar; (e. Ventilsteuerung:) push rod
Stoßstelle f joint
Stoßstrom m transient voltage
Stoßverbindung f joint

Stoßwelle f (= Druckwelle) shock wave
Strahl m *(geom.)* straight-line; *(opt.)* ray; *(radar)* beam; (Dampf, Luft, Wasser:) jet, stream
Stahlantrieb m jet propulsion
Strahldüse f jet
Strahleinspritzmotor m solid-injection engine
strahlen v.t. (mit Sand:) blast; – v.i. (Licht:) shine; (Wärme:) radiate
strahlen mit Sand, sandblast
strahlen mit Stahlkies, shot-blast
Strahlenbrechung f refraction
Strahlenbündel n beam of rays, bundle of rays
strahlend a. radiant
Strahlenfilter m ray filter
Strahlenkegel m ray cone
Strahlenschutz m protection against radiation
Strahler m radiator
Strahlläppmaschine f vapor-blast liquid lapping machine
Strahlpumpe f injector pump
Strahlrohrbeheizung f (e. Ofens:) radiant-tube heating
Strahlung f radiation
STRAHLUNGS ~, (~dichte, ~dosis, ~empfänger, ~energie, ~feld, ~frequenz, ~heizung, ~intensität, ~leistung, ~quelle, ~trockner, ~verlust, ~wärme, ~widerstand) radiation ~
Strahlungspyrometer n radiation pyrometer
Strahlvortrieb m jet propulsion
Strahlwasserschutz m (Motor:) hoseproof enclosure
Strang m (Kette, Seil:) strand; (Schienen:) track

Stranggießanlage f continuous casting plant
Stranggußverfahren n continuous casting method
Strangpresse f extrusion press, extruder
strangpressen v.t. *(plastics)* extrude
Strangpreßform f extrusion die
Strangpreßprofil n extruded section
Strangpreßteil n *(plastics)* extruded article
Strangpreßverfahren n extrusion process, power-press extrusion
Strangpreßwerkzeug n *(plastics)* extrusion die
Straße f *(civ. eng.)* road; street; *(rolling mill)* mill train, rolling train
Straßenbau m road building, road construction, road-making
Straßenbauer m road builder
Straßenbauingenieur m road engineer
Straßenbaumaschine f road-building machine
Straßenbautechnik f highway engineering
Straßenbauunternehmer m road contractor
Straßenbeleuchtung f steel lighting
Straßenbeschotterung f road gravelling
Straßenbrücke f road bridge
Straßendecke f pavement
Straßenfahrzeug n road vehicle
Straßenfahrzeugwaage f road vehicle balance
Straßengabelung f *(auto.)* road junction, bifurcation
Straßenkreuzer m *(auto.)* turnpike cruiser
Straßenkreuzung f crossing, road junction
Straßenlage f *(auto.)* roadability
Straßenschlepper m industrial wheel tractor, wheeled tractor

Straßenverkehr m road traffic
Straßenverkehrsordnung f Highway Code
Straßenwalze f road roller
Straßenzugmaschine f road tractor
Strebe f brace, strut; cross beam
Strebenfachwerk n strut frame
Strebepfeiler m buttress
Strecke f distance; length; *(auto.)* route, track; *(railway)* line; *(rolling mill)* rolling train; *(math.)* straight line
strecken v.t. stretch, extend, lengthen, elongate, draw; (Flüssigkeiten:) dilute; (Walzgut:) rough down; (Blechstürze:) break down
Streckgerüst n break-down stand
Streckgesenk n fuller
Streckgrenze f yield point
Streckgrenzenverhältnis n elastic limit-tensile strength ratio
Streckkaliber n breaking down pass
Streckmetall n expanded metal
Streckmittel n extender
Streckspannung f yield stress
Streckwalze f breaking down roll
Streckwalzen n roughing, breaking down; blooming; *(UK)* cogging
streckziehen v.t. stretch-form
Streckziehpresse f stretching press
strehlen v.t. chase
Strehler m chaser
Strehlvorrichtung f thread-chasing attachment
Streichasphalt m mastic asphalt
streichbar a. suitable for painting
Streichbarkeit f brushability
Streichlack m brushing lacquer
Streichmaß n scratch gage
Streifen m band, strip; (Rohrherstellung:) skelp; *(metallo.)* band, streak; *(recording)* chart
Streifenblech n strip steel
Streifengefüge n banded structure
Streifenwagen m patrol car
Streifenwalzwerk n strip rolling mill
strengflüssig a. viscous
Streubereich m range of variation; scattering range
Streustrahlung f scattered radiation
Streustrom m leakage current; stray current
Streuung f scattering, spreading, variation; *(magn.)* dispersion
Streuwert m erratic value
Strich m line, dash; (Feile, Magnet:) stroke
stricheln v.t. dot
Strichendmaß n hairline gage
Strichmarke f size line
Strichmaß n rule, straightedge
Strichraster m *(telev.)* bar pattern
Strichteilung f line graduation
Strichzeichnung f line drawing
Strichziehmaschine f road-marking machine
Stripperkran m stripper crane
Stripperlaufkran m travelling stripper crane
Strippertorkran m stripper gantry
Strom m *(electr.)* current; *(electron.)* flow; (≈ Strömung:) stream, flow
STROM ~, (~amplitude, ~bahn, ~bauch, ~bedarf, ~dichte, ~durchfluß, ~durchgang, ~fluß, ~gleichrichter, ~knoten, ~kurve, ~modulation, ~pfad, ~phase, ~prüfer, ~regler, ~richter, ~rückkopplung, ~schwankung, ~spule, ~stärke, ~stoß, ~transformator, ~überlastung,

~unterbrechung, ~verbrauch, ~versorgung, ~verteilung, ~wandler, ~wärme, ~weg, ~zähler) current ~

Stromabnehmer m current collector, commutator
Stromart f type of current
Stromaufnahme f power input
Stromausfall m power breakdown
Stromauslöser m circuit-breaker
Stromerzeuger m generator
stromführend a. current-carrying, live
Stromkreis m circuit
Stromlaufbild n wiring diagram, circuit diagram
Stromleiter m conductor
Stromleitung f circuit line
stromlinienförmig a. streamlined
Stromlinienwagen m streamlined car
Strommangelsicherung f power failure safety system
Strommesser m ammeter
Strompfeiler m river pier
Stromrelais n relay
Stromschiene f bus-bar
Stromverbraucher m power consumer
Stromversorgungsnetz n power supply main
Stromwächter m automatic controller
Stromwender m commutator
Stromzufuhr f power supply
Stromzuführungskabel n power feed cable
Stuckarbeit f stucco work
Stückerz n lump ore
Stückkalk m (steelmaking) lump lime
Stückkohle f lump coal
Stückkoks m lump coke
Stückkosten pl. piece-production cos
Stückliste f parts list

Stückzeichnung f detail drawing
Stückzeit f (work study) piece-handling time
Stufe f stage; step; phase; (Drehzahlen:) range; rate; (Getriebe:) speed; (Stufenscheibe:) cone; (Gegentakt:) stage
Stufenantrieb m variable-speed drive
Stufengetriebe n variable-speed transmission
Stufenhärtung f hot tempering
Stufenkompensator m (electr.) deflection potentiometer
stufenlos a. (Getriebe) infinitely variable, stepless
stufenloses Getriebe, infinitely variable speed gear drive
stufenlos regeln v.t. vary infinitely
Stufenpresse f (= Mehrstempelpresse) multiple-die press, multiple plunger press
Stufenrad n (gearing) cluster gear
Stufenrädergetriebe n variable speed gear drive
Stufenschalter m multiple-contact switch
Stufenschalten n variable-speed control
Stufenschaltgetriebe n automobile gear transmission
Stufenschaltung f (electr.) cascade connection
Stufenscheibe f cone pulley
Stufenscheibenantrieb m cone pulley drive
Stufensprung m (gearing) progressive ratio
stufenweise a. gradual
Stufenzahn m (gears) staggered tooth
Stufung f (von Drehzahlen:) grading, progression

stumpf *a.* blunt; dull; (Winkel:) obtuse
Stumpflasche *f (welding)* butt strap
Stumpfnaht *f (welding)* butt weld, butt seam
Stumpfnaht mit Bördelvorbereitung, butt weld with edges prepared by flanging
Stumpfnaht mit Fugenvorbereitung, butt weld with edge preparation
Stumpfnahtschweißung *f* butt seam weld
Stumpfschweißung *f* butt welding
Stumpfstoß *m (welding)* butt joint
Stumpfverzahnung *f* stub-tooth gearing
Stundenleistung *f* output per hour
Stundenlohn *m* pay per hour
Sturzenglühofen *m* pack annealing furnace
Sturzenglühung *f* pack annealing
Sturzenwalzung *f* pack-rolling
Sturzgußverfahren *n (founding)* slip casting
Sturzmesser *m (auto.)* camber gage
Sturzwalzen *n* pack rolling
Stützauflage *f* support rest
Stütze *f* support
stützen *v.t.* support
Stutzen *m* connection piece; (Rohre:) fitting
Stützlager *n* supporting bearing
Stützpfeiler *m* supporting column
Stützwalze *f* backing roll, back-up roll
Stützweite *f* span
Submission *f* submission (of competitive tenders)
Suchantenne *f* search antenna *(or aerial)*
Sucher *m (auto.)* spotlight; *(opt.)* view finder
Suchscheinwerfer *m (auto.)* spotlight
Summenfehler *m* cumulative error
Summenspannung *f (electr.)* summated voltage
Summer *m* buzzer, hummer
Summton *m (tel.)* buzzer signal
Sumpferz *n* bog ore
Sumpfgas *n* marsh gas, methane
Support *m (lathe)* slide rest; saddle; *(planer)* toolhead; *(vertical turret lathe)* turret head
Supportschieber *m (planer)* crossrail slide, toolhead slide
Symmetrieschaltung *f* balanced circuit
synchron *a.* synchronous
SYNCHRON ~, (~abtastung, ~drehzahl, ~generator, ~kondensator, ~maschine, ~motor, ~phasenschieber) synchronous ~
Synchrongetriebe *n (auto.)* synchromesh gear
synchronisieren *v.t.* synchronize
Synchronuhr *f* synchronous timer
System *n* system

T

Tabelle *f* table; list; chart; diagram; *(mach.)* (für Drehzahlen, Geschwindigkeiten usw.:) plate; (Gleisbild:) diagram

Tachometer *m* tachometer, speedometer

Tachometer mit Kilometerzähler, *(auto.)* speedometer-odometer

Tachometerleuchte *f (auto.)* speedometer light

Tafel *f* (≈ Tabelle:) table; chart; (Blech:) sheet; panel; plate; (Holz:) panel board; (Leim:) tablet; *(plastics)* slab

Tafelglas *n* sheet glass

Tafelleim *m* tablet glue

Tafelschere *f* plate shears, lever shears

Tagebau *m* open-pit mining

Tagesschicht *f* dayshift

Takt *m (techn.)* cycle

taktieren *v.t.* (Maschinen:) index; time

Taktierung *f (mach.)* timing

Taktschaltwerk *n* timing gear

Taktzeit *f* cycle time

Talkum *n* talc, talcum

Tandemwalzwerk *n* tandem mill

Tangentenbussole *f* tangent galvanometer

Tangentialkraft *f* tangential force

tanken *v.t. (auto.)* refuel, refill

Tankstelle *f* service station, filling station

Tankwart *m* fuelling attendant

tanzen *v.i.* (Lichtbogen:) scatter, be wild, be unsteady

Tapetenkleister *m* wallpaper adhesive

tapezieren *v.t.* paper

Tapezierer *m* paper hanger

Tariflohn *m* official wage

Taschenlampe *f* pocket lamp

Taschenschieblehre *f* pocket caliper square

Tastameter *m* inside thread caliper

Taste *f* key

Tasteinrichtung *f (copying)* tracer unit

tasten *v.t.* key

Tastenplatte *f (programming)* keyboard

Tastenschnelltelegraf *m* keyboard printing telegraph

Taster *m* (Meßzeug:) machinists' caliper; (≈ Tasterlehre:) snap gage; *(copying)* tracer

Tastereinstellung *f* caliper setting

Tastermeßuhr *f* dial gage caliper

Tastgerät *n* minimeter type instrument

Tastuhr *f* dial indicator

Tastzirkel *m* hermaphrodite caliper

Tauchbadschmierung *f* oil bath lubrication

Tauchelektrode *f* dipped electrode, washed electrode; (mit e. Flußmittel:) fluxed electrode

Tauchemaillelack *m* dipping enamel

tauchen *v.t.* immerse, dip, submerge

tauchfräsen *v.t.* plunge-mill

Tauchfräsmaschine *f* plunge-cut milling machine

tauchhärten *v.t.* dip-harden

Tauchhartlötung *f* dip brazing

Tauchkolben *m* plunger piston

Tauchkolbenpumpe *f* plunger pump

Tauchlack *m* dipping varnish

Tauchlöten *n* dip brazing

Tauchpatentieren *n* batch patenting

Tauchschleuderschmierung f splash lubrication
Tauchsiedegerät n immersion heater
Tauchveredlung f hot dipping process, hot dip process
Taumelfehler m (threading) drunkenness
taumeln v.i. (gears) wobble
Taupunkt m dewpoint
Taupunktmeßanlage f dewpoint metering unit
Technik f engineering; practice; technique; technology
Techniker m technician
technische Frequenz, industrial frequency
technische Lieferbedingung, technical delivery condition
technische Messe, industrial fair
technische Messung, engineering measurement
technischer Überwachungsverein, test code association
Teer m tar
Teerabscheidung f tar separation, tar extraction, detarring
Teerfarbstoff m coal-tar dye
Teermagnesit m tar-bonded magnesite
Teeröl n tar oil
Teerpappe f tar-board
Teersplitt m tarred chippings
Teerung f tarring
Teil m & n part, piece, portion; (Bauteil:) component, member; (Abschnitt:) section
Teilansicht f partial view
Teilapparat m indexing attachment, dividing attachment
Teilchen n particle, corpuscle

Teildrehung f fractional turn, fractional rotation
Teileliste f parts list
teilen v.t. (mach.) index, divide; graduate, scale
Teiler m (Frequenzen:) divider
Teilflankenwinkel m flank angle
Teilfuge f joint; parting line; (e. Gesenkes:) die-parting line
Teilgerät n indexing attachment
Teilhärtung f selective hardening
Teilkegel m (gears) pitch cone
Teilkopf m indexing head
Teilkreis m pitch circle
Teilmontage f sub-assembly
Teilmontagezeichnung f exploded view
Teilnehmerleitung f (tel.) subscriber's line
Teilscheibe f index plate
Teilstrich m division line
Teilung f (gearing, threading) pitch; (Teilarbeiten:) indexing, dividing; (math.) division; (≈ Einteilung:) graduation; (in gleiche Abstände:) spacing
Teilzahl f (math.) quotient
Teilzeichnung f component drawing, detail drawing
Telefon n telephone
TELEFON ~, (~amt, ~anschluß, ~apparat, ~buch, ~gespräch, ~nummer) telephone ~
Telefonie f telephony
telefonieren v.t. telephone, phone; call
TELEGRAFEN ~, (~amt, ~kabel, ~leitung, ~mast, ~relais, ~stange, ~technik) telegraph ~
Telegrafenzange f linemen's plier
Telegrafie f telegraphy
Tellerfeder f plate spring

Tellerrad *n (US)* rim gear; *(UK)* crown wheel, crown gear
Tellerventil *n* plate valve
Temperatur *f* temperature
TEMPERATUR ~, (~abfall, ~abhängigkeit, ~bereich, ~gefälle, ~gleichgewicht, ~meßgerät, ~messung, ~schreiber, ~schwankung, ~steuerung, ~zunahme) temperature ~
Temperaturregler *m* thermostat
Tempereisen *n* malleable cast iron
Tempererz *n* annealing ore
Tempergießerei *f* malleable iron foundry
Temperglühofen *m* malleable annealing furnace
Temperguß *m* malleable cast iron
Tempergußstück *n* malleable casting
tempern *v.t.* anneal, malleablize
Temperroheisen *n* malleable pig iron
Temperrohguß *m* malleable hard iron
Tempertopf *m* annealing pot
Temperung *f* annealing
Terpentin *n* turpentine
Terpentinöl *n* turpentine oil, rosin oil
Terrain *n (building)* site
Terrazzo *m* terrazzo
Testbenzin *n* paint thinner
Tetrachlorkohlenstoff *m* carbon tetrachloride
Thermalhärtung *f* hot quenching
Thermit-Gießschweißen *n* thermit fusion welding
Thermitschweißung *f* thermit welding
Thermoelement *n* thermocouple
Thermometer *n* thermometer
thermonuklear *a.* thermonuclear
Thermophysik *f* thermophysics
Thermoregler *m* thermostat
Thermosäule *f* thermopile

Thomasbirne *f (US)* basic Bessemer converter; *(UK)* Thomas converter
Thomaseisen *n* basic pig iron
Thomasflußstahl *m* basic converter steel; (*UK* auch:) Thomas steel
Thomasroheisen *n* basic Bessemer pig [iron], basic converter pig iron
Thomasschlacke *f* basic slag
Thomasschlackenmehl *m* ground basic slag
Thomasstahl *m* basic Bessemer steel, basic converter steel
Thomasverfahren *n (US)* basic converter process; *(UK)* Thomas process
Tiefbaggerung *f* deep cut digging
Tiefbau *m* structure below ground; civil engineering
Tiefbauunternehmer *m* civil engineering contractor
Tiefbettanhänger *m* flat-bed trailer
Tiefbettfelge *f (auto.)* drop-base rim, well base rim
Tiefbohren *n* deep-hole drilling; deep-hole boring; *s.a.* bohren
Tiefe *f* depth
Tiefenmaß *n* depth gage
Tiefenmessung *f* depth measurement
Tiefenschieblehre *f* rule depth gage
Tiefenvorschub *m (metal cutting)* vertical feed, downfeed
Tiefladeanhänger *m* deep-loading trailer
Tieflader *m* low-bed trailer
Tieflauf *m (rolling mill)* sloping loop channel
Tieflochbohrmaschine *f* deep-hole drilling machine; deep-hole boring machine
Tiefofen *m* soaking pit, pit-type furnace
Tiefofenkran *m* pit furnace crane

Tieftemperaturverkokung f low-temperature carbonization
Tiefungsprobe f cupping test
Tiefungsversuch m cupping test, deep drawing test
TIEFZIEH ~, (~blech, ~fähigkeit, ~presse, ~qualität, ~stahl, ~werkzeug) deep drawing ~
tiefziehen v.t. deep draw; (Rohre:) hot draw; (Hohlkörper:) cup; (Räder:) cone
Tiefziehprobe f cupping test
Tiefzug m deep drawing
Tiegel m *(steelmaking)* crucible
Tiegelgußstahl m crucible steel, crucible cast steel
Tiegelofen m crucible furnace
Tiegelschmelzofen m crucible melting furnace
Tiegelstahl m crucible steel
Tierleim m animal glue
Tippbetätigung f jogging
tippen v.t. *(auto.)* tickle
Tippschalter m inching switch
Tippschaltung f inching control
Tipptaste f inching push button
Tischbohrmaschine f bench drilling machine
Tischempfänger m *(telev.)* table set
Tischexzenterpresse f bench press
Tischgewindebohrmaschine f bench tapping machine
Tischhobelmaschine f (= Langtischhobelmaschine) planer
Tischkreissäge f circular saw bench; (als Handwerkzeug:) variety saw
Tischler m joiner; carpenter
Tischlerarbeit f joinery work
Tischlerplatte f blockboard

Tischrücklauf m table return
Tischschlitten m *(gear planer)* table saddle; *(miller)* table slide
Tischschwenkbohrmaschine f bench-type radial
Tischselbstgang m *(lathe)* table power traverse
Tischumkehr f table reversal
Tischvorlauf m table forward stroke
T-Nut f T-slot
Toleranz f tolerance; permissible variation
Toleranzanzeiger m limit indicator
Toleranzlehre f limit gage
Toleranztasterlehre f limit snap gage
tolerieren v.t. tolerance
Toluol n toluol, toluene
Ton m clay
 gebrannter ~, fire-clay
Tonabnehmer m pick-up
Tonband n recording tape
Tondinasstein m *(US)* quartzite-brick; *(UK)* ganister
tönen v.t. *(painting)* tint; *(photo.)* tone; v.i. *(acoust.)* sound
Tonerde f alumina
Tonfrequenz f audio frequency
Tonfrequenztelefonie f voice-frequency telephony
Tonfrequenzverstärker m audio amplifier
tonhaltig a. aluminous
tonig a. argillaceous, clayey
Tonmörtel m clay mortar
Tonnenblech n arched plate
Tonnenlager n spherical roller bearing
Tonwiedergabe f sound reproduction
Topf m (e. Auspuffes:) muffler
Topfglühofen m pot annealing furnace
Topfschleifscheibe f cup wheel
Torsiograph m torsion recorder

Torsiometer *m* torsion meter
Torstahl *m* twisted steel bars of deformed rounds
tote Körnerspitze, *(lathe)* dead center, tail center
töten *v.t.* (Flotationsschaumbildung:) kill, deaden
totes Kaliber, dummy pass
Totgangausgleich *m (miller)* backlash eliminator
totgebrannt *a.* dead-burned
Totlage *f* dead center position
Totpunktzündung *f (auto.)* dead-center ignition
Totzeit *f* dead time; downtime; non-productive time; ($\approx$ Verlustzeit:) lost time
Tourenwagen *m* touring car
Tourenzähler *m* speedometer, speed indicator, revolution counter
T-Profil *n* T-section; – *pl.* Tees
Tragachse *f* supporting axle
tragbar *a.* portable
tragbarer Empfänger, portable receiver
träge *a.* inert
Träger *m* girder; joist; *(chem.)* carrier, vehicle; (Kosten:) bearer; *(civ.eng.)* girder; (~ Balken) beam
Trägerfrequenz *f* carrier frequency
Trägerwalzwerk *n* beam rolling mill
Trägerwerkstoff *m* parent metal
Tragfähigkeit *f* load capacity
Trägheitsmoment *n* moment of inertia
Tragkranz *m (blast furnace)* lintel ring, mantle ring
Traglänge *f* bearing length
Tragseil *n* carrying rope
Traktor *m* tractor
Traktorenanhänger *m* tractor trailer
tränken *v.t.* soak

Transfermaschine *f* transfer machine
Transferpressen *n (plastics)* transfer molding
Transferstraße *f* transfer line
Transformator *m* transformer
TRANSFORMATOREN ~, (~betrieb, ~blech, ~kern, ~klemme, ~kreis, ~öl, ~schalter, ~spannung, ~spule, ~station, ~strom, ~wicklung, ~zelle) transformer ~
transformieren *v.t.* transform
Transistor *m* transistor
Transmission *f* lineshaft
Transmissionsantrieb *m* lineshaft drive
Transmissionsvorgelege *n* overhead lineshaft
Transmissionswelle *f* transmission shaft
Transportband *n* conveyor belt
transportierbar *a.* portable
Transportmischer *m (road building)* truck mixer
Transportpfanne *f (founding)* bull ladle, transfer ladle
Transportschnecke *f* conveyor worm
Transportunternehmer *m* hauling contractor
Transportwagen *m* transporter
Trapezgewinde *n* Acme thread; *(UK)* trapezoidal thread
Trassierung *f (road building)* route selection
Traufblech *n* eaves sheet
Traufe *f* gutter
Traverse *f* cross beam
treiben *v.t. (met.)* cupel; – *v.i.* float, drift; (Kokskuchen:) expand; (Steine beim Brennen:) swell
Treibkeil *m* drift key
Treibofen *m* cupelling furnace

Treiböl n fuel oil; *(hydr.)* hydraulic oil
Treibrolle f driving pulley
Treibsitz m drive fit
Treibstoff m power fuel
Trennbandsäge f resaw
trennen v.t. separate; cut-off; *(electr.)* disconnect; *(woodworking)* rip
Trennfuge f parting line
Trennkontakt m break contact
Trennkreissäge f circular cut-off saw
Trennsäge f metal slitting disc
Trennschalter m isolating switch, disconnecting switch, isolator
trennscharf a. selective
Trennschärfe f selectivity
Trennscheibe f abrasive cutting wheel; circular friction saw; *(Diamant:)* cutoff wheel
Trennschleifen n abrasive cutting
Trennschleifmaschine f abrasive cutting machine
Trenn- und Abkürz-Kreissäge f ripping and cutting-off saw
Trenn- und Besäumkreissäge f rip and edging saw
Trennungsbruch m rupture
Trennwand f partition wall
Treppenstufe f stair tread
Tretanlasser m *(auto.)* kick-starter
Trichter m funnel; (Gichtverschluß:) cone, hopper; (Rohrschweißung:) bell; *(founding)* gate
Trieb m [gear] drive
Triebachse f driving axle
Triebling m pinion
Triebwagen m motor-coach
Triebwerk n driving mechanism, gear mechnism
Trinkwasser n drinking water

Trio n three-high rolling mill, three-high mill
TRIO ~, (~blechwalzwerk, ~fertigstraße, ~gerüst, ~straße, ~universalwalzwerk, ~walzgerüst, ~walzstraße, ~walzwerk) three-high ~
Triplexverfahren n *(met.)* three furnace process
Trittbrett n *(auto.)* running board
Trockenbagger m excavator
Trockenbatterie f dry cell battery
Trockenelement n dry cell
Trockenfarbe f pigment
Trockengestell n drying rack
Trockengleichrichter m dry disc rectifier, dry plate rectifier, metal rectifier
Trockenguß m dry sand casting
Trockenkammer f baking oven, drying chamber
Trockenmahlung f dry grinding
Trockenofen m (Holz:) seasoning kiln; baking oven, drying furnace, drying kiln
Trockenscheibenkupplung f dry-plate clutch
Trockenwäsche f dry scrubbing
trocknen v.t. dry; (Kerne:) bake; (Holz:) season; (Steine:) kiln
Trockner m dryer
Trogband n trough belt
Troggurt m trough belt
Trogmischer m *(concrete)* open-pan mixer
Trolleybus m trolley bus
Trommel f drum; (Kabel:) reel
Trommelbohrwerk n drum-type boring machine
Trommelfräsmaschine f drum miller
Trommelkonverter m barrel converter
Trommelkurve f barrel cam
Trommelmagazin n drum magazine
Trommelmantel m drum barrel

Trommelmischer *m (concrete)* rotary drum mixer
trommeln *v.t.* (Guß:) tumble, barrel
Trommelrevolverkopf *m* drum turret
Trommeltrockner *m (road building)* rotary drier
tropenfest *a.* tropicallized
Tropfbrett *n* drain board
Tropföler *m* drip oiler
Tropfpunkt *m* drop point
Tropfpunktmeßgerät *n* drop point tester
Tropfschmierung *f* drip oil lubrication
Tropfwasserschutz *m* drip-proof enclosure
Trübe *f (met.)* pulp; dross; sludge, slurry, slime
Trübungsmesser *m* opacimeter
Trum *m & n* (= Trumm) (e. Kette:) strand
T-Stoß *m (welding)* Tee-joint
T-Träger *m* T-beam; T-girder; – *pl.* Tees
Tuffstein *m* tuff
Tulpennaht *f* single-U butt weld

Tünche *f* whitewash
tünchen *v.t.* limewash
Tunnel *m* (Kabel:) tunnel, duct
Tunnelofen *m* tunnel kiln, continuous-type furnace
Tupfer *m (auto.)* carburetor tickler
Tür *f* door
Turbine *f* turbine
TURBINEN ~, (~düse, ~gehäuse, ~leistung, ~motor, ~regler, ~schaufel, ~welle) turbine ~
TURBO ~, (~dynamo, ~gebläse, ~generator, ~kompressor, ~lader, ~pumpe) turbo-~
Türfüllung *f* door panel
Turmdrehkran *m* tower crane
Turmkraftwagen *m* tower wagon
Türpfosten *m* door post, door jamb
Türschwelle *f* (e. Ofens:) sill
Tusche *f* Chinese ink
Tuschierlineal *n* levelling straightedge
Tuschierplatte *f* cast iron surface plate
typisieren *v.t.* standardize

U

Überbau m superstructure
überbeanspruchen v.t. overstress; overload
überbelasten v.t. overload
Überbelastung f overload
überbelichten v.t. over-expose
überblasen v.t. (met.) over-blow
überbrücken v.t. (electr.) by-pass; shunt; (building) span
ÜBERBRÜCKUNGS ~, (~kondensator, ~schalter, ~schaltung) bypass ~
überdecken v.t. cover; overlap; mask
Überdeckungsgrad m (gears) engagement factor; (e. Profils:) overlap
überdimensionieren v.t. overdimension
überdrehen v.t. finish-turn; (Schrauben:) overturn; (Motor:) overspeed
Überdruck m excess pressure, pressure above the atmosphere, overpressure
Überdruckturbine f reaction turbine
Überdruckventil n pressure relief valve
Übereckmaß n across corner dimension
übereinanderlagern v.i. superpose
übereinanderschweißen v.t. lap-weld
übererregen v.t. (electr.) over-excite
Übererregung f (electr.) over-excitation
übereutektoider Stahl, hypereutectoid steel
überfahren v.t. overtravel, overrun
Überfall m (hydr. eng.) overflow, weir
Überfallkommando n flying squad
überführen v.t. (railw.) overbridge; (conveying) transfer, transport
Überführung f (railw.) viaduct
Übergabewalzwerk n pass-over mill

Übergang m change-over; transition; (railw.) crossing
Übergangseisen n off-grade iron
Übergangsrohr n reducing pipe
Übergangszone f (welding) heat-affected zone
übergarer Stahl, overblown steel
übergares Kupfer, dry copper
Übergröße f oversize
Überhang m overhang, projection
überhängen v.i. overhang
überhängend p.a. overhung
Überhebewalzwerk n pull-over mill
überhitzen v.t. overheat, superheat
Überhitzer m superheater
ÜBERHITZER ~, (~anlage, ~rohr, ~schlange, ~ventil) superheater ~
Überhöhung f camber; (road building) superelevation
überholen v.t. (Drehzahlen:) overtake; (Maschinen:) overhaul
Überholmelder m overtaking signal
Überholspur f overtaking lane
Überholungsgetriebe n reduction gearing
Überholungskupplung f overtaking clutch
Überkopfschweißung f overhead welding
Überkorn n (Aufbereitung:) oversize
überkragen v.i. overhang, project
überlagern v.t. superpose, superimpose; overlap
Überlagerung f superposition; overlapping
ÜBERLAGERUNGS ~, (~frequenz, ~kreis, ~schaltung, ~telegrafie) superimposed ~
Überlagerungsempfänger m superheterodyne receiver, superhet

Überlandleitung f *(electr.)* transmission line

überlappen v.t. & v.i. overlap

Überlappnaht f lap weld

Überlappnahtschweißung f lapped seam welding

Überlappschweißung f overlap welding

Überlappstoß m *(riveting, welding)* lap joint

überlappter Stoß, lap-joint

überlappt schweißen, lap weld

Überlappung f overlapping; *(forging, rolling)* lap seam; (Gießfehler:) lap

Überlappungsnietung f lap-joint riveting

Überlappungsschweißung f lap welding

Überlast f overload

Überlastbarkeit f overload capacity

Überlastdrehzahl f overload speed

Überlastschutz m overload protection

Überlastung f overload

Überlastungsauslöser m overload release

Überlastungskupplung f overload clutch

Überlauf m *(hydr. eng.)* overflow

überlaufen v.i. overflow

Überlaufrohr n overflow-pipe

Übermaß n oversize; (Passungen:) interference

überprüfen v.t. check, control; examine; inspect

Überschall m supersonics

ÜBERSCHALL ~, (~frequenz, ~geschwindigkeit, ~kanal, ~welle) supersonic ~

Überschlag m *(electr.)* flashover, spark-over; *(com.)* estimate

Überschlagfunken m jump spark

Überschlagsrechnung f estimate

Überschlagsspannung f spark-over voltage

Überschmiedung f lap seam

überschreiten v.t. exceed

Überschuß m *(electron.)* excess

Überschußgas n surplus gas

überschüssig a. excess, surplus

übersetzen v.t. *(mech.)* transmit

Übersetzung f *(mech.)* gear transmission ratio

Übersetzungsgetriebe n transmission gearing

übersichtlich a. well-ordered, clear

Überspannung f overvoltage

Überspannungsschutz m overvoltage protection

Überstand m excess length

ÜBERSTROM ~, (~auslöser, ~auslösung, ~relais) overcurrent ~

Überströmventil n by-pass valve

Überstundenzuschlag m overtime premium

übertragen v.t. (Kraft:) transmit; (Maße:) transfer; *(copying)* reproduce; *(radio)* broadcast

Übertragung f transmission; communication; (Maße:) transfer

ÜBERTRAGUNGS ~, (~bereich, ~faktor, ~geschwindigkeit, ~leitung, ~mittel, ~netz, ~organ, ~pegel, ~spannung, ~verlust) transmission ~

Übertragungsgerät n *(opt.)* transfer equipment

Überverbrauchszähler m *(electr.)* excess meter

überwachen v.t. supervise

Überwalzung f (Oberflächenfehler:) seam

Überwurfmutter f cap nut

überziehen v.t. (galvanisch:) plate; (Tauchveredlung:) coat
Überzug m coat, coating
Überzugslack m coating varnish
Überzugsmaterial n (met.) cladding material
Überzugsmetall n (= Auflagemetall) cladding material
Uhr f (metrol.) dial indicator
UKW-Bereich m (radio) VHF range
UKW-Rundfunk m VHF broadcasting
Ultrakurzwelle f ultrashort wave
Ultrakurzwellensender m ultrashort wave transmitter
Ultrarotstrahlung f infrared radiation
Ultraschall m supersonics, ultrasonics
ULTRASCHALL ~, (~bohren, ~echolotung, ~frequenz, ~geschwindigkeit, ~prüfung, ~schwingung, ~sender, ~welle) supersonic ~
Ultraviolettstrahlung f ultra-violet radiation
Umbau m reconstruction
umbauen v.t. redesign; reconstruct
umbördeln v.t. border, flange; (mit Drahteinlage:) wire
Umdrehung f revolution, rotation
Umdrehungszahl f number of revolutions
Umdrehungszähler m revolution counter
umfallen v.i. (Walzgut:) tilt over, turn over
umfalzen v.t. bead, crimp over
Umfang m periphery; circumference; extent
Umfangsgeschwindigkeit f peripheral speed, surface speed
Umfangskraft f tangential force
umflechten v.t. (Schläuche:) braid
Umformarbeit f (forging) deformation work
umformen v.t. form, shape; deform

Umformen n (explosive metal-forming) plastic deformation
Umformer m (electr.) converter; (welding) motor generator
Umformerstation f power converter station
Umformgeschwindigkeit f (forging) strain rate, rate of deformation
Umformkraft f (forging) deformation force; forming force
Umformung f (electr.) conversion, transformation; deformation; shaping
Umformverfahren n forming method
umführen v.t. (Walzgut:) loop
Umführen n (Walzgut:) loop mill rolling
Umführung f (rolling mill) repeater
Umfüllpfanne f transfer ladle
umgestalten v.t. redesign
Umkehr f reversal
UMKEHR ~, (~bewegung, ~blechgerüst, ~getriebe, ~hebel, ~motor, ~reibkupplung, ~schalter, ~schütz, ~walzwerk) reversing ~
umkehren v.t. reverse
Umkehrspülung f (auto.) reverse scavenging
Umkehrstromrichter m current converter
Umkehrung f reversal
umkonstruieren v.t. reconstruct
Umlage f (Kosten:) allocation
Umlauf m circulation; rotation
Umlaufdurchmesser m (lathe) swing
umlaufen v.i. rotate, revolve; circulate
umlaufende Schere, rotary shear
Umlauffräsen n profile milling
Umlaufgenauigkeit f concentricity
Umlaufgeschwindigkeit f speed [of rotation]; (≈ Umfangsgeschwindigkeit:) peripheral speed, surface speed

Umlaufgetriebe *n* epicyclic transmission; (≈ Planetengetriebe:) planetary gearing
Umlauföl *n* circulation oil
Umlaufpumpe *f* circulation pump, rotary pump
Umlaufrad *n* planet gear
Umlaufschmierung *f* circulation oiling
Umlaufschrott *m* home scrap
Umlaufzähler *m* revolution counter
umlegen *v.t.* (Blech:) fold over; (Riemen:) shift; (≈ sicken:) bead
umleiten *v.t. (traffic)* divert
Umleitung *f (traffic)* diversion
Umlenkrolle *f (belt drive)* guide pulley
Umlenkspiegel *m (opt.)* deflection mirror
ummantelte Elektrode, sheathed electrode
ummantelter Schlauch, sheathed hose
Ummantelung *f* jacketing, encasing; sheathing
umnummern *v.t.* re-number
umpolen *v.t. (electr.)* reverse polarity
umrichten *v.t.* (e. Werkzeugmaschine:) re-set
Umrichter *m (electr.)* frequency converter
Umriß *m* contour, outline
Umrißfräsmaschine *f* contour milling machine
Umrißfühler *m* profile tracer
Umrißkopieren *n* copying, contouring
Umroll-Rüttelformmaschine *f* jolt rollover machine
umschalten *v.t. (electr.)* switch over; (Hebel:) change over; (Spindel, Tisch:) reverse; (Riemen:) shift; (Revolverkopf:) index
Umschalter *m* change-over switch
Umschaltgetriebe *n* reversing mechanism
Umschaltkupplung *f* reversing clutch

Umschaltung *f (electr., radio, tel.)* commutation; switching
Umschaltventil *n* change-over valve
Umschlag *m (milling)* shift; (Farben:) change
Umschlagbohren *n* shift drilling
Umschmelzeisen *n* remelt iron
umschmelzen *v.t.* remelt; Zink:) redistil
Umschmelzofen *m* remelting furnace
Umschmelzverfahren *n* remelting process
Umsetzer *m* (Frequenzen:) changer
umspannen, *v.t.* (Werkstücke:) rechuck; (Werkzeuge:) re-set, reset; (Riemen:) shift; *(electr.)* transform
Umspanner *m (electr.)* transformer
Umspinnung *f* (Draht:) covering
Umsteckrad *n* (für Drehzahlenwechsel:) pickoff gear
Umsteckwalzwerk *n* looping mill
umstellen *v.t.* change over (to); (e. Maschine:) reset; (Meißelanordnung:) re-tool
Umstellzeit *f (mach.)* re-setting time
Umsteuergetriebe *n* reversing gear
Umsteuermotor *m* reversing motor
umsteuern *v.t.* reverse
umwälzen *v.t.* circulate
Umwälzflügelrad *n* circulating impeller
Umwälzpumpe *f* circulating pump
umwandeln *v.t.* convert; (Gewinde:) transpose
Umwandlung *f* conversion, transformation, change, modification
Umwandlungsbereich *m* transformation range
Umwandlungsgetriebe *n* reversing gears; *(screwcutting)* transposing gears

Umwandlungspunkt *m* thermal critical point, transformation point
Umwandlungstemperatur *f (met.)* transformation temperature
Unachtsamkeit *f* carelessness
U-Naht *f (welding)* single-U-butt weld
unbelasteter Zustand, no-load condition
unberuhigter Stahl, rimming steel, rimmed steel
unbeweglich *a.* fixed, stationary
unbewehrter Beton, unreinforced concrete, plain concrete
undicht *a.* leaky
Undichtheit *f* leakiness
undurchlässig *a.* impervious; *(opt.)* opaque
undurchsichtig *a.* opaque
Undurchsichtigkeit *f* opacity
uneben *a.* out of flat; uneven; rough
Unebenheit *f* out-of-flatness; unevenness
unerreicht *a.* unrivalled, unexcelled
Unfall *m* accident
Unfallgefahr *f* risk of accident
Unfallhilfskraftfahrzeug *n* accident ambulance
unfallsicher *a.* accident-proof
Unfallstation *f* ambulance station
Unfallverhütung *f* accident prevention
Unfallverhütungsvorschrift *f* safety regulation
Unfallversicherung *f* accident insurance
Unfallwagen *m* ambulance car
ungelöschter Kalk, quick-lime
ungerade *a.* out-of-straight; uneven
ungleichmäßig *a.* non-uniform, heterogeneous
Ungleichmäßigkeit *f* non-uniformity, heterogeneity
unhandlich *a.* unwieldy

UNIVERSAL ~, (~bohrmaschine, ~drehmaschine, ~fräsmaschine, ~futter, ~gelenk, ~kreissäge, ~läppmaschine, ~motor, ~planscheibe, ~räummaschine, ~schleifmaschine, ~schraubenschlüssel, ~stoßmaschine, ~teilgerät, ~tiefenlehre, ~teilkopf, ~tischkreissäge, ~walzwerk, ~werkzeughalter, ~windeisen, ~winkelmesser, ~zange) universal ~
Universalspannfutter *n* scroll chuck
Universalstahl *m* universal mill plate
unlegiert *a.* unalloyed
unlegierter Kohlenstoffstahl, plain carbon steel
unlegierter Werkzeugstahl, plain carbon tool steel
unmittelbares Teilen, direct indexing
unruhig vergossener Block, rimmed steel ingot
unrund *a.* out of round, untrue
Unsauberkeit *f (surface finish)* (sandige Oberfläche:) scabbiness
Unterbau *m (railway)* substructure
unterbrechen *v.t.* interrupt; stop; *(electr.)* disconnect
Unterbrecher *m (electr.)* interrupter, contact breaker
Unterbrecherkontakt *m* interrupter contact
Unterbrecherschalter *m* interrupter switch
Unterbrechung *f* interruption; *(electr.)* break, disconnection
unterbringen *v.t.* accommodate
unterbrochene Härtung, broken hardening
Unterdruck *m* underpressure; vacuum
Unterdruckbremse *f* vacuum brake

Unterdruckmanometer n vacuum gage
Unterdruckservobremse f vacuum servo brake
untereutektoid a. hypoeutectoid
Unterfangung f (civ. eng.) underpinning
Unterflurebene f below ground level
Unterflurmotor m (auto.) underfloor engine
Unterflurofen m underground furnace
Unterführung f (railway) subway crossing, underpass, underbridge
Untergesenk n lower die, bottom die
Untergestell n (e. Maschine:) base frame
untergießen v.t. (mit Zement:) grout
Untergrund m (building) sub-floor; subsoil
Untergurt m bottom boom, bottom chord
Unterklotzung f underpinning
Unterkorn n (Aufbereitung:) undersize
Unterkühlung f subcooling, undercooling
Unterlagblech n shim
Unterlage f pad, padding
unterlegen v.t. (mit Blech:) shim
Unterlegscheibe f washer; (≈ Abstandscheibe:) spacer
Untermaß n undersize
Unternehmer m (building) contractor
Unterpulverschweißen n submerged-arc welding
Unterputz m (building) scratch coat
Unterputzsteckdose f flush socket
Unterputzverlegung f underplaster installation
Untersatz m base
Unterschienenschweißung f firecracker welding

Unterschlitten m (lathe) bottom slide; (miller) saddle; (planer) base slide
unterschneiden v.t. (metal cutting) undercut
Untersetzungsgetriebe n (= Reduziergetriebe) speed reduction gear
Unterspannung f under-voltage
Unterspannungsauslöser m undervoltage trip
Unterspannungsschalter m undervoltage circuit-breaker
unterspülen v.t. (building) scour, underwash
Unterstempel m lower die, bottom die
unterstützen v.t. support
untersuchen v.t. investigate, examine, test, inspect, study
Unterteil n (mach.) base
Unterwaschung f (building) scouring, undermining, underwashing, subsurface erosion
Unterwasserschweißung f underwater welding
Unterwind m down-draft
Unterzug m (road building) cross girder
unveränderlich a. invariable
unverseifbar a. unsaponifiable
Unwucht f out-of-balance, unbalance
unzugänglich a. inaccessible
unzweckmäßig a. impractical
Uranpechblende f pitchblende
Urlehre f master gage
US-Schweißen n firecracker welding
U-Stahl m channels pl.
UST-Gewinde n unified screw thread
U-Stoß m (welding) single-U-butt joint
Utensilien pl. utensils

V

Vakuum *n* vacuum
VAKUUM ~, (~bremse, ~lampe, ~messer, ~pumpe, ~röhre, ~schalter, ~trokkenofen) vacuum ~
V-Bahn *f (lathe)* V-track
Vektordiagramm *n* vector diagram
Vektormodell *n (nucl.)* vector model
Vektorwinkel *m* vectorial angle
Ventil *n* valve
VENTIL ~, (~einstellehre, ~gehäuse, ~hub, ~kegel, ~klappe, ~körper, ~nadel, ~schaft, ~sitz) valve ~
Ventilation *f* ventilation
Ventilator *m* ventilator, fan; blower
Ventilatorflügel *m* fan blade
Ventilatorkühlung *f* forced-draught cooling
Ventilatorriemen *m* fan belt
Ventileinstellschlüssel *m (auto.)* tappet wrench
Ventileinstellung *f* valve timing
Ventilkegelschleifmaschine *f* valve refacer
ventilloser Motor, sleeve-type engine
Ventilschleifmaschine *f* valve refacer
Ventilsitzfräser *m* valve seat reamer
Ventilsitzring *m (auto.)* valve seat insert
Ventilsitzschleifmaschine *f* valve seat grinder
Ventilspiel *n (auto.)* tappet clearance
Ventilspindel *f* valve stem
Ventilspülung *f (auto.)* valve scavenging
Ventilsteuerung *f* valve timing; (als Bauteil:) valve gear
Ventilteller *m* valve disc
veraluminieren *v.t.* aluminize

veränderlich *a.* variable
verändern *v.t.* vary, alter, change
verankern *v.t.* anchor; tie; stay, guy
Verankerung *f* anchorage
Verankerungsbolzen *m* tie-bolt, anchor bolt
Verarbeitbarkeit *f* workability
verarbeiten *v.t.* work, treat; process, manufacture; use up, work up
Verarbeiter *m* processor
Verarbeitung *f* working, forming, shaping; machining; fabrication
Verarbeitungsbetrieb *m* fabricating shop
Verarbeitungskostenstandard *m* operating cost standard
veraschen *v.t.* ash
Veraschung *f* incineration
Verband *m* (Steine:) bond
Verbandputz *m (building)* bond plastering
verbiegen *v.t.* deform, distort; bend
verbinden *v.t.* connect; joint; combine; unite; tie together; *(chem.)* combine
Verbinder *m* (Riemen:) fastener, lacer; (Kabel:) connector
Verbindung *f* connection; combination; junction; union; *(chem.)* composition; *(electr.)* connection; *(mech.)* joint; *(met.)* fusion; amalgamation; *(railway)* communication; (Kabel:) joint; *(radio)* communication
Verbindungsdose *f* junction box
Verbindungsgleis *n* junction track
Verbindungsglied *n* link
Verbindungskabel *n* junction cable
Verbindungsklemme *f* fastener; *(electr.)* terminal

Verbindungsleitung f *(electr.)* connection line
Verbindungslöten n fusion brazing
Verbindungsmuffe f *(electr.)* cable joint
Verbindungsschalter m *(tel.)* position coupling key
Verbindungsschnur f flexible cord
Verbindungsschweißen n full-fusion welding
Verbindungsschweißung f joint weld
Verbindungsstecker m *(electr.)* connection plug
Verbindungsstelle f joint; junction
verblasen v.t. *(steelmaking)* blow
verbleien v.t. lead-coat
verblenden v.t. *(building)* face
Verblendstein m facing brick
verbolzen v.t. bolt; ($\approx$ verstiften) pin, insert a pin
Verbrauch m consumption
Verbraucher m consumer, user
Verbraucherleitung f *(electr.)* service cable
verbreiten v.t. spread; propagate
verbreitern v.t. widen; extend; enlarge
verbrennbar a. combustible
Verbrennbarkeit f combustibility
verbrennen v.t. burn
Verbrennung f combustion; burning; oxidation
VERBRENNUNGS ~, (~druck, ~formel, ~gase, ~kammer, ~luft, ~motor, ~raum, ~rohr, ~rückstand, ~zone) combustion ~
Verbrennungskraftmaschine f internal combustion engine
Verbund m *(building)* bond
VERBUND ~, (~block, ~dampfmaschine, ~dynamo, ~erregung, ~kokille, ~maschine, ~motor, ~pumpe, ~träger, ~transformator, ~turbine, ~wicklung) compound ~
Verbundbetrieb m *(electr.)* interconnection of power systems; *(met.)* compound plant
Verbundguß m composite casting, compound casting
Verbundmaterial n *(plastics)* sandwich material
Verbundmetall n composite metal
Verbundsicherheitsglas n laminated safety glass
Verbundstahl m composite steel
Verbundverfahren n *(met.)* duplexing process
verchromen v.t. chrome-plate; chromize
Verchromung f chrome-plating, chromium plating
verdampfen v.i. vaporize; evaporate
Verdampfung f vaporization, evaporation
Verdeck n *(auto.)* top, roof; (Riemen:) guard
verdecken v.t. cover; mask
verdeckt p.a. hidden, concealed
verdecktes Lichtbogenschweißen, shielded arc welding
verdichten v.t. *(mech.)* compress; condense; *(concrete)* compact
Verdichter m compressor; condenser
Verdichtung f compression; condensation; concentration; (Beton:) compaction; (Formsand:) pucking
VERDICHTUNGS ~, (~druck, ~grad, ~hub, ~verhältnis, ~zündung) compression ~
verdicken v.t. thicken; concentrate; (Rohre:) bulge out
verdoppeln v.f. double

Verdopplung f (Frequenzen:)) doubling
verdrahten v.t. wire
Verdrahtungsplan m wirng diagram
verdrallen v.t. twist
Verdrängerkolben m *(power press)* plunger
Verdrängung f displacement
verdrehen v.t. twist, turn round
Verdrehfestigkeit f torsional strength
Verdrehmoment n torsional moment
Verdrehung f torsion, twist
VERDREHUNG ~, (~beanspruchung, ~festigkeit, ~kraft, ~messer, ~modul, ~moment, ~prüfmaschine, ~versuch, ~widerstand) torsional ~, torsion ~
verdübeln v.t. dowel
verdünnen v.t. dilute, thin; (Gas:) rarefy
Verdünnung f dilution, thinning
Verdünnungsmittel n thinner, diluting media, diluent
verdunsten v.i. evaporate
Verdunstung f evaporation
veredeln v.t. (Stahl:) refine, improve; (Leichtmetall:) age-harden
vereinfachen v.t. simplify
verfahren v.t. (e. Maschinentisch:) traverse, move; proceed
Verfahren n process; method; procedure
Verfahrenstechnik f process engineering, processing technology
verfahrenstechnisch a. operational
verfärben v.r. decolorize
Verfärbung f discoloration
verfestigen v.t. reinforce; stiffen; strengthen
Verfestigung f strengthening, reinforcement; (durch Kaltverformung;) strain hardening, work hardening; (e. Klebstoffes:) solidifying process

verformbar a. workable, deformable
Verformbarkeit f deformability, forming property
verformen v.t. form, deform, work; shape; (≈ deformieren:) deform
Verformung f deformation; shaping, etc.; cf. verformen
Verformungsbruch m fracture
verfugen v.t. join
verfüllen v.t. (civ.eng.) fill
vergasen v.t. gasify
Vergaser m *(US)* carburetor; *(UK)* carburetter
VERGASER ~, (~düse, ~einstellung, ~luft, ~schwimmer) carburetor ~
Vergaserklappe f *(auto.)* choke
Vergaserluftklappe f *(auto.)* choke
Vergasermotor m carburetor engine, gasoline engine, petrol engine
Vergießbarkeit f *(founding)* castability, pourability
vergießen v.t. *(met.)* pour; cast; teem; *(concrete)* grout in)
vergilben v.i. yellow
vergipsen v.t. plaster
verglasen v.t. glaze
vergleichen v.t. compare
Vergleichsmaß n reference gage
Vergleichsmessung f comparative measurement
Vergleichsprüfung f comparison test
vergrößern v.t. enlarge; increase; *(opt.)* magnify
Vergrößerung f enlargement; increase; *(opt.)* magnification
Vergußmasse f *(plastics)* sealing compound; (Zement:) grout
vergütbar a. heat treatable; (Stahl:) susceptible to hardening and tempering;

Vergütbarkeit (Al-Legierungen:) agehardenable

Vergütbarkeit f (Stahl:) heat treating property, tempering quality; (Al-Legierungen:) quench-aging property

vergüten v.t. (Stahl:) quench and temper; (Al-Legierungen:) age-harden

Vergüteofen m quenching and tempering furnace

Vergütung f (Stahl:) quenching and tempering; (auf Bainit:) austempering; (Al-Legierungen:) age-hardening

Vergütungsstahl m heat treatable steel; (Erzeugnis:) quenched and tempered steel

Vergütungszustand m (von Stahl:) quenched and tempered condition; temper

Verhalten n behaviour; characteristics

Verhältnis n ratio; proportion; relation

verharren v.i. dwell

verharzen v.i. gum

verhütten v.t. smelt

verjüngen v.r. taper

verkadmieren v.t. cadmium-plate

verkanten v.t. tilt, tip; cant

Verkehr m traffic

VERKEHRS ~, (~ader, ~ampel, ~anlage, ~regelung, ~stockung, ~zeichen) traffic ~

Verkehrslast f live load

Verkehrsteilnehmer m (auto.) road user

Verkehrswert m trade-in value

verkehrt konischer Block, big-end-up ingot

verkeilen v.t. key; wedge

verketten v.t. interlink

verkitten v.t. stop with putty; cement; lute; seal; (Formmasken:) stick

verkleben v.t. bond; glue together

verkleiden v.t. panel, sheath; board; line; jacket; *(building)* case

verkleinern v.t. reduce; decrease

Verkleinerung f reduction; decrease

verklemmen v.t. lock, bind; – v.r. seize, jam; block

verkohlen v.t. & v.i. carbonize, coke

verkoken v.t. coke; carbonize; distil

Verkokungskammer f coking chamber

verkröpfen v.t. offset; crank

Verkröpfgesenk n snaker

verkrümmen v.t. & v.i. warp, distort

verkupfern v.t. copper-plate; copper

Verladeanlage f handling equipment

Verladebrücke f handling bridge, loading bridge

Verladebühne f loading ramp

Verladekran m material handling crane

Verladerampe f loading ramp

verlagern v.t. displace, dislocate; – v.r. misalign

Verlagerung f ($\approx$ Lageänderung:) displacement; dislocation; (aus der Fluchtebene:) misalignment

verlängern v.t. lengthen, extend; elongate

Verlängerung f extension, elongation

verlaschen v.t. fishplate; (Holz, Schienen:) fish

Verlauf m course, run; (Kosten:) flow

verlegen v.t. (Leitungen:) lay, install; ($\approx$ verschieben:) displace

Verlegung f (railw.) laying

verleimen v.t. cement, glue together

Verlustfaktor-Meßgerät n dielectric loss factor meter

Verlustwinkel m *(electr.)* loss angle

Verlustzeit f *(work study)* lost time

vermauern v.t. lay bricks, mason

vermessen *v.t.* measure; survey
Vermessung *f* survey
Vermessungskunde *f* surveying
Vermittlung *f (tel.)* exchange
Vermittlungsamt *n* central exchange
vernageln *v.t.* nail
vernichten *v.t.* destroy, ruin, spoil; – *v.r.* annihilate
Vernichtung *f (phys.)* annihilation; (Abfall:) removal
vernickeln *v.t.* nickel; (galvanisch:) nickel-plate
vernieten *v.t.* rivet
verölt *a.* oily
Verputz *m* plaster work
verputzen *v.t.* (Guß:) dress, clean; (Wände:) plaster
Verputzerei *f (founding)* dressing room; *(building)* plastering
verriegeln *v.t.* lock; latch; arrest
Verriegelung *f* locking, bolting, interlocking; (als Bauelement:) locking mechanism
verrippen *v.t.* rib
verrohren *v.t.* pipe
verrosten *v.i.* rust, corrode, get rusty
verrußen *v.i.* soot
versagen *v.i.* fail
Versand *m* delivery, conveyance; shipment
Versatz *m* (e. Gesenkes:) die shift; (≈ Fluchtung:) misalignment
verschalen *v.t.* encase, case, board; jacket; *(building)* sheet, shutter
Verschaltung *f (electr.)* faulty connection
Verschalung *f* encasing; panelling, boarding, planking, shuttering, facing, sheeting
Verschalungsbrett *n* shuttering board

verschiebbar *a.* movable, displaceable
Verschiebeankermotor *m* armature shifting motor
verschieben *v.t.* move, shift; displace; (≈ verfahren:) traverse; (zeitlich:) postpone
Verschieberädergetriebe *n* sliding gear drive
Verschiebezahnrad *n* sliding gear
Verschiebung *f (mech.)* shifting, moving, movement, displacement; *(electr.)* shift, shifting
verschlacken *v.t.* slag, flux, scorify; – *v.i.* sinter, scale
Verschlackungsperiode *f* (Birnenbetrieb) slag-forming period
Verschlackungsvermögen *n* fluxing power
Verschleiß *m* wear; abrasion; erosion
verschleißen *v.i.* wear out; abrade; erode
verschleißfest *a.* wear-resistant
Verschleißfläche *f* wearing surface
Verschleißmulde *f (metal cutting)* crater
Verschleißteil *n* wearing part
Verschleißwiderstand *m* wear resistance
verschlissen *p.p.* worn out
Verschluß *m* closure; lock; seal; *(photo.)* shutter
Verschlußstopfen *m* stopper
verschmutzen *v.i.* contaminate, soil
Verschmutzung *f* contamination; dirt; pollution; (e. Ventils:) fouling
verschneiden *v.t.* (Flüssigkeiten:) thin, mix; adulterate
verschrammen *v.t.* scratch, score
verschränken *v.t.* (Riemen:) cross; (Sägen:) set
verschrauben *v.t.* screw; bolt

Verschraubung f screw joint; screwing; bolting

verschrotten v.t. scrap

verschweißen v.t. weld; – v.i. weld together

verschwelen v.t. carbonize under vacuum at a low temperature

Verschwelung f *(coking)* low-temperature distillation

verseifbar a. saponifiable

verseifen v.t. saponify

Verseifungszahl f saponification value

verseilen v.t. twist

versenken v.t. *(mach.)* countersink; (zylindrisch:) counterbore

versetzen v.t. (≈ verschieben:) shift, displace; (≈ umsetzen:) relocate; (auf Luke:) stagger; (≈ kröpfen:) offset

versiegeln v.t. seal

versorgen v.t. supply

VERSORGUNGS ~, (~anlage, ~gebiet, ~leitung, ~netz) supply ~

verspannen v.t. fasten, brace; stay, guy; tension; (falsch:) grip faulty; – v.r. distort

versplinten v.t. cotter

verstählen v.t. steel face

verstärken v.t. reinforce; *(radio)* magnify, amplify

Verstärker m *(electr.)* amplifier

Verstärkeramt n repeater station

Verstärkerröhre f amplifying valve, amplifier tube

Verstärkung f *(mech.)* reinforcement; *(radio, telec.)* amplification, gain; *(photo.)* intensification

versteifen v.t. stiffen; (≈ verstärken:) reinforce; (≈ verstreben:) brace

Versteifung f reinforcement, stiffening

verstellbar a. adjustable

verstellbarer Nutenfräser, interlocking slotting cutter

verstellbares Strichmaß, slide caliper rule

Verstellbereich m range of adjustment

verstellen v.t. adjust; displace; (Drehzahlen, Vorschübe:) vary; (Hebel:) shift; (senkrecht:) raise and lower

Verstellmotor m brush-shifting motor; (zum Verfahren von Supporten, Maschinentischen:) traversing motor

Verstellung f adjustment; displacement; shifting

verstemmen v.t. calk; *(UK)* caulk

verstiften v.t. pin; dowel

verstopfen v.t. & v.r. clog

verstreben v.t. strut, brace

Verstreichmörtel m rendering mortar

Versuch m test; experiment; trial

versuchen v.t. test; try out; experiment

Versuchsmodell n test model, pilot model

Versuchsreihe f test series

Versuchsstadium n experimental stage

verteilen v.t. distribute; divide; proportion

Verteiler m *(tel.)* distribution frame; *(auto.)* distributor

Verteilerfeld n distribution panel

Verteilergetriebe n *(auto.)* power take-off gear; *(mach.)* distributor drive

Verteilerkasten m *(electr.)* distribution box

Verteilerklemme f *(electr.)* terminal

Verteilerkopf m *(auto.)* distributor head

Verteilerrohr n *(auto.)* manifold

Verteilgesenk n edger

Verteilung f distribution

Verteilungskasten m (für Kabel:) distribution box

Verteilungsnetz n *(telec.)* distributing network
Verteilungsschalttafel f distribution board
Verteilungstafel f *(electr.)* distribution board
Vertiefung f cavity; depression; indentation; (Oberflächenfehler:) impression
VERTIKAL ~, (~antenne, ~bohrmaschine, ~fräsmaschine, ~frequenz, ~hub, ~räummaschine, ~revolverdrehmaschine, ~ständer, ~vorschub) vertical ~; *s.a.* Senkrecht ~
Vertikalgattersäge f mill saw
verunreinigen v.t. soil, pollute, contaminate
Verunreinigung f impurity; contamination; (Gewässer:) pollution; (e. Ventils:) fouling; clogging
vervielfachen v.t. multiply
Vervielfacher m *(electron.)* multiplier
verwackeln v.t. *(photo., telev.)* blur
Verwaltungskostenstelle f administrative cost center
verwandeln v.t. change, convert, transform
verweilen v.i. dwell
Verweilzeit f (Ofen:) holding time; (Maschinentisch:) dwell idling time
verwerfen v.r. warp, distort
Verwerfung f (von Gleisen:) distortion; *(radio)* shift
verwinden v.t. twist
Verwindung f torsion, twist; distortion
verwindungssteif a torsion-resistant
verzahnen v.t. cut teeth; cut gears
Verzahnen n gear cutting; (ballig:) convex tooth cutting, crowning
Verzahnmaschine f gear cutting machine

Verzahnung f [gear] tooth system; gear production; gear design; (als Arbeitsvorgang:) gear cutting
Verzahnungsfräser m tooth milling cutter
Verzahnungswälzfräser m gear hob
verzapfen v.t. *(woodworking)* mortise
verzerren v.t. distort
Verzerrung f distortion
verziehen v.r. distort, warp
Verzimmerung f timbering
verzinken v.t. galvanize; (feuerverzinken:) hot-galvanize; (galvanisch:) cold-galvanize, electroplate
Verzinkerei f galvanizing plant
Verzinkungsanlage f galvanizing plant
verzinnen v.t. tin; (feuerverzinnen:) tincoat; (galvanisch:) tin-plate
Verzögerer m *(nucl.)* moderator
verzögern v.t. delay, retard; *(speeds)* decelerate
Verzögerung f retardation, delay; deceleration
Verzögerungslinse f *(opt.)* cutoff lens
Verzögerungsrelais n time-delay relay
Verzögerungsspannung f *(electr.)* delay-action voltage; *(telev.)* threshold voltage
Verzug m *(mech.)* distortion, warping, warpage; *(electr.)* delay, lag
verzundern v.i. scale
Verzweiger m *(telegr.)* branch joint
Verzweigung f branching; bifurcation
Verzweigungspunkt m junction
V-Getriebe n V-transmission
Viadukt m viaduct
Vibration f vibration
VIBRATIONS ~, (~dämpfung, ~galvanometer, ~schwelle) vibration ~
Vibrationsmeßgerät n vibrating reed instrument

Vibrationsregler *m* Tirrill regulator
Vibrationsschweißung *f* pulsation welding
Videofrequenz *f* (telev.) video frequency
Videosignal *n* (telev.) video signal
vieladrig *a.* multi-core
vielatomig *a.* polyatomic
Vieleck *n* polygon
vielfach *a.* multiple, multi-
VIELFACH ~, (~abtastung, ~beschleuniger, ~betrieb, ~empfang, ~feld, ~leiter, ~zerlegung) multiple ~
Vielfachsteckdose *f* multiple-contact jack
vielgliederig *a.* (chem.) polymerous; (math.) polyminal
Vielhärtungsriß *m* hot crack
Vielkant *m* polyhedron, polygon
Vielkeilwelle *f* multiple-spline shaft
vielphasig *a.* multiphase, polyphase
vielpolig *a.* multipole, multi-contact
Vielschnittdrehmaschine *f* multi-cut lathe
vielseitig *a.* versatile
Vielseitigkeit *f* versatility
Vielspindelautomat *m* multi-spindle automatic
Vielspindelbohrmaschine *f* multiple-spindle drilling machine
vielstufig *a.* multi-stage
vielstufiges Zahnradgetriebe, multiple-geared transmission
Vielwalzengerüst *n* cluster mill
Vielwalzenwerk *n* cluster mill
Vielzweckmaschine *f* multi-purpose machine
Vierbackenfutter *n* four-jaw chuck
Viereck *n* square; rectangle
viereckig *a.* square; rectangular
Viererleitung *f* (electr.) phantom circuit

Vielfachmeißelhalter *m* four-way tool block
Vierfachspaltung *f* (nucl.) quadruple fission
Vierganggetriebe *n* (auto.) four-speed drive; four-speed transmission
Vierkant *m* (geom.) tetrahedral; (techn.) square
Vierkantblock *m* (met.) square ingot
Vierkantfeile *f* square file
Vierkantholzschraube *f* (UK) square head coach screw
Vierkantkopfschraube *f* square head bolt
Vierkantlineal *n* knife edge
Vierkantmaterial *n* square stock
Vierkantmeißel *m* box tool
Vierkantrevolverkopf *m* square turret
Vierkantschraubenschlüssel *m* square box wrench
Vierkantstahl *m* square steel
Vierkantstiftschlüssel *m* square socket screw wrench
Vierleiterkabel *n* four-core cable
Vierpolröhre *f* tetrode
Vierpunktaufhängung *f* (auto.) four-point suspension
Vierradantrieb *m* (auto.) four-wheel drive
Vierradbremse *f* (auto.) four-wheel brake
Vierrädergetriebe *n* four-gear drive
Viersäulen-Friktionsspindelpresse *f* four-column friction screw press
Vierspindelautomat *m* four-spindle automatic machine
Viertakt *m* four-stroke-cycle
Viertakter *m* four-stroke engine
Viertaktmotor *m* four-stroke engine
Vierwalzengerüst *n* four-high rolling stand
Vierwegebohrmaschine *f* four-way drilling machine; four-way boring machine

Vignolschiene f flat-bottom rail
Vincentpresse f Vincent friction screw press
V-Innengetriebe n V-type internal transmission
Visiereinrichtung f sighting device
visieren v.t. sight
Visiergerüst n (surveying) sight rail
Visierkreuz n cross wires
Visiermikroskop n locating microscope
Visierwinkel m angle of sight
Viskosität f viscosity, viscidity
V-Naht f single-V-butt weld
V-O-Außengetriebe n V-O-external transmission
V-O-Innengetriebe n V-O-internal transmission
voll a. full; complete; ($\approx$ massiv:) solid
Vollast f (Motor:) full load
VOLLAST ~, (~anlauf, ~drehzahl, ~leistung, ~leistungsfaktor, ~moment, ~schlupf, ~spannung) full-load ~
Vollautomat m fully automatic machine; (building) automatic block machine
vollautomatisch a. fully automatic, full-automatic
vollbohren v.t. drill
vollelektrisch a. all-electric
Vollgatter n multiple blade frame saw
Vollgummireifen m (UK) solid tyre; (US) solid tire
Vollguß m solid casting
vollhydraulische Maschine, fully hydraulic machine
völlig a. complete, full, quite; – adv. totally, perfectly
vollsaugen v.r. soak
Vollseil n solid rope
vollselbstätig a. fully automatic

vollständig a. total, complete
vollwandiger Träger, plate girder
Vollwandkonstruktion f solid contruction
Vollwelle f solid shaft
Voltmesser m voltmeter
VOR ~, (arbeitstechnisch:) (~arbeiten, ~bearbeiten, ~bohren, ~drehen, ~kopieren, ~läppen, ~planen, ~polieren, ~räumen, ~reiben, ~schleifen, ~schlichten, ~schmieden, ~schneiden, ~stoßen) rough ~
VOR ~, (Zeit, Folge:) (~bearbeiten, ~behandeln, ~belasten, ~einstellen, ~erwärmen, ~fertigen, ~heizen, ~kühlen, ~lochen, ~planen, ~mischen, ~montieren, ~spannen, ~trocknen, ~wärmen) pre-~
Voranschlag m (com.) estimate
Vorarbeit f preliminary work
Vorarbeiter m leading hand
Vorbau m extension
Vorbaumeißelhalter m adapter toolholder
Vorbehandlung f preliminary treatment, pretreatment; (plastics) conditioning treatment
Vorbeizen n (Weißblechfabrikation) black pickling, first pickling
vorbereiten v.t. prepare
Vorbereitung f preparation
vorblasen v.t. (met.) fore-blow, pre-blow
Vorblock m (met.) bloom; (UK) cogged ingot
vorblocken v.t. (met.) bloom; (UK) cog
Vorblockwalze f (US) blooming roll; (UK) cogging roll
VORDERACHS ~, (~antrieb, ~aufhängung, ~gehäuse, ~lager) front axle ~
Vorderachse f front axle
Vorderachsschenkel m (auto.) stub axle

Vorderansicht f front view
Vorderkipper m front dump truck
Vorderkotflügel m (auto.) front wing; front mudguard
Vorderlager n front bearing
Vorderradantrieb m (auto.) frontwheel drive
Vorderradaufhängung f (auto.) front-wheel suspension
Vorderradbremse f (auto.) front-wheel brake
Vorderseite f front
voreilen v.i. lead; (Walzgut:) slip forward
Voreilung f lead; (Walzgut:) forward slip
Voreinspritzung f (auto.) pilot injection
Vorfräser m stocking cutter
Vorgabebezeichnung f (cost accounting) budget determinant
Vorgabeleistung f (work study) standard performance
Vorgabewert m preset value
Vorgabezeit f standard time
Vorgabezeitermittlung f (time study) rate setting
Vorgang m operation; process; procedure
vorgeben v.t. (work study) allow
vorgegossenes Loch, core hole
Vorgelege n (Riemen:) countershaft; (Zahnräder:) backgears
Vorgelegerad n backgear
Vorgelegescheibe f (Riemen:) countershaft pulley
Vorgelegewelle f (Riemen:) countershaft; (Zahnräder:) back gear shaft
Vorgerüst n break-down stand, cogging stand, roughing stand
vorhalten v.t. (Niete:) back up
Vorhalter m (riveting) dolly
Vorherd m forehearth

Vorkaliber n (rolling mill) preceding pass
Vorkammer f (engine) pre-combustion chamber
Vorkammermotor m precombustion chamber engine
vorkommen v.i. occur
Vorlack m undercoating varnish
Vorlage f (Gasreinigung:) off-take main, collecting main; (welding) water seal
Vorlast f initial load, pre-load
Vorlauf m forward motion, advance, approach
vorlaufen v.i. move forward, advance, approach
Vorlaufgetriebe n forward gears
Vorlegscheibe f washer
Vormauerziegel m facing brick
Vormischung f (concrete) premixing
Vormontage f pre-assembly
Vorpreßwerkzeug n (plastics) preforming tool
Vorprobe f preliminary test
vorraffinieren v.t. (Werkblei:) (US) soften; (UK) improve
Vorraffinierofen m (US) softening furnace; (UK) improving furnace
Vorreibahle f roughing reamer
Vorrichtung f device; apparatus; facility; attachment; (als Spanner:) fixture; (als Schaltwerk:) mechanism; (boring, welding) jig
Vorrichtungsbau m design of jigs and fixtures
Vorrichtungsbohrmaschine f jig borer
vorschalten v.t. superpose; (electr.) connect in series
Vorschaltkondensator m series capacitor
Vorschaltwiderstand m series resistor; series resistance

vorschieben *v.t.* advance, move forward
Vorschlaghammer *m* blacksmiths' sledge hammer
Vorschmiedegesenk *n* blocker, rougher, blanker
Vorschmiedegravur *f* semi-finishing impression, blocking impression
Vorschneidezange *f* end cutting plier
Vorschrift *f* specification; instruction; code; regulation
Vorschub *m* feed [motion]
VORSCHUB ~, (~anschlag, ~antrieb, ~bereich, ~bewegung, ~druck, ~einstellung, ~geschwindigkeit, ~getrieberad, ~größe, ~hebel, ~kraft, ~kurve, ~motor, ~patrone, ~räderkasten, ~regelung, ~reihe, ~rohr, ~schalter, ~schlitten, ~tabelle, ~vorwähler, ~wechsel, ~wechselrad, ~welle, ~zange) feed ~
Vorschubgetriebe *n* feed gearbox, feed gear mechanism
Vorschubräderplatte *f* feed-gear quadrant
Vorschubschaltgetriebe *n* in-and-out feed gear mechanism
Vorschubselbstgang *m* power feed motion
Vorschubverfahren *n (grinding)* traverse grinding
Vorschubwendegetriebe *n* feed reversing mechanism
Vorsichtsmaßnahme *f* precautionary measure
Vorspannbatterie *f* biasing battery
vorspannen *v.t.* prestress, preload; *(electron.)* bias
Vorspannung *f* prestressing; preloading; *(electr.)* bias potential

vorspringend *p.a.* protruding
Vorsprung *m* projection; *(building)* jump
Vorspurwinkel *m (auto.)* toe-in
Vorsteckstift *m* stop pin, locking pin
vorstehen *v.i.* protrude, project
Vorstich *m (rolling mill)* roughing pass, breaking down pass
Vorstoßschiene *f (civ.eng.)* edging strip
Vorstraße *f (rolling mill) (US)* blooming train; breaking down train; *(UK)* roughing train, cogging train
vorstrecken *v.t. (rolling)* rough, rough down, cog; bloom; break down
Vorstreckgerüst *n* pony rougher
Vorsturzwalzwerk *n* mill for rolling breakdowns
Vortrieb *m (rocket)* propulsion; *(civ. eng.)* driving, heading
Vortriebsstelle *f (road building)* placement site
vorverdichten *v.t.* supercharge
Vorverdichter *m (engine)* supercharger
Vorverdichtung *f (auto.)* supercharging, precompression
Vorwahl *f* preselection
Vorwähler *m* preselector
Vorwählschaltung *f* preselection control
Vorwalze *f* roughing roll, cogging roll, blooming roll; break-down roll
vorwalzen *v.t.* rough down, cog down, bloom; (Blechstürze:) break down
Vorwalzgerüst *n* rougher, roughing stand, cogging stand, blooming stand; breaking-down stand
Vorwärmer *m* economizer
Vorwärmflamme *f* preheat flame
Vorwärmofen *m* preheating furnace
Vorwärmung *f* preliminary heating; preheating

Vorwärtsgang *m* forward speed, forward gear
Vorwärtsgeschwindigkeit *f* forward speed
Vorwärtsschweißung *f* rightward welding

Vorwiderstand *m* series resistor; series resistance
Vorzeichen *n* (math.) sign
V-Rad *n* V-gear
V-Stoß *m* (welding) single-Vee butt joint
Vulkanfiber *f* vulcanized fibre

W

Waage f balance, scales pl.
WAAGERECHT ~, (~bohrmaschine, ~bohr- und ~fräsmaschine, ~feinbohrmaschine, ~fräsmaschine, ~konsolfräsmaschine, ~räummaschine, ~revolverdrehmaschine) horizontal ~
Waagerechtstoßmaschine f shaper, shaping machine
Wabenkühler m (auto.) honeycomb radiator
Wachs n wax
Wachsausschmelzgießerei f investment foundry
Wachsausschmelzgießverfahren n investment-casting process
Wachsausschmelzguß m investment casting
Wachsausschmelzverfahren n lostwax process, investment casting process
wachsen v.t. (Parkett:) wax; – v.i. grow
Wächter m (electr.) automatic controller
Wächteruhr f tell-tale watch
Wackelkontakt m intermittent contact, defective contact
Waffelblech n goffered plate
Wagen m (auto.) motorcar, passenger car, car; (railway) wagon, freight car; (ingot) car
Wagenbauprofil n car-building section
Wagenbauschraube f carriage bolt
Wagenherdofen m car bottom furnace
Wagenpark m (railway) rolling equipment
Wagenpflege f maintenance, servicing
Wagenschlosser m millwright
Wagenwaschanlage f (auto.) vehicle washing equipment

Wagenwinde f car jack
Waggon m cf. Wagen
Wahl f selection; (tel.) dialing
wählen v.t. select; (tel.) dial
Wähler m selector
Wählerfernamt n automatic toll exchange
Wahlschalter m selector switch
Wählscheibe f selector dial; (tel.) calling dial
Wählzeichen n (tel.) dial signal
Walzbacke f (threading) rolling die
Walzblech n tin plate
Walzblei n sheet-lead
Walzblock m (US) bloom; (UK) cog
Walzdorn m roll mandrel
Walzdraht m wire rod
Walze f roll; (recording) drum
walzen v.t. (met.) roll; (≈ walzfräsen:) slab-mill
wälzen v.t. hob; generate
Walzenanstellung f adjustment of rolls
Walzenballen m roll body, barrel of a roll
Walzendrehmaschine f roll turning lathe
Walzenfräser m plain milling cutter
Walzenkalibriermaschine f roll drafting machine
Walzenschalter m drum switch, barrel controller
Walzenständer m roll housing, roll standard
Walzenstirnfräser m shell end mill
Walzenstraße f mill train
Walzenzapfen m neck of a roll, journal of a roll
Walzenzunder m mill scale
Walzer m roller

Walzerzeugnis *n* rolled product
Wälzfräsautomat *m* automatic hobbing machine
walzfräsen *v.t.* slab-mill
wälzfräsen *v.t.* hob; generate
Walzfräser *m* slab milling cutter
Wälzfräser *m* hob
Wälzfräsmaschine *f* hobbing machine; *(gear cutting)* gear generator
walzgerade *a.* straight in the as rolled condition
Walzgerüst *n* roll stand
Wälzgetriebe *n* rolling gear transmission
Walzgrat *m* sliver
Walzgut *n* rolling stock
Walzhaut *f* rolling skin
Wälzhobelmaschine *f* gear shaper
wälzhobeln *v.t.* plane by the generating method; (auf der Waagerechtstoßmaschine:) shape by the generating method
Walzkaliber *n* roll groove, roll pass
Walzkante *f* rolling edge
Wälzkegel *m* *(gear cutting)* pitch cone
Wälzkegelradhobelmaschine *f* bevel gear generator
Walzkessel *m* (Weißblechfabrikation) grease pot
Wälzkörper *m* (e. Lagers:) rolling element
Wälzkreis *m* *(gearing)* pitch circle
Wälzlager *n* *(US)* anti-friction bearing; *(UK)* rolling bearing; – *pl.* ball and parallel roller bearings
Wälzlinie *f* line of contact
Walzmaterial *n* rolling stock
Walzmessing *n* rolled brass
Walznaht *f* rolling fin, seam
Wälzpunkt *m* *(gearing)* point of contact

Wälzrad *n* rolling gear
Walzschlacke *f* roll scale, mill scale
wälzschleifen *v.t.* grind by the generating method
Wälzschleifmaschine *f* generating grinder
Walzschweißverfahren *n* sealed assembly rolling process
Walzsplitter *mpl.* slivers
wälzstoßen *v.t.* shape by the generating method
Wälzstoßmaschine *f* gear shaper
Walzstraße *f* rolling train
Walztoleranz *f* mill limits
Wälzung *f* rolling motion, roll motion, generating motion
wälzverzahnen *v.t.* generate gears
Wälzverzahnung *f* gear generation
Walzwerk *n* rolling mill, mill
Walzwerker *m* rolling mill engineer
Walzwerksbetrieb *m* rolling-mill practice
Wälzwerkzeug *n* generating tool; hobbing tool
Walzzapfen *m* neck of a roll
Walzzunder *m* mill scale, roll scale
Wandanschlußdose *f* *(electr.)* wall socket
Wandarm *m* wall bracket
Wandauslegerbohrmaschine *f* wall radial drill
Wandbohrmaschine *f* post drilling machine
wandern *v.i.* move, travel; wander; *(chem.)* diffuse, migrate
Wanderrost *m* travelling grate
Wanderrostfeuerung *f* travelling grate firing
Wanderwelle *f* *(electr.)* travelling wave, transient wave
Wandfaserplatte *f* wallboard
Wandfliese *f* wall tile

Wandkran *m* wall crane
Wandlager *n* bracket bearing
Wandler *m* *(electr.)* instrument transformer; converter, transducer
Wandradialbohrmaschine *f* wall radial drill
Wandschalter *m* wall-mounted switch
Wandschaltschrank *m* wall-mounting cabinet
Wandsteckdose *f* wall socket
Wandstecker *m* wall plug
Wange *f* (e. Kurbelwelle:) cheek; (e. Maschinenbettes:) shear; (e. Ständers:) front wall
Wanknutsäge *f* wobbling saw
Wanne *f* (Öl, Späne:) pan, tray, sump trough; reservoir
Waren *fpl.* (z. B. aus Blech:) ware
warmabgraten *v.t.* hot-trim
Warmabgratwerkzeug *n* hot trimming die
Warmarbeitsgesenk *n* hot forging die
Warmarbeitsstahl *m* hot working steel
Warmarbeitswerkzeug *n* hot working die
warmaufziehen *v.t.* shrink on
warmbadhärten *v.t.* hot-temper; (auf martensitisches Gefüge:) martemper
Warmband *n* hot rolled strip
Warmbandstraße *f* hot strip mill
warmbearbeiten *v.t.* hot work
Warmbearbeitung *f* hot work
warmbehandeln *v.t.* hot work; *s.a.* wärmebehandeln
Warmbehandlung *f* hot work, hot working; *s.a.* Wärmebehandlung
Warmbett *n* (rolling mill) hot bed
Warmbildsamkeit *f* (Kunststoffschweißen:) thermo-plasticity

Warmblasen *n* (Birnenbetrieb:) silicon blow (raising metal temperature by silicon oxidation)
warmbrüchig *a.* hot-short, hot brittle
Warmbrüchigkeit *f* hot-shortness, hot brittleness
warmdrücken *v.t.* hot spin
Wärme *f* heat
WÄRME ~, (~abfall, ~aufnahme, ~ausdehnung, ~austausch, ~behandlung, ~bilanz, ~durchgang, ~energie, ~fluß, ~inhalt, ~isolation, ~leiter, ~leitfähigkeit, ~leitung, ~riß, ~strahlung, ~übergang, ~übertragung, ~verlust, ~wirtschaft, ~zufuhr) heat ~
wärmeabgebend *p.a.* exothermic
wärmeaufnehmend *a* heat absorbing; endothermic
wärmebehandeln *v.t.* heat treat
Wärmebehandlungsofen *m* heat-treating furnace
Wärmebeständigkeit *f* heat-proof quality
Wärmedämmung *f* thermal insulation
Wärmedehnung *f* dilatation
Wärmedehnungsmesser *m* dilatometer
Wärmegefälle *n* temperature gradient
Wärmegrube *f* *(met.)* soaking pit
Wärmeimpulsschweißen *n* thermal impulse welding
Wärmemengenmesser *m* calorimeter
wärmen *v.t.* heat
Wärmeschrank *m* hot cabinet
Wärmespannung *f* thermal stress
Wärmetönung *f* evolution of heat
warmfest *a.* resistant to deformation at elevated temperatures
warmfester Stahl, high-temperature steel
Warmfestigkeit *f* high-temperature strength

warmformbare Preßmasse, thermoplastic compression molding material

Warmformgebung f hot forming, hot shaping, hot working

warmgewalzter Bandstahl, hot-rolled strip

warmhärtbare Preßmasse, thermosetting compression molding material

Warmhaube f (met.) hot dozzle, hot top

warmkleben v.t. heat-bond

Warmkleber m hot-setting adhesive

Warmlager n (rolling mill) hot bed

warmlaufen v.i. run hot

warmlochen v.t. hollow-forge

Warmluftheizung f hot-air heating

Wärmofen m (met.) reheating furnace

warmpressen v.t. press hot, hot-press; (plastics) heat-mold, hot-mold

Warmpreßform f hot pressing die

Warmpreßstahl m hot pressing steel

Warmpreßteil n hot-pressed part

Warmriß m heat crack, hot crack, thermal crack

Warmsäge f hot saw

Warmschermesser n hot shear blade

Warmschmiedegesenk n drop hammer die

warmschmieden v.t. hot-forge

Warmschrotmeißel m hot chisel

warmstauchen v.t. pressure-forge, hot-press, press-forge; (Bunde, Flanschen:) upset

Warmstauchmatrize f hot upsetting die, hot pressing die

Warmstreckgrenze f yield point at elevated temperatures

warmverformbar a. hot-workable

Warmverformbarkeit f hot forming property

warmverformen v.t. hot-work

warmwalzen v.t. hot-roll

Warmwalzwerk n hot rolling mill

Warmwasserbereiter m hot water heater

Warmzähigkeit f hot ductility

Warnlampe f (auto.) tell-tale light

Warnzeichen n (traffic) warning sign

Warte f (electr.) switchboard gallery

Wartezeit f (e. Maschine:) stand-by time

Wartung f upkeep; maintenance; service, servicing; routine maintenance

Wartungskosten pl. maintenance cost

Warzenblech n warted plate, pinned plate, studded plate

Warzenpunktschweißnaht f projection weld

Warzenpunktschweißung f projection welding, Woodpecker welding

Wäsche f (Benzol:) recovery

Wäscher m (Benzol:) washer, scrubber

Waschöl n wash oil

Waschpetroleum n kerosene

Waschprozeß m washing process, cleaning process, scrubbing process

WASSER ~, (~aufbereitung, ~aufnahme, ~bad, ~behälter, ~bremse, ~dampf, ~druck, ~dunst, ~filter, ~gas, ~glas, ~härtung, ~kraft, ~kraftwerk, ~kreislauf, ~kühler, ~kühlung, ~libelle, ~pumpe, ~rad, ~spiegel, ~stand, ~strahl, ~turbine, ~umlauf, ~versorgung) water ~

Wasserbau m hydraulix engineering

wasserbeständig a. waterproof

wasserdicht a. waterproof, watertight

wasserdurchlässig a. permeable to water

wasserfest a. waterproof

Wassergas-Preßschweißung f watergas pressure welding

Wasserhaltung f dewatering	**Wechselrichter** m *(electr.)* inverter
Wasserhärter m water hardening steel	**Wechselschalter** m change-over switch
Wasserhärtungsstahl m water hardening steel	**Wechselschaltung** f two-way wiring
	Wechselspannung f alternating voltage
Wasserkraftmaschine f hydraulic engine	**Wechselstrom** m alternating current
Wasserkraftwerk n hydro-electric station	**WECHSELSTROM** ~, (~empfänger, ~erzeuger, ~feld, ~gleichrichter, ~kreis, ~leistung, ~maschine, ~motor, ~netz, ~transformator, ~umformer, ~verstärker, ~widerstand) alternating-current ~
Wasserpumpenzange f water pump plier	
Wassermörtel m hydraulic lime mortar	
Wasserstoff m hydrogen	
Wasserstoff-Sauerstoffschweißung f oxy-hydrogen welding	
Wasserverdrängung f displacement of water	**Wechselversuch** m alternating stress test
	Wechsler m *(electr.)* changer, commutator
Wasserwaage f level, spirit level	**Weg** m path; way; travel; movement
Wattleistung f wattage	**Wegebohrmaschine** f way-drilling machine; way-boring machine
Wattstundenzähler m watt-hour meter	
Wechsel m change; variation; *(electr.)* alternation	**wegschwenken** v.t. swing out of the way
	wegspülen v.t. flush away
Wechselbrenner m *(welding)* variable head torch	**Wegspülung** f *(civ.eng.)* undermining, underwashing
Wechselfestigkeit f fatigue resistance, fatigue strength, endurance strength	**Wehr** n weir
	Weichblei n refined lead, soft lead
Wechselgetriebe n change gear mechanism	**Weiche** f rail point, rail switch
	Weicheiseninstrument n moving-iron instrument
Wechselklappe f butterfly valve	
Wechselkugellager n double-thrust ball bearing	**Weichenzungenhobelmaschine** f switch rail planer
Wechsellager n double-thrust bearing	**Weichfleckigkeit** f *(met.)* liability to [harden with] soft spots
wechseln v.t. & v.i. change, vary; *(electr.)* alternate	
	weichglühen v.t. soft anneal, spheroidize
Wechselrad n change gear	**Weichhaut** f soft skin
Wechselradantrieb m change gear drive	**Weichholz** n softwood
Wechselrädergetriebe n quick-change gear mechanism	**Weichlot** n [soft] solder
	weichlöten v.t. solder
Wechselräderschere f change gear quadrant	**Weichmacher** m softener, softening agent; *(plastics)* plasticizer
Wechselradgetriebe n quick-change gear drive	**Weißblech** n tinplate
	weißen v.t. *(building)* limewash
wechselrichten v.t. *(electr.)* invert	**Weißguß** m white malleable cast iron
	Weißkernguß m white heart malleable iron

Weißmetall *n* babbitt metal
Weiterbehandlung *f* subsequent treatment
weiterziehen *v.t. (cold work)* (e. Hohlkörper bei gleichbleibender Wandstärke ausstrecken:) redraw
Weitstrahler *m (auto.)* distance light
Wellblech *n* corrugated sheet steel
Wellblechwalzwerk *n* corrugating rolling mill
Welldrahtglas *n* corrugated safety glass
Welle *f* (Transmission:) shaft; (Kreissäge:) spindle, arbor; (Oberfläche:) undulation; (Riementrieb:) countershaft; *(opt., phys., electr. radio)* wave
WELLEN ~, (~band, ~bereich, ~frequenz, ~länge, ~leitwert, ~messung, ~strahlung) wave ~
Wellenbund *m* shaft collar
Wellenkupplung *f* coupling
Wellenleitung *f* transmission line, lineshaft
Wellenmesser *m (electr.)* wavemeter, cynometer
Wellenstrang *m* lineshaft
Wellenstumpf *m* shaft extension, shaft stub
Wellenzapfen *m* journal
Wellrohr *n* corrugated tube
Wellziegel *m* corrugated tile
Wendeformmaschine *f* turnover molding machine
Wendegetriebe *n* reversing gear mechanism
Wendeherz *n* reverse plate
Wendel *m* coil; helix; spiral
Wendelspan *m* coil chip
Wendeplatte *f* (e. Formmaschine:) turnover plate

Wendeplattenformmaschine *f* turnover table molding machine
Wendeplattenrüttler *m* jolt turnover machine
Wendepol *m* commutating pole
Wendepolumschalter *m* pole-changing switch
Wendepolwicklung *f* commutating pole winding
Wender *m* (für Walzgut:) manipulator
Wendeschalter *m* reversing switch
Wendeschütz *n* reversing contactor
Wendezapfen *m* (e. Konverters:) trunnion
Wendigkeit *f (auto.)* manoeuvrability
Wendung *f* turn
Werg *n* tow, oakum
Werk *n* plant; factory; works; mill; shop
Werksbescheinigung *f* works' test certificate
Werksprüfung *f* inspection test
Werkstatt *f* workshop, shop
Werkstattabnahmelehre *f* inspection gage
Werkstattmessung *f* inspection test, shop test
Werkstattpraxis *f* workshop practice
Werkstattwinkel *m* common try square
Werkstattzeichnung *f* workshop drawing
Werkstein *m* (Beton:) cast stone
Werkstoff *m* material; (für die laufende Fertigung:) stock
WERKSTOFF ~, (~abnahme, ~durchmesser, ~führungsrohr, ~halter, ~ständer, ~vorschub, ~zuführung) stock ~
Werkstoffprüfung *f* testing of materials
Werkstoffstange *f automatic lathe)* bar stock
Werkstück *n* workpiece, workpart, work

WERKSTÜCK ~, (~auflagenbock, ~aufspannung, ~ausrichtung, ~auswerfer, ~befestigung, ~drehzahl, ~durchbiegung, ~durchmesser, ~mitnehmer, ~muster, ~rutsche, ~spindel, ~stoff, ~support, ~zuführung) work ~

Werkstückaufnahmevorrichtung f workholding fixture

WERKZEUG ~, (~auflage, ~ausrüstung, ~bau, ~einrichter, ~einrichtung, ~einstellung, ~fertigung, ~fräsmaschine, ~futter, ~gestaltung, ~halter, ~härterei, ~instandhaltung, ~kasten, ~konstrukteur, ~kopf, ~loch, ~macher, ~schaft, ~schleifmaschine, ~schlitten, ~schmied, ~schrank, ~spindel, ~stahl, ~standzeit, ~stoff, ~überhang, ~verschleiß, ~vorschub, ~winkel, ~zeichnung, ~zustellung) tool ~

Werkzeuganordnung f tooling setup, tooling system

Werkzeugfräsmaschine f toolroom milling machine

Werkzeugmacherei f toolroom

Werkzeugmaschine f machine tool

Werkzeugsaal m toolroom

Werkzeugschlosser m toolmaker

Werkzeugspanner m toolholder

Wert m value

Wetterführung f (mining) system of ventilation

Wetterkorrosion f atmospheric corrosion

Wetterprüfung f weather exposure test

Wichte f volumetric weight; specific gravity

Wickelkondensator m roller type capacitor

wickeln v.t. (e. Feder:) wind, coil

Wickelnaht f (electr.) pitch

Wicklung f (electr.) winding

Widerlager n (civ.eng.) abutment

Widerstand m (electr., mech.) resistance; (als Bauelement:) resistor
magnetischer ~, reluctance

Widerstand zwischen Stöpseln, resistance between plug electrodes

WIDERSTANDS ~, (~nahtschweißung, ~ofen, ~preßschweißung, ~schmelzschweißung, ~schweißung, ~stumpfschweißung, ~thermometer) resistance ~

Widerstandsfähigkeit f resistivity

WIEDER ~, (~anlaufen, ~anstellen, ~aufbauen, ~aufkohlen, ~einfangen, ~einführen, ~einrücken, ~einschalten, ~erhitzen, ~festziehen, ~verwenden) re-~

Wiederaufbau m reconstruction

Wiedergabe f (copying, radio) reproduction; (Frequenzen:) response

Wiedergabetreue f (radio, telev.) fidelity [of reproduction]

Wiedergewinnung f regeneration, recovery

wiederholen v.t. repeat; – v.r. recur

Wiederholung f repetition, recurrence; (programming) rerun

Wiege f rocker

Wind m (Schmelzbetrieb:) blast, airblast

Winddruck m (met.) blast-pressure

Winddüse f tuyere

Winde f winch; ($\approx$ Schraubenwinde:) jack

Windeisen n tap wrench

Winderhitzer m (met.) hot blast stove

Windform f tuyere

Windfrischen n air blowing, air refining

Windfrischstahl m air-refined steel

Windkasten m (Kupolofen:) wind box; (Konverter:) blast box, air box

Windlast f wind pressure
Windleitung f blast main, blast pipe
Windschutzscheibe f *(auto.)* windscreen; windshield
Windung f turn: (e. Feder:) coil
Winkel m *(geom.)* angle; *(tool)* square
WINKEL ~, (~abweichung, ~ausschlag, ~beschleunigung, ~bewegung, ~fräsen, ~frequenz, ~geschwindigkeit, ~grad, ~maß, ~minute, ~trieb, ~verschiebung, ~verstellung) angular ~
Winkelaufspanntisch m angle plate table
Winkelendmaß n angle gage
Winkelfräser m angle milling cutter
Winkelgelenk n toggle joint
Winkelgenauigkeit f squareness
Winkeligkeit f angularity
Winkelkonsole f knee-table
Winkellibelle f protractor bevel
Winkelmaß n *(carp.)* bevel steel square; *(metrol.)* bevel protractor
Winkelmeßokular n gomiometer ocular
Winkelminute f angular minute
Winkelnaht f *(welding)* corner joint
winkelrecht a. square
winkelschleifen v.t. *(woodworking)* angle
Winkelschleifer m angle grinder
Winkelstahl m angle steel
Winkelstellung f angularity
Winkelstoß m corner joint
Winkelstreichmaß n square marking gage
Winkelteilungsprüfer m theodolite
Winkeltisch m angle plate; (≈ Konsol:) knee
Winker m *(auto.)* direction indicator, trafficator
Winkerschalter m *(auto.)* direction indicator switch

Wippe f (e. Motors:) rocker, hinged plate; *(rolling mill)* tilting table
Wipper m tippler
Wipptisch m tilting table
Wirbel m (Kette:) swivel; *(hydrodyn.)* vortex
Wirbelhaken m swivel hook, shackle hook
Wirbelkammer f (Motor:) turbulence chamber
Wirbelmaschine f (Gewinde:) whirling machine
wirbeln v.t. *(threading)* whirl
Wirbelstrom m *(electr.)* eddy current
Wirbelstromläufer m squirrel cage rotor
Wirbelströmung f turbulence
Wirbelstromverlust m eddy current loss
Wirkdruckgeber m active pressure indicator
Wirkleistung f *(electr.)* effective power, real power
Wirkleistungsmesser m active power meter
Wirkleitwert m conductance
Wirkspannung f *(electr.)* useful voltage
Wirkstrom m active current, wattful current
Wirkungsgrad m efficiency
Wirkverbrauchszähler m *(electr.)* watt-hour meter
Wirkwinkel m (e. Meißels:) working angle
Wirrspan m snarly chip
Wischer m *(auto.)* windscreen wiper
Wischerarm m *(auto.)* wiper arm
Wohnhaus n dwelling house
Wohnung f apartment
Wohnungsbau m home building, house building
Wohnwagen m caravan
Wohnwagenanhänger m caravan trailer

Wölbnaht f convex fillet weld
Wölbung f curvature, convexity; crowning
Wolfram n tungsten
Wolfram-Inertschweißen n inert gas arc welding with non-consumable electrodes
Wolle f wool
Wrasen m water vapor
Wulst m bulge, bulb; *(welding)* reinforcement
Wulstnaht f *(welding)* reinforced seam, upset butt weld

Wulstrand m beaded edge
Wulstreifen m *(auto.)* bead tire
Wulstwinkelstahl m bulb angle steel
Würfelfestigkeit f *(concrete)* cube compression strength
Wurzeleinbrand m root fusion zone
Wurzellage f root layer, bottom run
Wurzelraupe f *(welding)* root bead
Wurzelschweißung f root weld
Wurzelüberhöhung f *(welding)* penetration bead
Wurzelverschweißung f root penetration

Z

zäh *a.* tough; ductile; (Flüssigkeiten:) viscous
zähfest *a.* tenacious
Zähfestigkeit *f* toughness, tenacity
zähflüssige Schlacke, sticky slag, viscous slag
Zähflüssigkeit *f* viscosity
Zähigkeit *f* toughness; tenacity
Zahl *f* number, cipher; digit; figure
Zähler *m* (*metrol.*) counter; (*electr.*) meter; (*math.*) numerator
Zählerschaltuhr *f* meter change-over clock, time switch
Zählrohr *n* counting tube, Geiger-Müller tube
Zählwaage *f* counting balance
Zählwerk *n* counter mechanism
Zahn *m* tooth; (e. Keilwelle:) splinetooth; (e. Kerbverzahnung:) serration; (e. Messerkopfes:) blade
Zahnanlage *f* tooth contact
Zahneingriff *m* tooth engagement
zahnen *v.t.* (*gears*) cut teeth; ($\approx$ kerbverzahnen:) serrate
Zahnflanke *f* tooth flank
Zahnflankenlinie *f* tooth trace
Zahnform *f* tooth profile
Zahnformfehler *m* tooth profile error
Zahnformfräser *m* gear milling cutter
Zahnfuß *m* (*gears*) dedendum; ($\approx$ Zahngrund:) root
Zahnfußausrundung *f* fillet
Zahnfußkreis *m* root circle
Zahngesperre *n* ratchet and pawl [mechanism]
Zahngrund *m* root

Zahnhöhe *f* depth of tooth
Zahnkette *f* sprocket chain; silent chain
Zahnkettentrieb *m* silent chain drive
Zahnkopf *m* addendum; top of the tooth, tip of the tooth
Zahnkopfabrundung *f* top radius
Zahnkranz *m* gear rim, gear ring; (e. Planscheibe:) rim gear; (e. Planfutters:) scroll; (*auto.*) rim gear, ring gear
Zahnkupplung *f* jaw clutch coupling, dog clutch
Zahnlücke *f* tooth gap, gash; (e. Säge:) chip pocket
Zahnmeßlehre *f* tooth measuring gage
Zahnmeß-Schieblehre *f* gear tooth vernier caliper
Zahnmeßschraublehre *f* gear tooth micrometer
Zahnprofil *n* tooth profile, tooth form, tooth contour
Zahnrad *n* gear, gear wheel
ZAHNRAD ~, (~antrieb, ~bearbeitungsmaschine, ~fertigung, ~fräser, ~fräsmaschine, ~getriebe, ~körper, ~läppmaschine, ~prüfgerät, ~rohling, ~schleifmaschine, ~übersetzung, ~untersetzung, ~vorschub, ~wälzfräsmaschine, ~welle, ~zahn) gear ~
Zahnradhobelmaschine *f* gear planer; (Stoßverfahren:) gear shaper; (Wälzverfahren:) gear generator
Zahnradpumpe *f* geared pump
Zahnradstoßmaschine *f* gear slotter; (Abwälzverfahren:) gear generator
Zahnscheibe *f* ratchet wheel; (als Unterlegscheibe:) serrated lock washer

Zahnsegment n gear segment
Zahnspiel n tooth clearance
Zahnstange f gear rack
Zahnstangenantrieb m rack and pinion drive
Zahnstangenfräser m rack tooth cutter
Zahnstangengetriebe n rack gearing
Zahnstangenritzel n rack pinion
Zahnstollen m radial face
Zahntrieb m gear drive
Zange f pincers; tongs; (Draht:) plier, cutter; (Rohr:) wrench; (welding) electrode holder; (≈ Spannzange:) collet [chuck]; (rolling mill) (≈ Blockzange:) stirrup, tongs, dog
Zangenkran m (met.) dogging crane
Zangenspannfutter n spring collet chuck
Zangenwandler m (= Anlegewandler) split-core type transformer
Zapfen m stud, pin; (≈ Drehzapfen:) pivot; (≈ Zapfenlager:) journal; (≈ Führungszapfen:) spigot; (≈ Kurbelzapfen:) crankpin; (e. Konverters:) trunnion; (e. Revolverkopfes:) stud; (e. Schraube:) full dog point
zapfenausschneiden v.t. (woodworking) relish
Zapfenlager n pivot bearing
Zapfenloch n (e. Pressenstößels:) punch stem opening, tool shank hole
Zapfensäge f dovetail saw
Zapfenschlüssel m pin spanner
zapfenschneiden v.t. (woodworking) tenon
Zapfensenker m spot facer
Zapfenspannloch n (power press) punch stem opening, tool shank hole
Zapfen- und Nutenschneidmaschine f tenoner and gainer

Zapfhahn m faucet, tap
Zapfsäule f (auto.) petrol pump pillar
Zapfstelle f (auto.) filling station; (US) service station
Zapfwelle f (auto.) power take-off shaft
Zaponlack m zapon lacquer
Zarge f frame
Zaun m fence
Zebrastreifen m (auto.) zebra crossing
Zeche f (techn.) mine; colliery
Zechenkohle f mine coal
Zechenkoks m coke-oven coke
Zechenteer m coke tar
Zeichen n sign, mark; symbol; signal
ZEICHEN ~, (~brett, ~büro, ~gerät, ~tusche, ~utensilien, ~winkelmesser) drawing ~
Zeichenpegel m (radio) signal level
Zeichenspannung f (radio) signal voltage
Zeichner m (US) draftsman; (UK) draughtsman
Zeiger m (e. Meßgerätes:) index; pointer
Zeilenabtastung f (telev.) line scanning
Zeilenfräsen n line-by-line milling
Zeilenfrequenz f line frequency
Zeilenraster m (telev.) line raster
Zeilenstruktur f banded structure
Zeilenvorschub m straight feed
Zeit f time; period
ZEIT ~, (~auslösung, ~einstellung, ~konstante, ~schalter, ~studium, ~waage) time ~
Zeitdehner m (photo.) slow-motion camera
Zeitdehngrenze f time-yield limit
Zeitdehnung f time yield
Zeitfestigkeit f creep strength depending on time
Zeitfließgrenze f time yield
Zeitlohn m wages per unit of time

Zeitlupenaufnahme f slow-motion picture
Zeitmesser m chronometer, timer
Zeitmessung f timing
Zeitnehmer m time study man, time-keeper
Zeitraffung f quick-motion effect
Zeitschaltwerk n timing gear
Zeitstandard m *(cost accounting)* production standard
Zeitstandfestigkeit f creep strength depending on time
Zeitstandwert m stress-rupture property
Zeitstreckgrenze f time yield
Zeit-Wegschreiber m *(auto.)* time-mileage recorder
Zeitzähler m *(electr.)* hour meter
Zeitzündung f delay action firing
Zellenbeton m cellular-expanded concrete
Zellstoff m cellulose
Zellstoffwatte f cellulose wadding
Zelluloselack m cellulose varnish
Zellwolle f viscose rayon
Zement m cement
Zementbeton m cement concrete
Zementbetondecke f concrete pavement
Zementdrehofen m rotary cement kiln
Zementestrich m cement flooring
zementieren v.t. *(building)* cement
Zementkitt m cement adhesive
Zementklinker m cement clinker
Zementmörtel m cement mortar
Zentralamt n *(tel.)* central exchange
Zentrale f *(electr.)* power station; *(tel.)* exchange
Zentralschaltpult n central control desk
Zentralschmierpumpe f one-shot pump
Zentralschmierung f centralized lubrication
Zentrierbohrer m center drill

zentrieren v.t. center; *(UK)* centre
Zentrierwerkzeug n centering tool
Zentrierwinkel m *(tool)* center square
ZENTRIFUGAL ~, (~kraft, ~pumpe, ~regler, ~schalter) centrifugal ~
Zentrifugalanlasser m centrifugal starting switch
zentrisch a. centric, central; concentric
Zerfall m disintegration; decay
zerfallen v.i. disintegrate
Zerfallprodukt n disintegration product
Zerfallschlacke f disintegrating slag, slaking slag
Zerfallzeit f decay time, time of disintegration
Zerhacker m *(electr.)* vibrator, chopper
zerkleinern v.t. crush
Zerkleinerungsanlage f crushing plant
Zerkleinerungsmaschine f crushing machine, crusher; pulverizer
zerlegbar a. *(mech.)* detachable; resolvable, decomposable; *(nucl.)* fissionable
zerlegen v.t. detach, disassemble, take apart; decompose
Zerlegung f (Farben:) decomposition; *(nucl.)* decay
Zerreißbelastung f ultimate tensile stress, breaking load
zerreißen v.t. & v.i. tear, disrupt, break, rupture
Zerreißfestigkeit f ultimate strength, tensile strength
Zerreißmaschine f tensile testing machine, pull test machine
Zerreißprobe f tensile test specimen
Zerreißprüfung f cf. Zugversuch
Zerreißspannung f tensile load, tensile stress
Zerreißstab m tensile test bar

Zerreißversuch *m* tension test, tensile test, pull test

Zerschmiedungsriß *m* forging crack; star crack; shadow crack; shatter crack

zersetzen *v.r.* decay, decompose; dissociate

Zersetzung *f* decomposition, disintegration

Zersetzungsprodukt *n* decomposition

zerspanbar *a.* machinable

Zerspanbarkeit *f* machinability

zerspanen *v.t.* machine, cut

Zerspanung *f* machining, metal cutting

Zerspanungsarbeit *f* metal-cutting work

Zerspanungsbedingungen *fpl.* machining requirements

Zerspanungseigenschaft *f* cutting property

Zerspanungsleistung *f* metal-removing capacity

Zerspanungsmaschine *f* metal-cutting machine

Zerspanungszeit *f* machining time

Zerstäuber *m* atomizer

zerstörungsfreie Prüfung, non-destructive test[ing]

Zertrümmerung *f (phys.)* fission

Zickzacknaht *f (welding)* staggered seam

Zickzacknietung *f* (= versetzte Nietung) staggered riveting

Zickzackpunktschweißung *f* staggered spot weld

Zickzackschaltung *f* zig-zag connection

Zickzacktrio *n* cross-country mill

Ziegel *m* brick; (Dach:) tile

Ziegelei *f* brickworks

Ziegelofen *m* brick kiln

Ziegelpflaster *n* brick paving

Ziegelsplittbeton *m* crushed brick concrete

Ziehbank *f* drawing bench, draw bench

Zieheisen *n* draw plate

ziehen *v.t. (cold work)* (Draht:) draw; (Rohre:) sink

Zieherei *f* drawing mill

Ziehfähigkeit *f* drawing quality

Ziehherd *m* furnace discharge end

Ziehkeil *m* diving key

Ziehkeilgetriebe *n* diving key transmission

Ziehkissen *n (power press)* die cushion

Ziehmatrize *f* drawing die

Ziehnadel *f* broach

Ziehpresse *f* drawing press

Ziehräummaschine *f* pull-type broaching machine

Ziehräumnadel *f* pull broach

Ziehring *m* drawing block

Ziehschleifeinrichtung *f* honing equipment

ziehschleifen *v.t.* hone

Ziehschleifmaschine *f* honing machine

Ziehstein *m* wire-drawing die

Ziehstempel *m (power press)* draw punch

Ziehtiefe *f* depth of draw

Zielscheibe *f* target

Zierkappe *f (auto.)* ornamental hub cap

Zierleiste *f (auto.)* belt molding

Zimmerer *m* carpenter

Zimmererhandwerk *n* carpentry

Zimmermann *m* carpenter

Zimmermannsbeil *n* carpenters' hatchet

Zimmerung *f* timbering

Zink *n* zinc

Zinkblech *n* galvanizing sheet; galvanized sheet

Zinkdestillierofen *m* zinc distilling furnace

Zinkelektrolyse *f* electrolytic zinc process

zinken v.t. (woodworking) dovetail
zinkenfräsen v.t. dovetail
Zinkgewinnung f zinc extraction
Zinkhütte f zinc smelting plant
Zinkoxid n zinc oxide
Zinkweiß n zinc white, zinc oxide
Zinnblech n tin metal sheet
Zinnfolie f tin foil
Zinnhütte f tin smelting plant
Zinnlot n tin solder
Zoll m inch
Zollgewinde n English thread
Zollmaß n (als Maßstab:) inch rule; (Maßsystem:) English system
Zollstock m inch rule
Z-Profil n Z-section; pl. Zees
Zubringerfirma f sub-contractor
Zufuhr f supply
Zuführeinrichtung f (automatic lathe) feeding attachment
zuführen v.t. feed, supply
Zuführung f supply, delivery; (Werkstoffe:) feeding, feed; (Vorrichtung:) feeding attachment
Zuführungskabel n supply cable
Zug m tension; pull; (cold work) draw; (Kette, Riemen:) pull; (Draht:) (tool) drawing block; (railw.) train; (gears) train
Zugabe f (mach.) allowance
Zugang m access
zugänglich a. accessible
Zugänglichkeit f accessibility
Zuganker m tie rod
Zugbeanspruchung f tensile stress
Zugbelastung f tensile load, tension load
Zugbrücke f draw bridge
Zug-Ermüdungsversuch m tensile fatigue test
Zugfestigkeit f tensile strength

Zughaken m (auto.) tow hook
zügig a. consistent
Zugkraft f tractive power; pulling power; (magn.) attraction force
Zugmaschine f truck-tractor
Zugmaschinenanhänger m tractor-trailer
Zugmesser m traction dynamometer
Zugprüfung f tensile test, pull test
Zugraupe f [crawler] tractor
Zugschalter m pull switch
Zugschraube f draw-in bolt
Zugseil n hauling rope, traction rope, hoisting cable
Zugspannung f (mat. test.) tensile stress
Zugspindel f (lathe) feed rod
Zugspindeldrehmaschine f regular-type engine lathe
Zugstange f drawbar
Zug- und Leitspindeldrehmaschine f engine lathe
Zugversuch m tensile test
Zugvorrichtung f (auto.) towing gear
zulässig a. allowable, admissible; permissible
zulässige Beanspruchung, safe stress
zulässige Belastung, safe load
zulässige Spannung, permissible stress
Zulassung f permit
Zulassungsnummer f (auto.) licence number, registration number
Zulegestoff m (welding) metallic filler
Zuleitung f (electr.) lead; (radio) feeder
Zulieferer m sub-supplier
Zündanlaßschalter m (auto.) ignition switch
zündbar a. inflammable
Zündbarkeit f inflammability
Zündbatterie f (auto.) ignition battery

Zündeinstellung f *(auto.)* ignition timing, spark setting
zünden v.t. (Lichtbogen:) strike
Zunder m scale
zunderbeständiger Stahl, non-scaling steel
Zunderbeständigkeit f resistance to scaling
zundern v.i. scale
Zündgeschwindigkeit f *(welding)* rate of flame propagation; ignition rate
Zündkabel n *(civ. eng.)* shot-firing cable
Zündkabelschuh m *(auto.)* ignition cable end fitting
Zündkerze f sparking plug, spark plug
Zündkerzenreiniger m sparking plug cleaner
Zündkerzenschlüssel m sparking plug wrench
Zündkopf m *(engine)* hot bulb
Zündmagnet m magneto
Zündmaschine f *(civ. eng.)* blasting machine
Zündschalter m ignition switch
Zündschloß n *(auto.)* ignition lock
Zündschlüssel m *(auto.)* ignition switch key
Zündspannung f *(welding)* striking voltage, firing voltage
Zündung f *(auto.)* ignition, firing
Zündversteller m *(auto.)* automatic timing advance
Zündverteiler m *(auto.)* ignition distributor
Zündverteilerkopf m *(auto.)* distributor head
Zündverzug m *(auto.)* ignition lag
Zungenfrequenzmesser m vibrating reed instrument, reed frequency meter
Zungenweiche f tongue switch

zurichten v.t. recondition, refit; dress
Zurichterei f dressing shop
zurückknallen v.i. *(welding)* backfire
zurückschalten v.t. (Drehzahlen:) reduce
zurückwerfen v.t. *(opt.)* reflect
zurückziehen v.t. draw back, pull back, retract, withdraw
Zusammenbau m assembly
zusammenbauen v.t. assemble; fit
zusammendrücken v.t. compress
zusammenfügen v.t. (Paßteile:) mate; assemble
zusammenklappbar a. collapsible
Zusammenschnürung f (e. Probestabes:) necking, bottling
zusammensetzen v.t. assemble; *(chem.)* compose
Zusatz m addition, admixture; (Öl:) additive
Zusatzdraht m *(welding)* filler wire
Zusatzgetriebe n supplementary gear transmission
Zusatzmetall n *(welding)* filler metal
Zusatzpumpe f *(auto.)* booster pump
Zusatzstichmaß n micrometer extension rod
Zusatztransformator m booster transformer
Zusatzwerkstoff m *(welding)* filler metal
Zuschlag m *(time study)* allowance
Zuschläge mpl. *(met.)* additions
zuschneiden v.t. cut to size
Zuschub m *(metal cutting)* feed
zustellen v.t. (= justieren) adjust; (Meißel:) set; *(grinding)* feed [in]
Zustellung f *(mach.)* adjustment; *(grinding)*; *(met.)* (Ofen:) lining
Zutritt m access

zwangläufig *a.* positive
Zweckentfremdung *f* diversion
zweckmäßig *a. practical*
zweiatomig *a.* diatomic
Zweibackenfutter *n* two-jaw chuck
Zweifachstecker *m (electr.)* bipolar plug
Zweiflammenbrenner *m* twin-jet blowpipe
Zweiklanghorn *n (auto.)* dual-tone horn, klaxon
Zweilagenblech *n* two-layer steel sheet
zweiphasiger Wechselstrom, two-phase alternating current
zweipolig *a.* bipolar; double-pole
zweipoliger Stecker, two-pin plug
Zweiradsattelschlepper *m* two-wheel tractor
zweireihiges Lager, double-row bearing
Zweiröhrenempfänger *m* two-valve receiver
Zweisäulen-Friktionsspindelpresse *f* two-column friction screw press
Zweiständerfräsmaschine *f* double-column milling machine, planer-miller
Zweiständerhobelmaschine *f* double-housing planer
Zweiständerkarusseldrehmaschine *f* double-column vertical boring mill
Zweiständerräderziehpresse *f* straight-sided reducing press
Zweistofflegierung *f* binary alloy
Zweistufenstellgetriebe *n* two-speed adjusting gears
Zweistufenwendegetriebe *n* two-speed reversing mechanism
Zweitakter *m* two-stroke engine
Zweitaktmotor *m* two-stroke engine
Zweitluft *f* secondary air
Zweiwalzengerüst *n* two-high mill

Zweiwegebohrmaschine *f* two-way boring and drilling machine
Zweiwegemaschine *f* two-way machine
Zweizylinderboxermotor *m* flat-twin engine
Zweizylindermotor *m (auto.)* two-cylinder engine
Zwiemetall *n* bimetal
Zwillingsbereifung *f* dual tires
Zwillingsräummaschine *f* dual-ram broaching machine
Zwillingsreifen *m (auto.)* dual tire
Zwillingszahnradfräsmaschine *f* twin-head gear cutting machine
Zwischenformung *f (forging)* pre-forging, semi-forging
Zwischenfrequenz *f* intermediate frequency
Zwischenglühung *f* intermediate anneal
Zwischengröße *f* fractional size
Zwischenmaß *n* fractional size
Zwischenpfanne *f* tundish
zwischenschalten *v.t.* interpose; interconnect
Zwischenstufengefüge *n* intermediate structure, bainite
zwischenstufenvergüten *v.t.* austemper
Zwischenstufenvergütung *f* austempering
Zwischenverteiler *m (tel.)* intermediate board; *(electr.)* intermediate distributing frame
Zwischenwelle *f* intermediate shaft
Zyansalzbadhärtung *f* cyaniding
Zylinderblockaufbohrmaschine *f* cylinder block boring machine
Zylinderbohrmaschine *f* cylinder block boring machine
Zylinderbuchse *f* cylinder liner

Zylinderdichtungsring *m* cylinder-head gasket
Zylindereinsatz *m* [cylinder] liner
Zylinderkopf *m* (Schraube:) cheese head
Zylinderkurbelgehäuse *n* (auto.) crankcase
Zylinderlaufbuchse *f* cylinder liner
Zylindermantel *m* cylinder jacket
Zylinderrollenlager *n* cylindrical roller bearing
Zylinderschraube *f* fillister head cap screw
zylindrisch *a.* cylindrical; (Loch:) straight round

Wichtige Fachwörterbücher im Bereich Technik:

Henry G. Freeman

Technisches Taschenwörterbuch E/D

(Hueber-Nr. 6213)

Taschenwörterbuch Kraftfahrzeugtechnik D/E

(Hueber-Nr. 6270)

Hans Heidrich

Englischer Allgemeinwortschatz Naturwissenschaften

(Hueber-Nr. 2196)

Theo M. Herrmann

Electricity/Electronics: Minimum Wordage englisch-deutsch

(Hueber-Nr. 2.9301)

MAX HUEBER VERLAG · D-8045 ISMANING

Albert Schmitz
Englisch Grundkurs Technik
Dieser Kurs ermöglicht dem Anfänger ohne Vorkenntnisse den Einstieg in das technische Englisch ohne den Umweg über das »normale« Englisch. Für Facharbeiter, Ingenieure, Techniker sowie alle, die im Berufsleben mit technischem Englisch in Verbindung kommen.

Lehrbuch, 268 Seiten, mit vielen Zeichnungen und Abbildungen, kart. (Hueber-Nr. 2181)

Handbuch mit Schlüssel zu den Übungen, 84 Seiten, kart. (Hueber-Nr. 2.2181)

Arbeitsbuch, 128 Seiten, kart. (Hueber-Nr. 6.2181)

Außerdem:
Lernwörterbuch (Hueber-Nr. 8.2181), Cassetten (Hueber-Nr. 5.2181) mit Nachsprechpausen (Hueber-Nr. 9.2181)

Albert Schmitz
Englisch Aufbaukurs Technik
Dieses Lehrwerk vermittelt technisches Englisch für den fortgeschrittenen Lernenden. Vorkenntnisse werden vorausgesetzt, wie sie mit dem Grundkurs Technik, aber auch mit anderen Lehrwerken erworben werden.

Lehrbuch, 200 Seiten, mit Fotos und Zeichnungen, kart. (Hueber-Nr. 2189)

Handbuch mit Schlüssel zu den Übungen, 92 Seiten, kart. (Hueber-Nr. 2.2189)

Arbeitsbuch (Hueber-Nr. 6.2189), Lernwörterbuch (Hueber-Nr. 8.2189), Cassetten (Hueber-Nr. 5.2189) mit Nachsprechpausen (Hueber-Nr. 9.2189)

MAX HUEBER VERLAG · D-8045 ISMANING